Amphibian Evolution

Books in the **Topics in Paleobiology** series will feature key fossil groups, key events, and analytical methods, with emphasis on paleobiology, large-scale macroevolutionary studies, and the latest phylogenetic debates.

The books will provide a summary of the current state of knowledge and a trusted route into the primary literature, and will act as pointers for future directions for research. As well as volumes on individual groups, the Series will also deal with topics that have a cross-cutting relevance, such as the evolution of significant ecosystems, particular key times and events in the history of life, climate change, and the application of new techniques such as molecular paleontology.

The books are written by leading international experts and will be pitched at a level suitable for advanced undergraduates, postgraduates, and researchers in both the paleontological and biological sciences.

The Series Editor is *Mike Benton*, Professor of Vertebrate Palaeontology in the School of Earth Sciences, University of Bristol.

The Series is a joint venture with the *Palaeontological Association*.

Previously published

Dinosaur Paleobiology
Stephen L. Brusatte
ISBN: 978-0-470-65658-7 Paperback; April 2012

Amphibian Evolution

The Life of Early Land Vertebrates

Rainer R. Schoch

WILEY Blackwell

Registered Office
John Wiley & Sons, Ltd, The Atrium, Southern Gate, Chichester, West Sussex, PO19 8SQ, UK

Editorial Offices
9600 Garsington Road, Oxford, OX4 2DQ, UK
The Atrium, Southern Gate, Chichester, West Sussex, PO19 8SQ, UK
111 River Street, Hoboken, NJ 07030-5774, USA

For details of our global editorial offices, for customer services and for information about how to apply for permission to reuse the copyright material in this book please see our website at www.wiley.com/wiley-blackwell.

Library of Congress Cataloging-in-Publication Data

Schoch, Rainer, 1970- author.
  Amphibian evolution : the life of early land vertebrates / Rainer R. Schoch.
     pages   cm
  Includes bibliographical references and index.
   ISBN 978-0-470-67177-1 (cloth) – ISBN 978-0-470-67178-8 (pbk.)   1. Amphibians, Fossil.
2. Paleobiology. 3. Amphibians–Evolution.   I. Title.
  QE867.S36 2014
  567'.8–dc23
                                        2013039743
A catalogue record for this book is available from the British Library.

Wiley also publishes its books in a variety of electronic formats. Some content that appears in print may not be available in electronic books.

Cover image: Stem-amphibian fossil. Skeleton of temnospondyl *Sclerocephalus* (Pennsylvanian, Germany). Courtesy of Rainer R. Schoch.
Cover design by Design Deluxe

Set in 9/11.5pt Trump Mediaeval by SPi Publishers, Pondicherry, India

1   2014

Contents

Preface

This book focuses on the first vertebrates to conquer the land, and on their long journey to become fully independent from the water. It will trace the origin of tetrapod features and try to explain how and why they transformed into organs that permit life on land. The classic idea of early land vertebrates is that they were similar to modern amphibians. Right or wrong, the vast majority of early tetrapods are therefore classified as amphibians (or more precisely their stem taxa). Accordingly, this book is centered on early amphibian evolution, a topic that effectively includes all early tetrapods, and it will also analyze facts and opinions on the origins of modern amphibians. The major part of the story covers events that occurred over the past 370 million years, but it is far from restricted to paleontology.

My own motivation to study the amphibian fossil record derives in large part from a fascination with the development, ecology, and evolution of their modern representatives. Therefore I consider many topics that can only be covered by examination of extant animals: features of the soft body, functions of organs that mediate breathing, feeding, hearing, and locomotion, the morphogenesis of body parts, larval development, metamorphosis, and ecology.

The aim is to achieve a comprehensive picture of amphibian evolution. This requires a walk through several dimensions, and I cannot claim to be an expert in all the fields to be covered. Nevertheless, I hope that the outcome will be worth reading, even though some data may become quickly outdated as new finds are made, and some concepts may change with new insights. The following research questions illustrate the central problems of amphibian evolution as understood here:

1. How did fishes evolve the necessary structures and organs to survive on land?
2. What was the life of early tetrapods like?
3. Are modern amphibians a good model to understand early tetrapods?
4. How did modern amphibians acquire their complex life cycles, encompassing an aquatic larva, drastic metamorphosis, and a terrestrial adult?
5. How diverse were early land vertebrates, and which evolutionary strategies did they employ?
6. What were the major factors of amphibian evolution, and how did mass extinctions affect them?

We should not expect to find equally complete or satisfactory answers to all of these questions, as the problems remain at very different stages of research. Research questions that involve examining many fossilized hard parts may be relatively easy to solve, while others require inference from extant taxa and will always remain more hypothetical. Yet other problems date back such a long time that the fossil record is too poor and ambiguous to permit decisive answers – and in such cases we shall have to consider alternative solutions and discuss their plausibility.

The diversity of questions relates also to the different research fields addressing them, and that leads me to the second major focus of the book, which is to consider the various current approaches and perspectives of paleobiology. The study of amphibian morphology and paleontology exemplifies many aspects of evolution. This topic offers a great opportunity to deepen our understanding of how organisms survive under the most diverse range of conditions, as both extant and extinct amphibians have been studied extensively. It sheds light on the pathways taken by evolution to alter developmental systems, phenotypes, and ecological relations in the amphibian world. Excellent fossils allow breathtaking insights into deep time: fishes with limb-like appendages, early tetrapods with gills and eight fingers, fossils of 1 cm larvae with bushy gills and dark eye pigments cast in stone, and spectacular skeletons of predators more than 5 m in length, with skulls exceeding a meter and hundreds of teeth in their jaws. Paleontology, zoology, developmental biology, histology, and evolutionary biology all meet in this area. The book outlines how these fields are integrated and how they come together to analyze aspects of early amphibian evolution.

Rainer Schoch
Stuttgart, Germany
October 20, 2013

Acknowledgments

Writing this book has been a joy, because it has allowed me to write down ideas that had accumulated over two decades. However, composing a book is also hard work, and it is never clear in the beginning what the manuscript will be like when the chain of thoughts has been laid down. In planning the framework of the book I have been influenced by many friends and colleagues, and although I have attempted to list them all below I suspect that some names may have been inadvertently omitted, for which I apologize.

My parents encouraged me to start the adventure that led to me becoming a paleontologist, and they contributed all they could. My family patiently accepted my devotion to the subject and always support me with their interest and critical appraisal. Hence, the present book is dedicated to my family.

When I was still at school, David Wake and Kevin Padian provided invaluable encouragement. Jürgen Boy was most supportive in bringing his critical eye and admirable experience to bear on my work when I wrote my first scientific paper. My academic teachers at Tübingen, Wolf-Ernst Reif (†) and Frank Westphal, were the most open-minded and supportive academics I ever met. Wolf, who became a close friend over the years, challenged my thoughts by asking the most unexpected and difficult questions. That he believed in me provided invaluable support. Dolf Seilacher was always ready to give my thoughts a new spin, and he would do this by asking the simplest questions. Andrew Milner has been a friend from early on; he was more ready than others to accept this would-be colleague, bringing order into my fuzzy thoughts and incomplete texts and dealing with my impatience whenever he worked with me – I owe him a lot.

I also owe much to Rupert Wild, who entrusted me with working on the valuable material he had excavated and curated for so long. Norbert Adorf and Isabell Rosin are more than the excellent preparators who make my work possible and exciting – they are friends. Achim Lehmkuhl and Marit Kamenz have also helped greatly with their preparatory skills. Johanna Eder and Reinhard Ziegler give me the necessary freedom to conduct my projects in a near-perfect working place, which I value highly.

Many friends and colleagues have contributed by discussions or helpful advice over the years, in alphabetical order: Jason Anderson, Gloria Arratia, Günter Bechly, Ronald Böttcher, Jürgen Boy, Michael Buchwitz, Ross Damiani, Dino Frey, Nadia Fröbisch, David Gower, Annalisa Gottmann-Quesada, Alexander Haas, Hans Hagdorn, Traugott Haubold, Hanna Hellrung, Axel Hungerbühler, Philippe Janvier, Farish Jenkins, Christian Klug, Klaus Krätschmer, Werner Kugler, George Lauder, Michel Laurin, Natalya Lebedkina, Kasia Lech, Hillary Maddin, Michael Maisch, Erin Maxwell, Andrew and Angela Milner, Markus Moser, Hendrik Müller, Johannes Müller, Lennart Olsson, Nadine Piekarski, Michael Rasser, Robert Reisz, Brigitte Rozynek, Martin Rücklin, Marcello Ruta, Sophie Sanchez, Marcelo Sanchez-Villagra, Thomas Schindler, Hans-Peter Schultze, Michail Shishkin, Neil Shubin, Sergej Smirnov, Jean-Sebastien Steyer, Hans-Dieter Sues, Tomasz Sulej, Thomas Tütken, Frank Ullmann, Peggy Vincent, David Wake, Anne Warren, Jaco Weinstock, Ingmar and Ralf Werneburg, and most of all Florian Witzmann.

The work of Ivan Schmalhausen, David Wake, and Jenny Clack has been most inspiring for this book project and beyond. More than anyone else, there were five people who helped me become a paleontologist: Jürgen Boy taught me how to observe, Dieter Korn how to work efficiently, Erich Weber how to argue, Andrew Milner how to write, and Wolf Reif how to structure my thoughts. If this book has any value, I owe them a huge debt. The flaws remain mine.

Finally, I am most grateful to Mike Benton for suggesting that I write this book, Delia Sandford for her professional and friendly guidance while planning the text and illustrations, and Andrew Milner, Hans-Dieter Sues, and Florian Witzmann for much helpful advice on earlier drafts of the manuscript. Comments by two anonymous reviewers were also very thoughtful and constructive, and Hugh Brazier has been of enormous help in improving the language.

1 Introduction

The study of amphibians – both extinct and extant – makes a significant contribution to our understanding of how organisms develop and evolve. Like few other vertebrate groups, amphibians have been studied extensively from an early historic phase until today. Their modern exemplars have made an essential contribution to our understanding of phenomena such as morphogenesis, plasticity, larvae, metamorphosis, heterochrony, viviparity, feeding, ecology, speciation and microevolution, and – most recently and sadly – extinction. Their rich fossil record provides unique insights into ontogeny and paleoecology, phylogeny and macroevolution. Hence, the knowledge of amphibian evolution holds a pivotal position in the study of vertebrates.

Admittedly, amphibians are neither the most speciose, nor particularly spectacular vertebrates. They are often sluggish and slow, with a cold and moist skin covered with mucous and venom glands. Most of them are not very large, and many species are so tiny that they are easily overlooked. At the same time, amphibians are often the preferred objects for studies in development, ecology, and evolution. What, then, makes them such prominent study taxa? Why should their evolutionary history be of such wide general interest to biologists? There are historical reasons, influenced by their ready availability for study and the relatively easy breeding conditions of some laboratory taxa. However, amphibians are also special among vertebrates in many ways, not least in their capacity to survive and propagate in unstable environments, as well as in their ability to change from one habitat to a profoundly different one. Some amphibians have mastered the regeneration of organs in a way unthinkable in most other vertebrates, and they have repeatedly evolved live-bearing species, each time with different features. Some amphibians breathe with lungs, others with gills, and yet others

Amphibian Evolution: The Life of Early Land Vertebrates, First Edition. Rainer R. Schoch.

through their skin – and many amphibians employ a combination of all these respiratory mechanisms. Finally, amphibians are a group whose evolutionary history dates back as far as the Early Carboniferous, a time span encompassing 330 million years of change and stasis, diversification and extinction, and fascinating examples of evolutionary innovation. It is the purpose of the present book to trace this history, seeking to understand features of amphibian evolution in the frameworks of development and ecology, the two major foci of modern evolutionary biology. It is the interdisciplinary questions that are the most fascinating in this field, and therefore the second major theme of the book is the question of how we conduct studies on the fossil record, development, ecology, and evolution of amphibians and beyond.

What is an amphibian? The phylogenetic definition that I will use is straightforward: any member of the three modern groups salamanders (Caudata), frogs (Anura), and caecilians (Gymnophiona) is an amphibian (Figure 1.1). The correct systematic name for that group is **Lissamphibia**, and all lissamphibians share a common ancestor that lived sometime in the Late Paleozoic (~330–290 myr).

There is a large gap between lissamphibians and the manifold Paleozoic and Mesozoic taxa commonly referred to as "amphibians." Some of these must rank among the ancestors of lissamphibians, but authors still debate which taxa fall into the lissamphibian stem-group. To avoid confusion, it is reasonable to distinguish between the lissamphibian relatives (phylogenetically called "stem-amphibians") and all other taxa. The others are referred to here as "early tetrapods" when their relationships to Lissamphibia and Amniota are uncertain, and as "stem-amniotes" if their affinity with amniotes can be made plausible. Here, I follow the majority view on the origin of Lissamphibia, which holds that **temnospondyls**, members of a speciose clade encompassing almost 300 species, form the stem-group of lissamphibians (Bolt 1969; Milner 1993; Ruta and Coates 2007; Sigurdsen and Green 2011; Maddin *et al.* 2012).

Therefore, when speaking of Paleozoic and Mesozoic amphibians, I refer to temnospondyls, and thus I employ a scheme in which Lissamphibia forms a subgroup within a larger clade Amphibia. The alternative views will be discussed in depth in Chapter 9 (phylogeny). Whereas this book deals mainly with

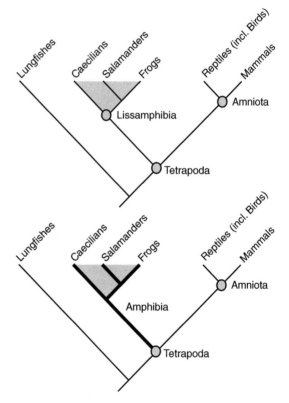

Figure 1.1 The relationships of extant tetrapods and their nearest relatives. Lissamphibians are probably a monophyletic group (clade), containing the limbless caecilians, salamanders, and frogs. Amphibia is a more inclusive name, here used to include all stem-group taxa, among which are many Paleozoic and Mesozoic forms ("early amphibians").

lissamphibians and amphibians, it also tackles many problems concerned with early tetrapods.

1.1 Changing paradigms in amphibian evolution

Amphibians bear a most appropriate name in several respects, and the scientist who coined the term was probably not aware of all of them. Literally meaning "living on both sides," the name points to the capacity to transform and adapt to divergent living conditions. In the narrow sense, the two sides are freshwater and land: the stereotyped amphibian life cycle includes the water-born newt or tadpole transforming into an adult land salamander or frog. Yet there are many other ways of amphibian existence, exemplified by the limbless caecilians, most of which live in the soil, the lungless and live-bearing salamanders, some of which ably climb trees, or the non-transforming axolotl, which is effectively a hypertrophied, sexually mature salamander larva. There are many more such cases, and on closer inspection one may even think there are as many different life cycles as there are species. These amazingly varied life histories differ far more than the slight variations in ontogeny known from other vertebrates. They often harbor built-in switches, responding to environmental inputs. Water conditions, temperature, food availability and properties, and oxygen form some of these factors, but there are many others, often confined to individual species or populations.

Amphibians are also peculiar because their fossil record is extraordinarily good. Although relatives of modern amphibians are often too small and delicate to be well preserved in most sediments, Paleozoic and early Mesozoic deposits yield a wealth of other, much larger amphibian fossils. These fossils tell us about a bizarre and alien world, playing in an exotic geographical setting and climate, and revealing highly unusual aspects of development and ecology. The abundance of early amphibians and their presence in numerous different deposits has made them preferred study objects for paleontologists ever since their first discovery in the 1820s. The most striking feature of these ancient forms is their huge size – ranging between 0.5 and 6 m. Compared with living amphibians, they had a very different morphology, many of them resembling modern crocodiles, while others reveal convergences to modern flatfishes, moray eels, giant salamanders, caecilians, and lizards.

In recent decades, discoveries of many new fossils have changed our view of early amphibians profoundly. Fossils are usually interpreted within the framework of phylogenetic hypotheses, spanned by well-known extant organisms. This procedure arrives at extant groups that give the best model for the understanding of the extinct group. In the case of amphibians and early tetrapods, the classic living model organisms were the modern salamanders, because of their apparently plesiomorphic appearance and the biphasic life cycle (larval–metamorphic). One might call this a central dogma in the study of tetrapod origins. Indeed, salamanders appeared to be perfect model organisms: their general body architecture, their "primitive" mode of locomotion on land, and the capacity of water-living larvae to transform into a terrestrial adult were seen as essential features of all early tetrapods. The central assumption was that the first tetrapods conquered land in the same way as many modern salamanders do it – namely, during metamorphosis.

Is the evolutionary conquest of land recapitulated in each baby salamander and frog? Formulations like that may be elegant, but have little to do with what really happened. There is no simple parallelism between ontogeny and phylogeny, let alone in such developmentally complex organisms as amphibians. The underlying processes are entirely different: stochastic selection on the evolutionary level, genetic and developmental mechanisms on the organism level. The whole issue of heterochrony, first triggered by such extraordinary cases as the axolotl, has become a multifaceted issue to analyze in recent years. New fossils, including those of Paleozoic baby amphibians, shed light on the life cycles of early amphibians (Boy 1974; Schoch 2009). These data amounted to the insight that metamorphosis was not shared by most of these early taxa, and that the salamander model is far from appropriate for the understanding of early tetrapods (Schoch 2002).

This model has also been challenged by many finds that indicate a more aquatic, fish-like habit of many early tetrapods (Coates and Clack 1990, 1991). These taxa (see Figure 1.2 for examples) retained lateral lines and gills as adults, and their skeletons were hardly capable of supporting longer excursions on land. The available evidence from fossil footprints confirms this, revealing that these animals were extremely slow when forced to cross dry land. They did not undergo a metamorphosis like modern amphibians. In many cases, adults are found in the same environments as their juveniles.

This touches the core of a second dogma on the fish–tetrapod transition, the ecological argument. The classic ecological scenario holds that tetrapods were attracted by food outside the water, that there must have been selection pressures driving their ancestors onto land. However, fossil evidence counters this idea by showing that early tetrapods and amphibians lived primarily in the water, retained many fish-like features and organs, and preyed on fish or other water-dwelling animals. New evidence from histology supports this conclusion, because many early tetrapods retained

Figure 1.2 Skulls of different Paleozoic taxa: (A) the stem-tetrapod *Acanthostega*; (B) the chroniosuchian *Chroniosaurus*; (C) the temnospondyl *Archegosaurus*; (D) the colosteid *Greererpeton*; (E) the dissorophoid *Cacops*.

calcified cartilage inside their long bones to make their bodies heavier, while others had lightly built bones, providing excellent swimming but very poor walking abilities. In all of these taxa, the internal structure of limbs was not adapted to meet torsional stress such as that caused by loco-motion on land (Sanchez *et al.* 2010). The old ideas of Alfred Sherwood Romer (1956, 1958), a pioneer in the study of early tetrapod evolution, are revived: then regarded as an oddity rather than mainstream opinion, his suggestion was that the origin of tetrapods took place under water, and that true land vertebrates appeared substantially later. Clearly, the salamander is not a reliable model for these long-extinct taxa. In turn, modern amphibians as a whole appear much more alien and interesting when these results are borne in mind. They form a separate, successive strategy to generate a land vertebrate, with many fascinating adaptations that were not features of early tetrapods, but evolved in the 330-million-year history of amphibian evolution after their split from the amniote ancestors. We are also more fully able now to trace some key aspects of this evolutionary pathway, although many problems are still unresolved.

The study of amphibian evolution – of extinct as well as extant taxa – reveals another very interesting aspect: ontogeny. In stark contrast to other groups of tetrapods, but similar to various fishes, amphibians are subject to profound ontogenetic change, reflecting a broad range of responses to environmental parameters. Although ancient taxa had very different ontogenies, they were sometimes as complex as modern ones. This reaches a stage at which it becomes necessary to consider the whole life cycle as a unit of taxonomy, phylogeny, ecology, and evolution. In paleontology, this concept has been put forward only recently. One outcome of these efforts is the present book, summarizing recent work and numerous still-unpublished observations. For paleontology, the life cycle concept means that single ontogenetic stages are not sufficient to trace evolutionary changes. Many problems in phylogenetic analyses result from the unsettled questions raised by ontogenies and developmental evolution. Fortunately, the preservation of different size classes in fossil amphibians provides insight into

this field, permitting detailed comparisons between extant and fossil ontogenies. The old and troubled concept of heterochrony comes into mind almost automatically here: neoteny, in its classic example of the axolotl as a sexually mature larva. Yet the new field of developmental evolution (evo-devo) is much more than the study of ontogeny and phylogeny. As pioneered by Ivan Ivanovich Schmalhausen and Conrad Hal Waddington, it focuses on the phenotype as an active player, responding to environmental changes, resisting perturbation from inside and outside, and being able to remain remarkably stable throughout evolution if required. However, the more obvious capacities of amphibian phenotypes are their flexibility and plasticity. This covers the important aspect of the *reaction norm*, a concept uniting development and ecology under the evolutionary umbrella.

The significance of fossil amphibians for the understanding of evolution is obviously manifold: their own evolutionary history is full of detailed stories, their relationship to modern amphibians is complex and reveals many perplexing convergences, their paleoecology has many unique features and provides insight into habitats, environments, and climates long ago, and the connection between evolution and development has been studied extensively in some Paleozoic and Mesozoic clades. This leads to the recognition of metamorphosis, a key feature of modern amphibians, as a life history strategy that evolved some 300 million years ago. Finally, the bearing of early tetrapod fossils on the fish–tetrapod transition is profound and has the potential to further shift the picture.

1.2 Paleobiology: data, methods, and time scales

Although there is *one* true history of early land vertebrates that needs to be found, only aspects of this story can be studied by any one approach at a time. Methods, time scales, and the data them-selves differ substantially between approaches. These are often complementary by nature – only when they are used in combination does a com-prehensive picture come within reach. Although efforts to make this picture clearer have met with tremendous success in the last few decades, there are inherent limitations and problems that will ensure that it remains forever incomplete. Understanding these problems is crucial for any successful contribution to this field.

Each of the research questions outlined in the Preface addresses complex and multifaceted problems. They require the integration of fossil data with those from embryology, genetics, physiology, developmental biology, and ecology. In concert, they form an inclusive research program of evolutionary biology, focused on early land vertebrates. The short list of questions leaves no doubt that different problems concerning the biology of early tetrapods require different research fields to be involved. But how this can be achieved is a far from trivial question, to be outlined as follows.

Despite their different problems and methods, scientists live in one world and want to grasp the whole story. To do that, interdisciplinary research is essential and inevitable. However, this often proves to be more difficult than it appears at first sight, especially when it concerns the integration of pattern- and process-focused disciplines. Paleontology and zoology are clearly centered on *patterns* – morphology, histology, embryology, and phylogeny dominate these fields. Description, statistics, and phylogenetic analysis are major approaches here, aimed at understanding the evolutionary history of the particular group. History, of course, is a sequence of unique events, it does not repeat itself in a predictable way, and has many causes. Consequently, zoology and paleontology are dominated by patterns that are historical, although it would be too simple to call them historical sciences.

On the other hand, genetics, developmental biology, ecology, and evolutionary biology study the *causes* of organismal structure and the reasons for its change. Genes and development are the domains where *mechanisms* of heredity act and the generation of organismal form takes place. These mechanisms are active within each and every organism, and they operate on microscopic scales of space and time. The actors in this play are cells, which gather in populations to coordinate movements, produce substances, and form tissues and hard parts. In the past two decades, genetics

and developmental biology have increasingly worked together to find unexpected levels of similarity between widely divergent taxa – referred to as deep homology. One facet of this very fruitful approach is that the new field of developmental genetics is able to bridge gaps between morphologically disjunct clades. It seems to hold one of the keys to *understand* major features of body plan evolution. The origin of tetrapod limbs from fish fins is one example where such novel approaches proved to be useful (Shubin *et al.* 1997, 2009). For instance, the tetrapod hand and foot have recently been found to be novel structures, without homologs among extant bony fishes (Clack 2009).

Conversely, ecology and evolutionary biology focus on markedly larger scales: the *processes* they study require much more time – from days to years in ecology, from years to thousands of millennia in evolution. The actors on this stage are not single individuals, but populations. Admittedly it is still not well understood how species are formed and what makes a population a species. After all, species are much more fuzzy and messy than atoms or molecules are in physics and chemistry. In sexually reproducing organisms, species boundaries are established (and maintained) by various mechanisms of reproductive isolation. In the long run, requiring at least 10^5–10^6 years, a given species transforms into a new one. This is the crucial gap between micro- and macroevolution. Rather than a principal difference, this gap is caused by the fact that our own time frame allows us to study the microscopic time scale of development, or the ecological time scale of predator–prey relationships, but not the evolutionary time scale at which species change.

How species form, by means of splitting (*cladogenetic*) or simple transformation within a lineage (*anagenetic*), is often unclear. Most probably, a broad range of modes exists, considering the enormous diversity of evolutionary rates and patterns known across the organismic world. Although paleontology cannot offer direct insight into processes, it reveals patterns of evolutionary transformation. However, it must be emphasized that it needs exceptional preservation, extraordinarily large samples, and a sequence of time slices that are not too distant in geological time, in order to permit evolutionary studies. Unfortunately, this reduces the number of possible cases, especially in vertebrate paleontology, to very few. Even then, it must be remembered that all we get is a sequence of snapshots of the evolutionary transformation of a given species, which cannot be compared to the data a developmental biologist or ecologist operates with. More than in other fields, evolutionary biology handles fragmentary data – and this is true not only in paleontology, which is so used to dealing with pieces of a puzzle.

In paleontology, a single exceptional deposit (*Lagerstätte*) often reveals more data on the ecology and microevolution of its fauna than dozens of other localities that yield only fragments. In the case of early amphibians, lake deposits rank first among such highly informative sites. When undisturbed by erosion, such lakes preserve hundreds to thousands of years of continued deposition, permitting the identification of changes on a small scale. Unfortunately, such lake deposits, even if preserved in close succession in the same area, are often separated by long time intervals undocumented or destroyed by erosion. When paleontologists put together data from the fossil record, they always have to consider how many sources of uncertainty remain.

To conclude, the study of evolutionary history – for instance, that of early land vertebrates – requires integration of data from various disciplines. This can only be achieved when (1) the nature and significance of data from each field are understood, (2) the strengths and limitations of the different methods are considered, and (3) the integration of results from different disciplines acknowledges the different levels (pattern versus process, time scales, levels of complexity).

1.3 Concepts and metaphors: how scientists "figure out" problems

"Words matter in science, because they often stand for concepts" (Wake 2009). Scientists need a theoretical platform on which to work and a framework of ideas and concepts into which they can fit their observations. In paleobiology this platform is evolution, a vast theoretical framework shared with other life sciences. While working on this platform, the developmental biologist, evolutionary biologist, ecologist, or paleontologist has to invent further concepts. These concepts build a framework within which problems are viewed and discussed. Such frameworks are essential for science, because they provide firm ground for hypotheses. The theory of evolution, with its constituent concepts of *natural selection* and *descent with modification*, provides the most general and stable pillars in the framework of modern life sciences.

An essential platform in evolutionary biology is the concept of homology (Hall 1994). First formulated by Richard Owen in 1840, it went through different phases of interpretation. First viewed as reflecting a divine body plan or *archetype*, it was then seen from the perspective of Darwin's theory of evolution. Shared features were now interpreted as based on *common ancestry*, whereas analogy was the outcome of independent evolution, highlighting the power of natural selection. The hands and feet of tetrapods go back to the last common ancestor of Tetrapoda, no matter how different they are in modern land vertebrates, or whether they have eventually disappeared, as in snakes or caecilians. More recently, the homology concept has been enriched by the addition of *homoplasy*, which embraces convergence, parallelism, and reversal. Originally, homology and homoplasy were viewed as a dichotomy. Today, the two are increasingly considered end points on a continuum (Hall 2007). After all, homology, reversal, and parallelism are just different evolutionary stages of common ancestry. A central theme of modern genetics and evolutionary biology is *deep homology*, or the observation that disparate organisms share fundamental genetic and regulatory similarities behind their divergent morphologies (Shubin *et al.* 2009). These new insights of developmental genetics, entirely unforeseen, have made an adjustment of the homology concept necessary. The historical transformation of this concept exemplifies the important point that scientific frameworks need to be sufficiently flexible to adjust to new ideas and changed paradigms.

The downside of scientific concepts is that they often employ metaphors – *descriptive images based on analogy*. Metaphors help researchers to figure out a complicated problem more clearly and in simple terms, but they may be easily over-stretched and overinterpreted. This is the point where the researcher has to perceive the difference between his metaphor and the process which it stands for – otherwise, the metaphor becomes the problem rather than the solution.

Like any science, paleobiology cannot work with-out metaphors, and knowing that one should always be aware of their existence and their limitations. It is appropriate to use the terms "homology," "selection," "genetic code," or "diversity" if we keep in mind that they represent much more complex phenomena than we are able to describe. In a complicated text, they may serve as handy abbreviations. Viewed in this sense, metaphors can be powerful tools, naming the unspeakable. They reduce a complex phenomenon of the biological world (which we often only know inadequately) to a situation resembling the human world. The crucial point is that we should never forget that – otherwise we might confuse description with reality.

1.4 Characters and phylogenies

Characters form the basis of any phylogenetic analysis, and thus play a crucial role in evolu-tionary biology. Cladistics treats characters as the "atoms" of phylogeny, but that requires an essential property: to become a useful character, a feature must be divisible into distinct character states. Here's why. A cladogram is a sequence of dichoto-mies or branching nodes. Each node is defined by at least one character that "supports" it. It forms the evidence that a given group has a common ancestor. Such evidence is provided only by exclu-sive (= derived) character states, the apomorphies.

What then makes a given feature a phylogenetic character? Although characters provide crucial evidence in the analysis of evolutionary history, they are still defined by researchers. It is quite common that newly published characters are disputed and their definition and coding subject to discussion and modification. In the long run, most proposed characters survive this test, albeit often with substantial reformulation and almost universally with recoding.

Reliable or "good" morphological characters are essential for phylogenetic analyses. But how can a character be recognized in an objective way? The reliability of morphological characters is difficult to assess because there are no objective, universally accepted criteria. The reality of characters itself is far from understood. Whereas it is undisputed that, for instance, a protein or cell really exists, there is no consensus on whether characters do. Organisms are modular, they fall into a nearly infinite number of units (Riedl 1978). Some units are obvious, but others can be very subtle and subject to scientific dispute (Wagner 2001). Some characters may be such modules, others are not. After all, characters are hypotheses of homology, not simple facts or undisputed building blocks of organisms.

Here are a few characters believed to be of some significance in early amphibian phylogeny (Figure 1.3):

- Presence of fingers and toes (yes/no).
- Number of fingers (8-7-6-5-4). This is a character that falls into more than two states.
- Shape of the occipital condyle. This character may be defined differently: either simple (one- or two-headed) or complex (contribution of basioccipital and surface area of facets). Depending on this, the character may have two states or be multistate.
- Length of ribs (short and straight/long and curved).

These four characters and their various states define major nodes in tetrapod phylogeny: (1) the limbed tetrapodomorphs, (2) the transition between limbed tetrapodomorphs and crown tetrapods, (3) the stem-group of modern amphibians, and (4) the stem-group of amniotes. These characters make most evolutionary considerations possible, thus forming the backbone of this book.

1.5 What's in a name?

There are two different ways to name monophyletic groups (clades), and despite much debate there is no consensus on which way should be preferred. Effectively, each author needs to make a decision which definition to use for a particular taxon

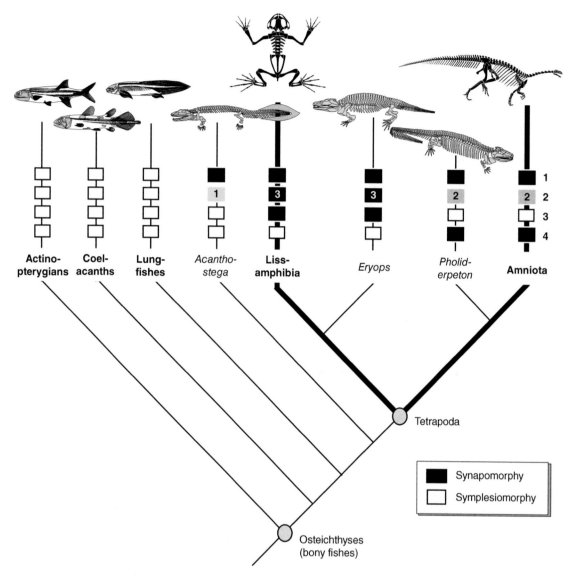

Figure 1.3 The importance of single morphological characters exemplified by early tetrapod phylogeny (see text). The *presence* of digits (1) is shared by some tetrapodomorphs. The *number* of digits varies from clade to clade: eight in *Acanthostega* (state 1) to five in stem-amniotes (state 2), and finally reduced to four in amphibians (state 3). The double occipital condyle (3) is a derived character of amphibians, whereas the long ribs characterizes amniotes and their stem (4).

name. It can only be hoped that in the long run authors will agree on a particular definition – but currently such agreement is not in sight. Without a clear statement by the author defining his/her use of taxa, much confusion can arise. The definitions of the names Amphibia and Lissamphibia have already been given. Here, I will briefly

explain the two alternative definitions as exemplified by the taxon Tetrapoda (land vertebrates), which includes Amphibia and Amniota.

The traditional way to define groups (predating cladistics) is to refer to key characters. It is called the *character-based concept*. Obviously, tetrapods have digits (fingers and toes) that their fish-like relatives lacked. This seems to be a perfect case, giving a clear-cut morphological definition that even corresponds to the meaning of the name Tetrapoda: four-footed animals (Greek: *tetra* = four; *pous*, *podos* = foot). In phylogenetic (cladistic) parlance, the presence of digits is a synapomorphy of all tetrapods, whereas "fishes" retain the plesiomorphic character state, the absence of fingers and toes. (In the case that digits evolved from radials, currently an alternative hypothesis, the distinction would be a functional one, highlighting the difference between radials in a fin and digits in a hand or foot.) Apart from the obvious advantage of referring a taxon to its most significant character, supporters of the character-based concept emphasize that it preserves the original meaning of taxon names better, upholding tradition and minimizing complicated nomenclatural changes.

The alternative way to define a taxon is *phylogenetic nomenclature*. This was introduced by Willi Hennig, the founder of phylogenetic systematics, who also first defined Tetrapoda in this new way. Here, taxa are defined entirely by the structure of the cladogram, and remain independent of particular characters (Figure 1.4). This is not such a bad idea, because our perception of characters often changes with new evidence, and sometimes characters are even abandoned when it is shown that they are ill-defined in principle. Without using characters, Tetrapoda can be defined as the group encompassing exclusively extant amphibians and amniotes. These two largest extant clades of land vertebrates form the two branches of modern tetrapods. All phylogenetic analyses, both morphological and molecular, agree on this. In this definition, fossil taxa fall either within this comb (in which case they are true tetrapods) or on the stem lineage (in which case they are stem-tetrapods).

Currently, the name Tetrapoda is used with divergent meanings by different authors. For instance, Ahlberg and Clack (1998), Anderson (2001), and Clack (2012) preferred the character-based definition. They speak of *Acanthostega* as a "basal tetrapod" because it has hand and foot skeletons, whereas *Tiktaalik* is considered a "fish-like sarcopterygian" because it lacks them. On the other hand, Laurin (1998, 2004) applied the

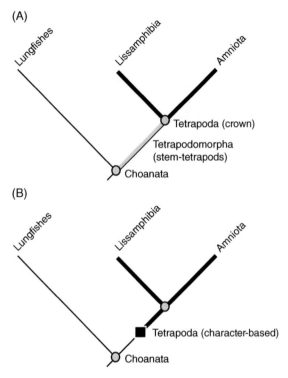

Figure 1.4 Two different ways to name a clade: (A) node-based versus (B) character-based.

phylogenetic nomenclature. This demands ranking both *Acanthostega* and *Tiktaalik* as stem-tetrapods (tetrapodomorphs). To acknowledge the presence of hand and foot skeletons in *Acanthostega*, Laurin (1998) has suggested naming all tetrapodomorphs with these features "stegocephalians." So far, this name has not been adopted by other authors because Laurin proposed a radically different phylogeny of lissamphibians which leaves numerous taxa traditionally regarded as crown tetrapods outside the Tetrapoda.

Throughout this book, I shall use phylogenetic definitions rather than those based on characters. My reasons for doing so are twofold: (1) my own experience has made me wary of character definitions, after even features long regarded as robust characters turned out (based on new evidence) to be poorly defined or, worse, impossible to define objectively; and (2) I agree with Hennig that there is a key difference between crown groups and other taxa in that extant species permit countless more traits to be studied than fossils. The constituent taxa of crown groups should therefore be much better known in the long run than fossil taxa will ever be. This is why crown groups – as one example of node-based phylogenetic definition – may serve as anchors for cladograms. The crown group Tetrapoda is a good example, as the monophyly of amniotes and lissamphibians is more robust than all taxa defined on the basis of extinct taxa. For those interested in the details of this debate, I recommend Laurin and Anderson's (2004) exchange of arguments for and against phylogenetic nomenclature.

References

Ahlberg, P.E. & Clack, J.A. (1998) Lower jaws, lower tetrapods ± a review based on the Devonian genus Acanthostega. *Transactions of the Royal Society of Edinburgh: Earth Sciences* **89**, 11–46.

Anderson, J.S. (2001) The phylogenetic trunk: maximal inclusion of taxa with missing data in an analysis of the Lepospondyli (Vertebrata, Tetrapoda). *Systematic Biology* **50**, 170–193.

Bolt, J.R. (1969) Lissamphibian origins: possible protolissamphibian from the Lower Permian of Oklahoma. *Science* **166**, 888–891.

Boy, J.A. (1974) Die Larven der rhachitomen Amphibien (Amphibia: Temnospondyli, Karbon-Trias). *Paläontologische Zeitschrift* **48**, 236–268.

Clack, J.A. (2009) The fin to limb transition: new data, intepretations, and hypotheses from paleontology and developmental biology. *Annual Reviews of Earth and Planetary Sciences* **37**, 163–179.

Clack, J.A. (2012) *Gaining Ground: the Origin and Evolution of Tetrapods*, 2nd edition. Bloomington: Indiana University Press.

Coates, M.I. & Clack, J.A. (1990) Polydactyly in the earliest known tetrapod limbs. *Nature* **347**, 66–69.

Coates, M.I. & Clack, J.A. (1991) Fish-like gills and breathing in the earliest known tetrapod. *Nature* **352**, 234–236.

Hall, B.K. (1994) *Homology: the Hierarchical Basis of Comparative Biology*. New York: Academic Press.

Hall, B.K. (2007) Homoplasy and homology: dichotomy or continuum? *Journal of Human Evolution* **52**, 473–479.

Laurin, M. (1998) The importance of global parsimony and historical bias in understanding tetrapod evolution. Part I. Systematics, middle ear evolution, and jaw suspension. *Annales des Sciences naturelles* **19**, 1–42.

Laurin, M. (2004) The evolution of body size, Cope's rule and the origin of amniotes. *Systematic Biology* **53**, 594–622.

Laurin, M. & Anderson, J.S. (2004) Meaning of the name Tetrapoda in the scientific literature: an exchange. *Systematic Biology* **53**, 68–80.

Maddin, H., Jenkins, F.A., & Anderson, J.S. (2012) The braincase of *Eocaecilia micropodia* (Lissamphibia, Gymnophiona) and the origin of caecilians. *PloS ONE* **7**, e50743.

Milner, A.R. (1993) The Paleozoic relatives of lissamphibians. In: D. Cannatella & D. Hillis (eds.), Amphibian relationships: phylogenetic analysis of morphology and molecules. *Herpetological Monographs* **7**, 8–27.

Riedl, R. (1978) *Order in Living Organisms*. New York: Wiley.

Romer, A.S. (1956) The early evolution of land vertebrates. *Proceedings of the American Philosophical Society* **100**, 157–167.

Romer, A.S. (1958) Tetrapod limbs and early tetrapod life. *Evolution* **12**, 365–369.

Ruta, M. & Coates, M.I. (2007) Dates, nodes and character conflict: addressing the lissamphibian origin problem. *Journal of Systematic Palaeontology* **5**, 69–122.

Sanchez, S., Germain, D., de Ricqlès, A., et al. (2010) Limb-bone histology of temnospondyls: implications for understanding the diversification of palaeoecologies and patterns of locomotion of Permo-Triassic tetrapods. *Journal of Evolutionary Biology* **23**, 2076– 2090.

Schoch, R.R. (2002) The evolution of metamorphosis in temnospondyls. *Lethaia* **35**, 309–327.

Schoch, R.R. (2009) The evolution of life cycles in early amphibians. *Annual Review of Earth and Planetary Sciences* **37**, 135–162.

Shubin, N.H., Tabin, C., & Carroll, S. (1997) Fossils, genes and the evolution of limbs. *Nature* **388**, 639–648.

Shubin, N.H., Tabin, C., & Carroll, S. (2009) Deep homology and the origins of evolutionary novelty. *Nature* **457**, 818–823.

Sigurdsen, T. & Green, D.M. (2011) The origin of modern amphibians: a re-evaluation. *Zoological Journal of the Linnean Society* **162**, 457–469.

Wagner, G.P. (2001) *The Character Concept in Evolutionary Biology*. New York: Academic Press.

Wake, D.B. (2009) What salamanders have taught us about evolution. *Annual Review of Ecology and Systematics* **40**, 333–352.

2 The Amphibian World: Now and Then

The amphibian world encompasses numerous groups of animals that evolved during the past 330 myr. Although most of them are long extinct, they played important roles in ancient ecosystems. The story begins with the first four-legged vertebrates (tetrapods), which were remarkably fish-like in many features – and only some of them fall within the ancestral lineage of modern amphibians. Lissamphibians and amniotes form the end points in an exciting sequence of early tetrapod evolution. Only the fossil record can shed light on how the extant groups formed and what the diversity of tetrapods was like in the Paleozoic, Mesozoic, and Cenozoic eras. The last few decades have produced many new and unexpected finds of these animals, and these discoveries have changed the big picture of early tetrapod evolution profoundly. Many of these taxa were radically different from all modern vertebrates, and there is no single extant model organism that can serve as a safe guide in understanding these animals. How was the fish skeleton modified to become that of a tetrapod? How many different tetrapod groups existed at a given time? How can they be identified, and what do we know about their evolution? Studying early tetrapods brings us face to face with fascinating and alien creatures whose reconstruction, life habits, development, and evolution pose major problems for paleobiology.

Amphibian Evolution: The Life of Early Land Vertebrates, First Edition. Rainer R. Schoch.
© 2014 Rainer R. Schoch. Published 2014 by John Wiley & Sons, Ltd.

2.1 Tetrapoda

The land vertebrates form the starting point of the present book, which in many respects deals as much with early tetrapods as it does with amphibians. The origin of tetrapods matters here because the understanding of amphibian evolution requires a deep knowledge of early tetrapod characters themselves. It is a major argument of this book that the traditional idea of modern amphibians as a guide to understanding extinct amphibians needs revision. This notion has emerged primarily from the study of fossil taxa themselves, but has been complemented by insights into the functional morphology, physiology, and developmental biology of lissamphibians. On closer inspection, early tetrapods appear stunningly different from both extant amphibians and amniotes. It is therefore important to approach the topic by setting a framework within which all further thoughts and discussions may be placed. The crown-group concept first advocated by Hennig (1966) is such a frame, and it will serve this purpose throughout the book. In the following sections, major features of the tetrapod skeleton will be described, followed by a discussion of the most important tetrapod characters.

2.1.1 The tetrapod skeleton

Tetrapod skeletons have evolved hundreds of very diverse forms, but they all share a common underlying architecture. This may be considered a coherent tetrapod *body plan* (a structuralist view) or it may be viewed as an *assemblage of characters* (a phylogenetic view). Either way, the hard parts of tetrapods are numerous and often highly complicated, but they all go back to a common ancestor. In turn, this ancestral stem-tetrapod inherited its bodily structure from bony fishes, that is, from aquatic vertebrates. Consequently, the tetrapod skeleton can be understood as a modification of the fish skeleton.

Skull structure. The tetrapod skull falls into three different units that can be defined under three entirely different aspects: embryology, phylogeny, and function. These include (1) the dermal skull ("outer skull"), (2) the endocranium ("inner skull"), and (3) the gill arches (visceral skeleton) (Figure 2.1). The different units are formed by cartilage or bone and serve many purposes: feeding,

---

BOX 2.1: VERTEBRATE PHYLOGENY AND RELATIONSHIPS

Gnathostomata: The jawed vertebrates include all fishes with jaws supported by a skeletal apparatus. They originated in the Early Silurian (~440 myr). Characters: (1) head skeleton composed of braincase, dermal skull, and gill arches plus jaws, (2) paired fins, (3) three unpaired fins (two dorsal, one anal), (4) teeth and bony scales. The gnathostomes include two large extant groups: the cartilaginous fishes (sharks, rays, and chimaeras) and bony fishes.

Osteichthyes: The bony vertebrates. The oldest osteichthyan fossils are from the Late Silurian (~428 myr). Characters: (1) lungs or swim bladder, (2) lepidotrichiae (bony fin-rays), (3) numerous new dermal bones in the skull and pectoral girdle. The Osteichthyes include two large branches, the ray-finned fishes (~30 000 extant species) and lobe-finned fishes and tetrapods (~24 000 extant species).

Actinopterygii: The ray-finned fishes, comprising more than 95% of living fishes. They are known from the Late Silurian onwards. Characters: (1) ganoid scales (containing the enamel-like substance ganoin), (2) crowns of teeth formed by acrodin, a transparent material, and (3) only one dorsal fin (which may be split to form two in some taxa). A plesiomorphic feature is the thin cross-section of the fins, in contrast to the lobe-finned fishes.

Sarcopterygii: The lobe-finned fishes have very few surviving aquatic taxa (only *Latimeria* and three genera of lungfishes), but they also include all living land vertebrates. They have been in existence since the latest Silurian (~420 myr). Characters: (1) strong paired fins or limbs with a single long axis, (2) teeth entirely covered by enamel, and (3) scleral eye ring with more than four plates.

Tetrapoda: The extant four-legged land vertebrates. They first appeared in the Early Carboniferous (~335 myr) and fall into two major clades: Lissamphibia (caecilians, salamanders, and frogs) and Amniota (mammals and reptiles, which include birds).

Figure 2.1 Essential units of the skull, exemplified by *Eusthenopteron*. Dermal bones in light grey, endoskeletal units in darker shades. All dermal bones marked in black were lost in tetrapods (opercular or gill-covering elements). ac, anocleithrum; bs, branchiostegal bones; es, extrascapular; op, operculum; pop, preoperculum; pot, posttemporal; sc, scapula; sop, suboperculum.

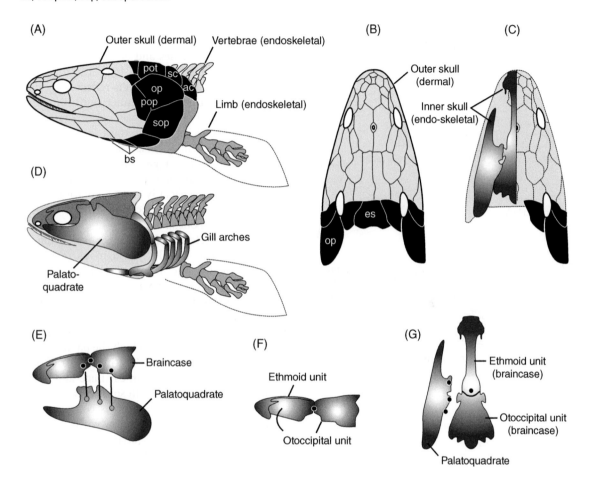

breathing, housing and protecting the brain and organs of sense, and the attachment of musculature, to name just the most obvious.

Although highly complex, the three units are found in all tetrapods, and indeed occur throughout vertebrates. Originally, each of these units was composed of numerous elements, but evolution has reduced the number and sometimes the complexity of elements in several major lineages. Comparing early tetrapods with modern amphibians reveals how far this reduction has gone: most salamanders retain just half of the skeletal elements possessed by the first tetrapods.

Structurally, the inner skull forms a cylindrical cover of the brain, while the outer skull in turn contains the inner skull – the two cranial skeletons are separated by thick sheets of musculature attaching at the jaws and eyeballs (Figure 2.2). The third unit is the gill arches, which form a basket primitively composed of five half-rings that contain the gills and permit the water to flow from the mouth through the gills; the gill openings are located between these half-rings. This basket is composed of rod-like elements formed of cartilage in the embryo, which may be replaced by bone in later life. Like the gill arches, the inner

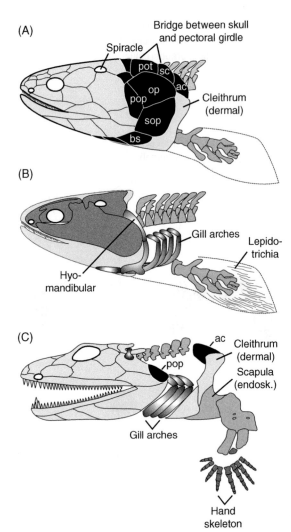

(A)

Bridge between skull and pectoral girdle

Spiracle

pot sc

op ac

pop

Cleithrum (dermal)

sop

bs

(B)

Gill arches

Lepido-trichia

Hyo-mandibular

(C)

ac

Cleithrum (dermal)

pop

Scapula (endosk.)

Gill arches

Hand skeleton

Figure 2.2 During the fish–tetrapod transition, the skull and forelimb underwent substantial modification: reduction of gill chamber, consolidation of skull, separation of pectoral girdle and forelimb, and the appearance of digits. (A, B) *Eusthenopteron*; (C) *Acanthostega*. Adapted from Jarvik (1980) and Clack (2002a). Abbreviations as in Figure 2.1.

lating the brain, whereas the endoskeletal jaws (palatoquadrate and mandible) form the paired upper and lower jaw halves, respectively. In bony fishes, the inner skull is moveable (kinetic) in itself: apart from the joint between upper and lower jaw, the upper jaw can also be moved against the braincase. This was already a functional property of early bony fishes and is retained in some extant bony fishes. Although long lost in most extant tetrapods, this kinetism is reflected by the embryonic patterning of the inner skull. In early tetrapods, the braincase was only partially replaced by bone, and only these portions are usually preserved in fossils. The most common bony portion is the region between the eyes (sphenethmoid), but the ear capsules may also be bony (otics), and especially the articulation with the first vertebra, originally composed of four elements (occipital bones).

The inner skeleton of the upper jaw (palatoquadrate) remains mostly cartilaginous in tetrapods, except for the jaw articulation, which ossifies as quadrate (upper jaw) and articular (lower jaw). Only in ancient lobe-finned fishes and early tetrapods, a second part of the palatoquadrate was bony: the epipterygoid. This element formed one of the joints by which the palatoquadrate hinged at the braincase. The inner part of the lower jaw is called Meckelian cartilage and ossifies only rarely and partially in some taxa.

Dermal skull. The outer skull is composed of numerous plate-like bones that grow within a more superficial layer of the skin (dermis). Referring to this developmental origin, they are called *dermal bones*. Figure 2.3 exemplifies the diversity of tetrapod skulls. Dermal bones are relatively thin but often form a complete shield, leaving only the openings for eyes (orbits) and nose (nares) uncovered. The dermal bones are often the only skeletal parts visible from outside, and they also bear the teeth, which also belong to the outer skeleton (Figure 2.4). The epidermis, or external layer of the skin, is never involved in bone formation and always covers the dermal bones. In bony fishes and their early tetrapod descendants, the dermal skull is composed of at least 43 elements, most of which occur in pairs. In modern salamanders, the number has been reduced to 21–23, in frogs to 19, and in gymnophionans to as few as 17.

skull originates as a cartilaginous structure in the embryo, but may be partially replaced by bone during later stages of ontogeny. In contrast, the dermal skull is bony from the start, it usually forms rather late, and cartilage is never involved.

Braincase and jaws. In the inner skull, the braincase forms an unpaired central unit encapsu-

Figure 2.3 Tetrapods then and now: (A) stem-amphibian *Sclerocephalus*; (B) stem-amniote *Seymouria*; (C) Jurassic salamander *Karaurus*; (D) extant giant salamander *Andrias*. B by courtesy of Thomas Martens, C of Ralf Werneburg.

Gill arches. The visceral skeleton is one of the most ancient structures of the vertebrate body plan. At closer look, the upper and lower jaws are consistent in many aspects with the gill arches and are referred to as the *mandibular arch*. Indeed, although much larger and more robust, the palato-quadrate and mandible are structurally similar to the gill arches, and embryologically form in a

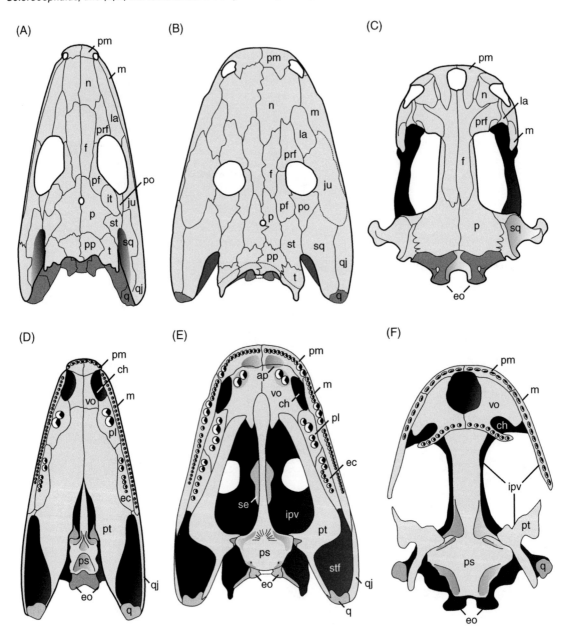

Figure 2.4 Tetrapod skull anatomy, exemplified by (A, D) the stem-amniote *Proterogyrinus*, (B, E) the stem-amphibian *Sclerocephalus*, and (C, F) the salamander *Dicamptodon*. A, D adapted from Holmes (1984).

similar way. Functionally, the movement of the jaws is consistent with that of the gill arches, which can be expanded and contracted. In addition, the gill arches also bear dermal elements with teeth, effectively forming a "pharyngeal jaw" that handles prey that has already been swallowed – a common feature in bony fishes. Otherwise the gill basket primarily manipulates the water current for the breathing cycle, a function in which the jaws are not involved. It was long believed that the enlargement and specialization of the mandibular arch is secondary to permit the formation of jaws for grasping and manipulating prey. Recent observations cast doubt on this scenario, suggesting that whereas jaws and gill arches are serial homologs, they need not have had a common functional origin in early vertebrates. Additional evidence is provided by the hyoid arch, which lies between the mandibular and gill arches proper and whose elements are not strictly homologous to those of the gill arches (Janvier 1996). In most fishes, the hyoid arch suspends the jaws rather than forms part of the gill basket, and it is associated with a gill cleft that extends in a different direction than the clefts of the gill arches: it is aligned dorsally rather than posterolaterally, ending in a slit-like opening in the skull, the spiracle. In conclusion, whereas the jaws are often grouped with the inner skull because of their tight articulation with the braincase, they are derived from the same embryonic source as the hyoid and gill arches, which is why they also considered part of the visceral skeleton. This highlights how recruitment of pre-existing elements for new functions has made skeletal parts more complex and difficult to group.

The gill region in bony fishes is covered by a series of dermal elements, the opercular bones. These articulate with the posterior margin of the skull by hinge joints, opening posteriorly to permit water flow out of the gill slits. The opercular bones are encircled by a rigid framework of dermal bones: the cheek, the pectoral girdle, and a series of connecting elements (extrascapulars and posttemporal) between the former two. In extant tetrapods, the connecting elements are absent, the opercular bones are absent, and the skull is completely free from the pectoral girdle.

Girdles. A common feature of all jawed vertebrates is the presence of two sets of paired appendages: the pectoral and pelvic fins. In fishes, the pectoral fin is firmly connected with the skull by means of the bony gill cover (opercular bones) and the pectoral girdle. In all extant tetrapods, the opercular bones are absent and the shoulder girdle and forelimb are separated from the skull. The pectoral girdle consist of both dermal and endoskeletal elements. The paired cleithrum and clavicle are of dermal origin, complemented by an unpaired interclavicle; these are all plesiomorphic features of bony fishes. Whereas in bony fishes the cleithrum is extensive, it was substantially smaller in early tetrapods and is lost in all extant taxa. The clavicles and interclavicle were large in many aquatic forms from the Paleozoic and Mesozoic, but are small or reduced in many modern tetrapods. In contrast to bony fishes, the endoskeletal elements are greatly enlarged and differentiated in tetrapods: these include the scapula and coracoid, which form the articular facet for the forelimb and may ossify as a single unit. The pelvic girdle of tetrapods is more extensive than in bony fishes and is three-rayed: a dorsal ilium connected to the vertebral column by means of enlarged sacral ribs, and blade-like ventral elements (pubis and ischium), which serve as attachments for limb and tail musculature.

Limbs. Throughout jawed vertebrates, the limbs arise from condensations of mesenchymous tissue. In lobe-finned bony fishes (sarcopterygians), the inner limb skeleton is segmental, forming in the embryo by successive splitting (bifurcation) of primordia (Clack 2009). Because of this common developmental process, fore- and hindlimb are generally of similar structure: the first element (humerus in the arm, femur in the leg) is long and single, followed by two elements (radius + ulna in the arm, tibia + fibula in the leg) (Figure 2.5). So far, these elements are present in all sarcopterygians. Primitively, bony fishes have numerous rod-like elements called radials that support the fins. In tetrapods, radials are absent, but there are digits – segmented and flexible outgrowths. Digits are not homologous to radials because their embryonic origin is different: radials develop from the anterior margin of the limb axis, digits from the posterior one (Clack 2009). Tetrapods primitively have five fingers in the hand (reduced to four or fewer in lissamphibians) and five toes in the foot. Further reduction of digits in tetrapods is common and occurred repeatedly, up to the complete loss of limbs (e.g., caecilians, snakes, and amphisbaenians). Apart from the radials, two additional elements of the fish fins are absent in tetrapods: the keratinous ceratotrichia and the bony lepidotrichia.

Figure 2.5 Tetrapod appendages and limbs share many features not found in other vertebrates, exemplified by (A) *Acanthostega*, (D) *Ichthyostega*, (B, E) *Sclerocephalus*, and (C, F) *Salamandra*.

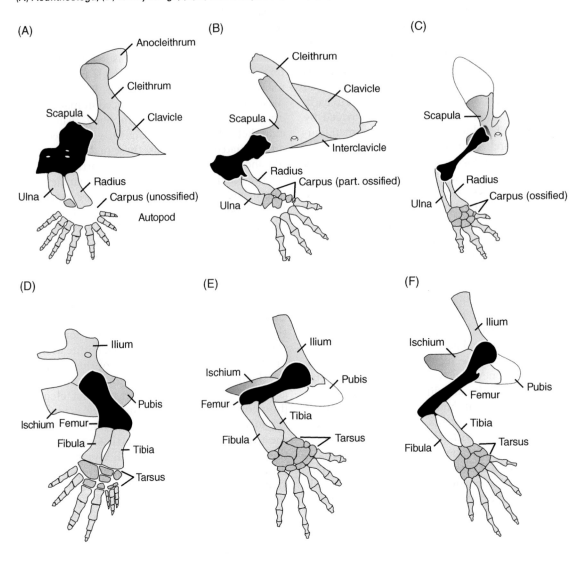

Vertebrae. In the vertebrate embryo, the main body axis is defined by the notochord, a liquid-filled rod that permits flexibility to move and stability to maintain the cylindrical body form at the same time. This is an essential requirement for fishes to swim, keeping the body length stable while the fins and trunk muscles are at work. The vertebral column develops around the notochord during later embryonic stages, while the notochord successively shrinks and disappears in many adult vertebrates. The vertebrae are part of the inner skeleton, formed first by cartilaginous elements that usually are replaced by bone later. The vertebral column encloses several vital organs that are aligned along the main body axis: the spinal cord, the embryonic notochord, and the dorsal ligament; in the tail, the vertebrae also enclose the aorta.

The adult vertebra of bony fishes is composed of a short disc (centrum) and a neural arch on top

Figure 2.6 Traditionally regarded as of high significance, the structure of vertebrae has received less attention recently, after numerous convergences have become known. The parallel reduction of the intercentrum is especially apparent (lissamphibians and amniotes). (A) *Ichthyostega*; (B) *Sclerocephalus*; (C) vertebral evolution mapped onto cladogram.

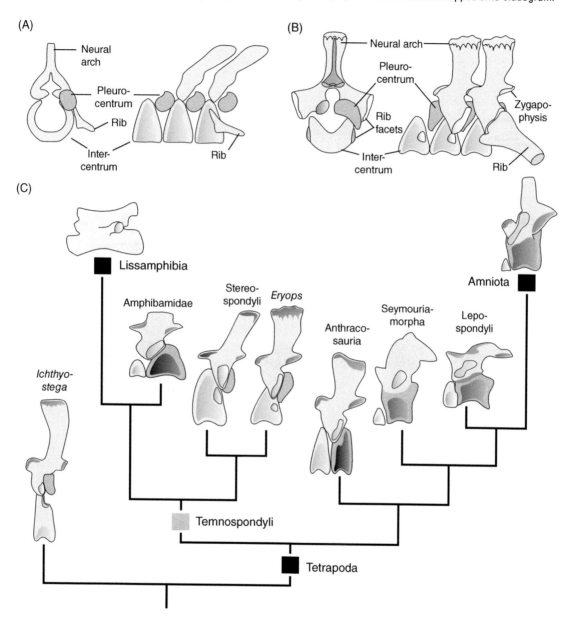

of it, which has an inverted Y shape. During ontogeny (bony fishes), the centrum develops from four components, which form two pairs of elements: two intercentra (ventral) and two pleurocentra (dorsal) (Figure 2.6). (Other names have been proposed for the cartilaginous precursors of these elements, but that is a different topic.) At any rate, the pleurocentra and intercentra often fuse in the midline to form half-rings, and in some bony fishes (*Amia*) and Paleozoic tetrapods

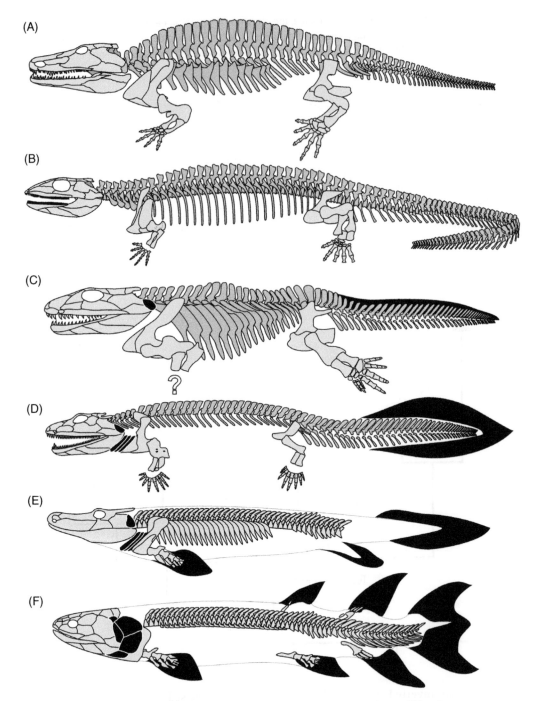

Figure 2.7 The changing tetrapodomorph skeleton as a whole: (A) temnospondyl *Eryops*, (B) anthracosaur *Proterogyrinus*, stem-tetrapods (C) *Ichthyostega* and (D) *Acanthostega*, and tetrapodomorph fishes (E) *Tiktaalik* and (F) *Eusthenopteron*. Fish-like features are marked in black.

(anthracosaurids) they form two complete discs per segment. In extant tetrapods, the pleurocentrum is the only remaining central element and forms an elongate cylinder, while the intercentrum has disappeared.

Ribs. There are several different elements referred to as "ribs" in bony fishes, but tetrapods retain only one type (Janvier 1996). The tetrapod ribs are part of the endoskeleton and develop within the horizontal septum, a sheet that divides muscle portions of the trunk. Ribs were short in the fish-like ancestors of tetrapods but elongated and strengthened in land vertebrates, where they originally had two heads articulating with both the vertebral centrum and the neural arch. In amniotes, the ribs are substantially longer than in lissamphibians, markedly curved, and ventrally attach to an unpaired cartilaginous or ossified element (sternum). Such a sternum is present in anurans and some salamanders, but there the ribs are short and straight rods. The ancient ribs of early tetrapods were of variable length, but usually their continuation by cartilage and attachment to a sternum remains unknown, because such elements are not preserved.

Bony scales. Gnathostomes are covered entirely by solid bony scales, which belong to the outer (dermal) skeleton. These are homologs of the scales present in sharks and other, more basal vertebrates (Janvier 1996). They are serially arranged, often overlap one another, and are originally composed of several layers of bone, dentine, and enamel. Histologically they are similar to vertebrate teeth, and like these are also formed in pockets; teeth and bony scales are therefore homologous structures (Janvier 1996). In most Paleozoic tetrapods, bony scales covered the belly, flanks, and back as well as the limbs and tail. They are simplified homologs of fish scales that have lost the enamel and dentine layers (Castanet *et al.* 2003).

Osteoderms. Like the bony scales, osteoderms develop in the dermis layer of the skin (Greek *osteon* = bone). They differ from scales in their embryology and histology, and are often ornamented in the same way as the skull bones. In contrast to bony scales, they do not form in pockets but like dermal bones simply ossify within the skin. Osteoderms may be isolated or arranged in rows, and sometimes form a carapace-like shield.

Their adaptational background is often unknown, and it is certainly too simple to view them as "armor." Apart from protection, they may provide attachment surface for muscles, such as for stabilization of the vertebral column during locomotion (Dilkes and Brown 2007). Other functions may include protection against desiccation or against skin abrasion in burrowing species.

Tetrapods and extant amphibians. A wide morphological gap separates early tetrapods and lissamphibians. Most notably, lissamphibians have open skulls with large windows in the cheek accommodating jaw-closing musculature (Figure 2.4). Associated with this is the absence of numerous cranial elements (jugal, postorbital, postfrontal, supratemporal, tabular, postparietal, ectopterygoid, supraoccipital, basisphenoid, basioccipital, and epipterygoid). In the postcranium, the cleithrum and interclavicle are always absent, and the coracoid and pubis are often not ossified (Figure 2.7). Bony scales in the dermis – typical features of Paleozoic amphibians – are absent in salamanders and frogs, but retained in caecilians (Zylberberg and Wake 1990). The complex vertebrae of early tetrapods also differ from the single-boned, cylindrical vertebrae of modern amphibians.

2.1.2 Tetrapod characters

Living tetrapods fall into widely divergent groups, each of which has gone through a long history of evolutionary changes. Despite numerous modifications there remains a wide range of tetrapod autapomorphies. This reflects how important and fundamental the transition to a terrestrial lifestyle was. The shared derived characters of tetrapods range from entirely novel structures – such as the hand and foot skeletons – to incremental but often functionally significant morphological and histological changes. Most common are functional complexes that were basically retained during the fish–tetrapod transition, but changed the functional context in which they were embedded, as exemplified by the tetrapod ear. The middle ear cavity of tetrapods was once a spiracular canal, and the middle ear ossicle (stapes) used to serve as a tightly integrated anchor for muscles of the jaws, gill basket, and shoulder girdle. The successive freeing of the stapes from its constituent functions eventually made this shift possible; the complex functional context will be discussed in Chapter 5, section 5.2. Once called "preadaptations," a term later dismissed for its teleological (goal-directed) implications, evolutionary changes involving integration of characters are now referred to as "exaptations" (Gould and Vrba 1982). It is clear that the origin of tetrapods must have involved numerous exaptations.

1. **Origin of hand and foot skeleton**. It is common knowledge that the paired fins of bony fishes – pectoral and pelvic – were remodeled into limbs. During the transformation of these fins into limbs, which required several dozens of millions of years, the shape of the constituent bones and their joints changed profoundly. In addition, the skeletal support of the distal ends of the fins was reduced – the endoskeletal radials disappeared and the dermal lepidotrichia were reduced. Likewise, the collagenous ceratotrichia are also absent in extant tetrapods. It is generally thought that these skeletal elements of fins had to be reduced before digits could evolve. Indeed, the last skeletal components of the tetrapod limb to appear in the fossil record are the carpals, tarsals, fingers, and toes. Embryology has revealed that these novel elements form in a different way and from a different primordial region of the developing limb. Hence, they cannot be homologs of radials and are considered *neomorphs* (new structures).

2. **Loss of unpaired fins**. The prominent fins on the back (two in most sarcopterygians) and in the anal region are absent in all extant tetrapods, as are the dorsal and ventral lobes of the tail fin. The only exception may be the tail fins of lissamphibian larvae, but their homology with the fins of adult fishes is not clear.

3. **Choanae**. In tetrapods, the choanae form a novel connection between the external nostril and the buccal cavity. They permit the flow of air from the nose to the lungs. In fishes, the nostril serves only the olfactory sense, and when air is taken outside the water it is swallowed through the mouth. Primitively, gnathostomes have two subdivided narial openings, an incurrent and excurrent nostril, permitting the water to flow through the nasal sack. In osteichthyans, these are completely separate. In the primarily aquatic fishes, the two nostrils are both located on the skull roof close to the jaw margin. In contrast, the tetrapod choana lies in the palate. It is generally agreed that the choana evolved from the posterior (excurrent) narial opening of osteichthyans by shifting the opening from the jaw margin into the palate (Janvier 1996). In lungfishes, the posterior nostril has also shifted into the palate, but fossils show that this condition arose independently from that in tetrapods.

4. **Endoskeletal part of pectoral girdle strengthened**. The strong forelimb musculature of tetrapods inserts to a larger extent along the endoskeletal part of the shoulder girdle than it does in fishes. This correlates with a larger and more differentiated scapulocoracoid element, which has partially replaced dermal bones of bony fishes (anocleithrum, cleithrum).

5. **Pelvis and sacrum**. The pelvic girdle is composed of three elements (ilium, pubis, and ischium), which are ventrally connected by a medial fusion and further articulate with the vertebral column by means of specialized sacral ribs. The ilium is always bony, whereas in the pubis and ischium larger portions of cartilage may persist. Ilium, pubis, and ischium together form the articular facet for the hindlimb (acetabulum, cleithrum).

6. **Vertebrae**. Tetrapod vertebrae are primitively composed of two complementary wedge-shaped centra per segment (pleurocentrum and intercentrum) and a neural arch. In both amniotes and lissamphibians, only a single cylindrical centrum exists. Whereas in amniotes this is certainly the pleurocentrum, in lissamphibians its homology is still debated. At any rate, the cylindrical centrum evolved convergently in lissamphibians and amniotes and thus is no autapomorphy of tetrapods *per se*. Throughout tetrapods, successive neural arches contact each other by means of specialized facets (zygapophyses).

7. **Ribs strengthened**. The ribs of bony fishes are thin rods, in contrast to tetrapod ribs, which are of variable length but have two well-defined heads that articulate with the vertebral centrum and neural arch.

8. **Eyes with lids**. The eyes are protected by skin folds, which house glands that keep the eyes moist. Superfluous secretions are taken away by the nasolacrimal duct, which is also a novel structure. This canal connects the periphery of the eye with the nasal sac.

9. **Internal gills and opercular bones**. Throughout bony fishes, the opercular bones open to release water that has passed the gills. In tetrapods, internal gills are entirely lost and the opercular elements are all absent. Paleozoic tetrapods preserve a range of stages that suggest how this apparatus was reduced; some of these taxa still retained internal gills (Schoch and Witzmann 2011).

10. **Middle ear**. The gill region has not entirely disappeared in tetrapods. Rather, the first gill slit (spiracle) was modified into an air-filled passage (middle ear cavity) that serves for the reception of airborne sound. The transmitter of vibrations is a bone that used to be part of the gill cover: the hyomandibula. This element is an elongate bone in bony fishes that articulates with the operculum – rotation of the hyomandibula opens the gill cover. *Parallel* to the hyomandibula runs the spiracle, which is usually water-filled in bony fishes. In tetrapods, the spiracle *contains* the hyomandibula, now called stapes or columella, which is freed from its former skeletal connections and swings freely in the air-filled cavity. The middle ear cavity is lost in salamanders and caecilians, but its consistent presence in frogs and amniotes indicates that it was present in the earliest crown-group tetrapods. By changing both their connections to other parts and their relation to one another, the hyomandibula and spiracle have been exapted from an old function (water release in the aquatic breathing cycle) to a new function (transmission of airborne vibrations to the fluid-filled inner ear).

11. **Jacobson's organ**. The olfactory sac has a ventral outgrowth in tetrapods that includes an additional sensory epithelium.

12. **Glands**. The tetrapod skin is rich in multicellular glands that serve various purposes. In lissamphibians, these include poison glands as well as mucous glands. It is unclear whether the lissamphibian condition is primitive for tetrapods, but their unkeratinized skin (unlike the scale-bearing epidermis of amniotes) is probably plesiomorphic. The lack of bony dermal scales in batrachians has sometimes been considered as evidence that lissamphibian glands were confined to that clade, but the co-occurrence of dermal scales and glands in caecilians counters this view.

Without paleontological evidence, numerous additional characters would be considered tetrapod autapomorphies, judging from their consistent presence in all extant tetrapods. Fossil taxa indicate that in reality these characters evolved convergently in lissamphibians and amniotes. In other words, some stem-amphibians and stem-amniotes do not have the derived state, and thus it must have evolved in parallel. Examples are (1) the single, cylindrical vertebral centrum, (2) the reduction of bones in the skull and pectoral girdle, (3) the widespread fenestration of the skull in the cheek region, and (4) the absence of external gills in adults. At least some of these convergences appear to correlate with an evolutionary pattern reported in both lissamphibian and amniote ancestors: miniaturization (see Chapter 10).

2.1.3 Stem-tetrapods (Tetrapodomorpha)

The tetrapod stem has been studied in great depth in recent decades, based on new finds and more detailed analysis of the iconic genera (*Ichthyostega*, *Acanthostega*) (Figure 2.8). These projects – which involve increasing numbers of researchers around the globe – have intensified the search for new fossils, new anatomical characters, and a better understanding of their evolution. Homology has become an important issue here, but also functional scenarios for organ change. Developmental biology and even genetics have started work on the fish–tetrapod transition, and collaboration between workers in the two fields has started.

Two researchers and their teams are in a pivotal position in this regard: Jenny Clack of the University of Cambridge (UK) and Neil Shubin at the University of Chicago (USA). Neither group was satisfied with simply analyzing existing material, and hence organized new field trips and

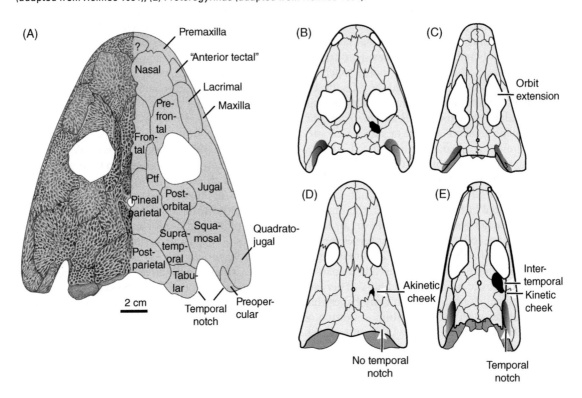

Figure 2.8 Skulls of Devonian and Carboniferous tetrapodomorphs: (A) *Ichthyostega* (adapted from Jarvik 1980); (B) *Whatcheeria* (adapted from Bolt and Lombard 2000); (C) *Baphetes* (adapted from Smithson 1982); (D) *Greererpeton* (adapted from Holmes 1984); (E) *Proterogyrinus* (adapted from Holmes 1984).

excavations. Jenny Clack built on the work of the Swedish School, returning to East Greenland. Her team found new specimens, excellently preserved, particularly of the poorly known *Acanthostega* (Clack 2012). These finds, analyzed by Clack, Coates, Ahlberg, and colleagues, have profoundly changed our perspective on stem-tetrapods. Although resembling many Paleozoic amphibians in bodily features, *Acanthostega* turned out on closer inspection to be a freakish animal: hands and feet with eight digits, a highly complex ear, internal gills, and a swimming tail with skeletal elements so far known only from bony fishes. Here was a tetrapod in the true sense of the term (four feet), but it had clearly always lived in the water.

Neil Shubin, in turn, decided to search for new localities in rocks of slightly older age than those bearing *Acanthostega*. He found these on Ellesmere Island, in the Canadian Arctic region of Nunavut, an autonomous territory of native Inuit people (Shubin 2008). Shubin and his colleagues Ted Daeschler and Farish Jenkins were most successful in discovering a new tetrapodomorph, *Tiktaalik*, named using the Inuit word for "big fish." Lacking hands and feet, retaining a partial bony gill cover, this was a fish in the traditional sense – but it already had a tetrapod skull. These two discoveries are only the most fascinating of many new finds that have made this field so attractive in recent years.

- **Eusthenopteron**. This iconic taxon was described in great detail by Andrews and Westoll (1970) and Jarvik (1954, 1980). It is the best-studied fish-like tetrapodomorph, based on material preserved three-dimensionally and with some rare soft-anatomical structures (internal gills). *Eusthenopteron* is found in a rich locality at Escuminac Bay (Quebec,

Canada), dating from the early Late Devonian (early Frasnian, 385–380 myr). It has a slender and deep skull that retains all elements common for bony fishes, especially the full complement of gill-cover bones and a firm connection between pectoral girdle and skull. The body outline is also that of a typical sarcopterygian fish, including two dorsal fins, one anal fin, and a trilobed tail fin. The deep, laterally compressed body was entirely covered by large and oval bony scales. The anatomy of *Eusthenopteron* was compared in detail with that of the extant ray-finned fishes *Amia* and *Polypterus*, which show numerous parallels interpreted as shared plesiomorphic features (Jarvik 1980). This helped to understand the complicated kinetic mechanism of the skull in *Eusthenopteron*: the endocranium was not only divided into braincase and upper jaws that hinged at three joints, but the braincase itself was subdivided, permitting the movement of the snout relative to the ear capsules and hindbrain. Apparently, this formed a unique mechanism to grasp large prey items. *Eusthenopteron* is placed in a Devonian clade called the Tristichopteridae, nesting above the other well-known tetrapodomorph *Osteolepis* (Ahlberg and Johannson 1998).

- **Panderichthys**. Known from Frasnian deposits in the Baltic States and Russia (~385 myr), this taxon constitutes a significant step towards the tetrapod condition (Vorobyeva and Schultze 1991). Most notable is the low body outline without the dorsal and anal fins. The only remaining fins are the paired appendages, the homologs of tetrapod fore- and hindlimbs.Only upper arm and forearm elements can be safely homologized between *Panderichthys* and tetrapods, whereas the further distal elements have long been believed not to match. Using CT scanning, however, Boisvert *et al.* (2008) found four irregular ossicles that might represent primitive versions of digits. If this is correct, then the origin of the hand skeleton would have to be predated to the *Panderichthys* node. As in tristichopterids, the snout of *Panderichthys* still houses a mosaic of numerous elements, and the cheek is still firmly connected with the pectoral girdle. Deep,

slit-like notches partially separate the skull table and cheek, which by analogy with *Polypterus* are thought to have housed the opening of the spiracle. *Panderichthys* has several close relatives, most of which are known only from incomplete material: *Elginerpeton* from Scotland, and *Obruchevichthys* and *Livoniana* from the East European Platform (Ahlberg *et al.* 2000).

- **Tiktaalik**. Mentioned briefly above, this taxon was discovered in 2005 (Daeschler *et al.* 2006; Shubin *et al.* 2006). *Tiktaalik* is known from Late Devonian stream deposits of Ellesmere Island (Middle Frasnian, ~380 myr). Most conspicuous are the flat, tetrapod-like skull and the complete absence of the opercular bone series. The absence of a bony gill cover does not imply a loss of gills; instead, their presence is indicated by grooved gill arch elements that bore branchial arteries (Daeschler *et al.* 2006). The absence of opercular bones and especially the posterior skull elements (extrascapulars, posttemporal) means that the skull and pectoral girdle were completely separate units. Remarkably, the tip of the snout forms a bulge as in many crocodiles, but the nares are placed at the lateral margin; the orbits are closely spaced and located on top of the skull. The forelimb of *Tiktaalik* still lacks fingers, but it is slightly more tetrapod-like in the presence of synovial joints in the distal elements, permitting a range of postures, including a substrate-supported stance as in tetrapod limbs. Despite its lack of true fingers, the pectoral appendage of *Tiktaalik* is functionally intermediate between a fin and a limb (Shubin *et al.* 2006). Originally considered a tetrapod because of its flat skull, *Elpistostege* from Escuminac Bay in Canada was a close relative of *Tiktaalik* (Schultze and Arsenault 1985).

- **Ventastega**. Between the almost completely preserved *Tiktaalik* and *Acanthostega*, there remains still a considerable gap. This is filled in part by *Ventastega*, a taxon from the latest Devonian (late Famennian, ~360 myr) of Latvia (Ahlberg *et al.* 1994). Its remains were first identified as belonging to a fish. Although the limbs remain unknown, *Ventastega* is more derived than *Tiktaalik* in its skull structure

Figure 2.9 Transformation of the skulls from fishes to land vertebrates, mapped onto a cladogram with important apomorphies.

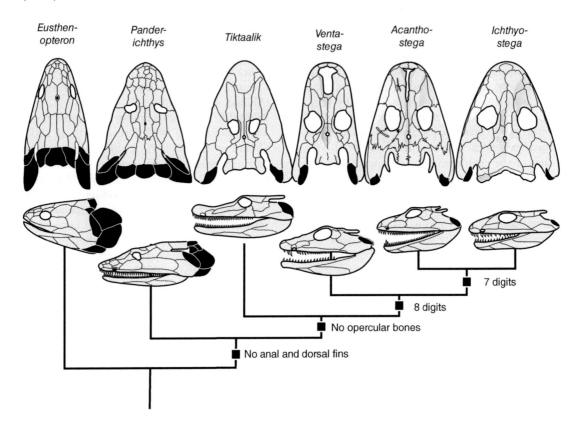

(Ahlberg *et al.* 2008). The orbits are large and the proportions of snout and skull table are more similar to *Acanthostega* than to finned tetrapodomorphs (Figure 2.9). In the snout, paired nasal bones are present, medially separated by a huge fontanelle. The "spiracular" notch in the cheek is much larger than in the previous taxa. In the cheek, both an intertemporal and a preopercular are present. The pectoral and pelvic girdle are consistent with those of *Acanthostega*.

The following taxa are usually ranked among tetrapods by authors using the character-based concept of classification (Ahlberg and Milner 1994; Anderson 2001; Anderson in Laurin and Anderson 2004; Clack 2012). In the phylogeny-based concept, they are still stem-tetrapods (tetrapodomorphs), because they fall outside

the crown taxon comprising Lissamphibia and Amniota.

- **Acanthostega**. Originally ranked second in importance after the iconic *Ichthyostega* (Jarvik 1980), *Acanthostega* was extensively studied in the 1990s by Clack and colleagues, who discovered a tremendous range of surprising features (Coates and Clack 1990, 1991; Clack 1994, 1998a; Coates 1996). By the nature of its completely known limbs, *Acanthostega* is the most primitive tetrapodomorph with fully developed hand and foot skeletons. Preparation of new material revealed that both fore- and hindlimb had eight digits, giving the hands and feet a wide, paddle-shaped structure. The skull is parabolically rounded with relatively large orbits and two separate notches in the cheek region (Figure 2.9). The lateral one of these is consistent with the squamosal embayment of many Paleozoic tetrapods, framed by the squamosal; the medial one is framed entirely by the tabular and unique, accommodating the stapes. The ear of *Acanthostega* is accordingly derived and very different from that of other taxa, notably *Ichthyostega*. The braincase is a single unit, unlike the kinetic structure of *Eusthenopteron* and more basal forms. The preopercular in the cheek and anocleithrum in the pectoral girdle are rudiments of the ancient bony bridge between the pectoral girdle and skull. The lateral-line organs were mostly enclosed in dermal bone, opening in so-called pit lines, which is a fish-like feature. The vertebrae are rhachitomous with a crescent-shaped ventral intercentrum and paired dorsal pleurocentra. The entire body was covered by bony scales, and the tail is long and deep, with substantial fin rays that include dermal elements typical of bony fishes (lepidotrichia). The skeleton was rather weak compared to most Paleozoic tetrapods and suggests that *Acanthostega* was aquatic throughout life, which is consistent with many other observations: the possession of internal gills, the structure and articulation of the limb elements, the lateral lines, the typical fish-eater dentition, and the swimming tail (Clack and Coates 1995).

- *Ichthyostega*. Familiar to schoolchildren like few other extinct animals, *Ichthyostega* was discovered by a Swedish expedition to East Greenland in the early 1930s (Säve-Söderbergh 1932). Unlike the gracile *Acanthostega*, this taxon has a robust skeleton with a heavy skull, massive limbs and girdles, and an overall stout appearance (Jarvik 1996; Ahlberg and Clack 2005). The skull is more similar to that of Paleozoic tetrapods in bone proportions, and has only one squamosal embayment (Figure 2.9). The ear is highly peculiar in that the stapes forms a huge blade. In the anterior trunk the ribs are relatively long with large uncinate processes. The foot has seven digits, whereas the structure of the hand remains unknown. Based on interesting parallels to modern elephant seals, Clack (2012) suggested that *Ichthyostega* had a similar mode of life:

the forelimbs are huge compared to the paddle-shaped hindlimbs, and probably served to drag the body. New data on the vertebrae show that the neural arches were regionally differentiated, possibly permitting a dorsoventral flexion in the posterior trunk and allowing a shuffling movement. All in all, *Ichthyostega* appears to represent an early but eventually unsuccessful lineage of tetrapodomorphs capable of locomotion along the shore, but not necessarily able to cover greater distances on land.

- *Tulerpeton*. Based on a partial skeleton from the late Devonian (Famennian) of Tula in Russia, *Tulerpeton* is clearly more derived than *Ichthyostega* in having only six digits in the hand and foot (Lebedev and Coates 1995). The radius and ulna are more similar to those of crown tetrapods, and the forelimb in general is more slender than in *Acanthostega* and *Ichthyostega*. The hindlimb, in turn, resembles that of *Ichthyostega* in its paddle-like shape and ankle construction. In the pectoral girdle an anocleithrum is retained, as in the Devonian taxa.

- *Stem-tetrapod tracks predate body fossils*. Until recently, the described phylogenetic sequence of taxa matched their stratigraphic occurrence quite well: tristichopterid fishes in the Middle to early Late Devonian, followed by *Tiktaalik*, these again slightly older than *Ventastega*, *Acanthostega*, *Ichthyostega*, and *Tulerpeton*, and finally more tetrapod-like taxa appearing in the Early Carboniferous. However, paleontology is famous (or infamous) for its discoveries of unexpected fossils that shake up conventional thinking. Such a case happened in 2010, when Niedźwiedzki *et al.* reported well-preserved tetrapodomorph tracks from marine deposits at Zachełmie in the Holy Cross Mountains (southern Poland). Had these tracks been discovered in Late Devonian deposits, they would have been readily assigned to *Acanthostega* or *Ichthyostega* because of their close resemblance to the hand and foot skeletons. However, these tracks were found in rocks of early Middle Devonian age, some 18 myr older than the oldest known body fossils of stem-tetrapods with limbs. The downside of this sensational find is that the

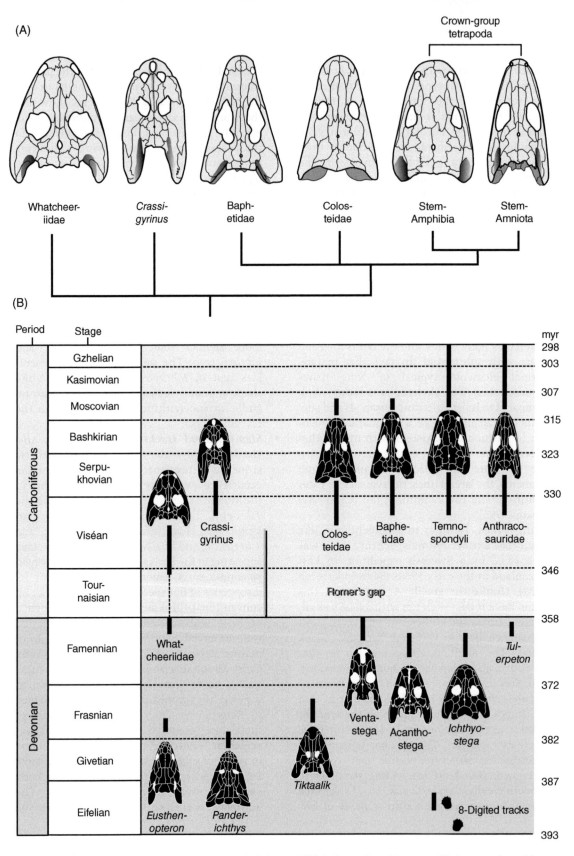

Figure 2.10 (A) Relationships of major stem-tetrapod clades and (B) their stratigraphic range. Phylogeny adapted from Ruta *et al.* (2003a), stratigraphy from Clack (2002a) and Carroll (2009).

fossil record has once again turned out to be much less reliable than expected – it turns out that the apparent good match between stratigraphic succession and hierarchically nested relationships of tetrapodomorphs was accidental. But the Zachełmie tracks have a second component that may be even more surprising: the track-makers lived in an undisputed marine habitat, a tidal flat close to a tropical coral reef lagoon. The origin of limbed taxa need not have taken place in freshwater, but may have occurred in estuaries. This had in fact been previously proposed, on the basis of functional and ecological considerations (Schultze 1997). That said, Pierce *et al.* (2012) have recently questioned this assignment on the basis of a functional model of locomotion in *Ichthyostega*.

2.1.4 Carboniferous tetrapods or tetrapodomorphs?

A wide range of Carboniferous taxa are more derived than *Ichthyostega* and *Tulerpeton*, but do not share any apparent synapomorphies with either amniotes or lissamphibians. Simply judging from this lack of crucial characters, it is impossible to place these taxa inside Tetrapoda (Clack and Carroll 2000), and consequently most phylogenetic analyses have found them nested below the crown-group tetrapods (Laurin and Reisz 1997; Anderson 2001; Ruta *et al.* 2003a, 2003b; Klembara and Ruta 2005; Clack 2012). This uncertainty will probably remain for a while, unless substantial new material is found soon.

- **Whatcheeriidae.** This is a small group of deep-skulled tetrapods with a length in the 1–1.2 m range. They form the oldest known well-preserved taxa after the Devonian tetrapodomorphs and are earliest Carboniferous in age (Tournaisian–Viséan, ~348–330 myr). Whatcheeriids have large skulls retaining a preopercular element in the cheek and pit lines instead of sensory grooves in the dermal skull bones (Figure 2.10). The Viséan *Whatcheeria* is known from a single locality in Iowa, USA, which has produced hundreds of skeletons (Lombard and Bolt 1995). This genus has an elongated trunk with 30 vertebrae and a long-stemmed interclavicle, two features resembling

the condition of stem-amniotes. Other plesiomorphic traits are the two-headed ilium and the retention of the intertemporal in the skull table. The ribs are moderately long with pronounced uncinate processes. The Tournaisian *Pederpes* is based on a single find from Scotland (Clack and Finney 2005). Finally, Daeschler *et al.* (2009) reported Late Devonian tetrapod material closely resembling *Pederpes*, indicating that Whatcheeriidae might turn out to be a grade. It has paddle-like, broad hands and feet, only 24 trunk vertebrae, robust limbs but no tarsal bones, and the coracoid and pubis are poorly ossified. All these traits indicate that whatcheeriids were aquatic, and the mass accumulation of *Whatcheeria* was probably formed in a small pond inhabited by the animals (Lombard and Bolt 1995).

- **Ossinodus.** This singular taxon is based on fragmentary remains from the Early Carboniferous of Australia (Warren 2007). Some features are shared with colosteids and temnospondyls, whereas the overall proportions resemble those of whatcheeriids. The ribs are moderately long with uncinate processes, the intertemporal is absent, and the wedge-shaped intercentra resemble the primitive condition of *Ichthyostega*; additional primitive features are the two-headed ilium, the ossified pubis, the massive humerus, and the palate.

- **Colosteidae.** This is a small clade of long-bodied, fully aquatic taxa with relatively small limbs (Smithson 1982; Hook 1983). Colosteids were a long-lived Carboniferous group, ranging from the Viséan through the Moscovian (~345–306 myr). Their skull is elongate with a moderate snout dominated by an extensive prefrontal, and the skull table includes a greatly enlarged postorbital and a rudimentary intertemporal, which may also be absent (Figure 2.10). Lateral lines are always present in adults, and gill arch elements are ossified, bearing elongate ossicles with up to 15 pharyngeal teeth (Hook 1983), indicating the existence of open gill slits. In the front of the snout and palate, the tusks are very large, and in contrast to anthracosaurs and whatcheeriids, the skull appears to have been essentially akinetic. The gastral scales are heavy and histologically similar to those of bony fishes (Witzmann 2007). The trunk includes 35–40 vertebrae of rhachitomous structure, with similar-sized intercentra and pleurocentra. The ribs are moderately long, and the scapula remained cartilaginous dorsally. In *Colosteus* the limbs were minute and gracile; in *Greererpeton* the very primitive humerus was more robust. The hand has only four fingers. It is possible that colosteids form part of the stem-group of temnospondyls–lissamphibians, but the numerous plesiomorphic characters raise some doubts. In contrast to many temnospondyls, the pubis and tarsals are ossified, but the single-headed ilium resembles that of temnospondyls. The ontogeny of the skull is known in *Greererpeton*, showing little proportional change except for a decrease in relative orbit size (Godfrey 1989). Colosteids were relatively large aquatic predators, which probably left the water only when forced to do so by environmental changes.

- *Crassigyrinus*. Because of its mixture of very primitive and derived embolomere characters, this single taxon has puzzled scientists ever since its discovery (Panchen 1973, 1985; Clack 1998b). Known from two different localities in Scotland, and a further record in West Virginia (Godfrey 1988), the finds of *Crassigyrinus* date around the boundary between Early and Late Carboniferous (~318 myr). The 1.5–2 m long animal was eel-shaped with a large, deep-sided skull, minute limbs, and feebly ossified vertebrae and girdles. The skull has a long depressed region in the midline of the snout, resembling the fontanelle of *Acanthostega* (Figure 2.10). The animals resemble large moray eels and were obviously aquatic.

- *Baphetidae*. Formerly called loxommatids, these distinctive forms encompass six genera and 13 species (Milner *et al*. 2009). The baphetids range from the late Viséan to the Moscovian (~330–306 myr). They are characterized by a keyhole-shaped orbit and a supratemporal extending well around the squamosal embayment (Figure 2.10). The skull is usually elongate, with the snout about double the length of the posterior skull table (measured without the triangular extension of the orbit,

which is clearly offset from the eye opening proper). The axial and limb skeleton is very poorly known but all data suggest that it was rather weakly ossified, as in aquatic forms. This is consistent with the presence of sensory grooves on the dermal skull bones, which are confined to the snout in most baphetids. *Baphetes*, *Kyrinion*, *Loxomma*, and *Megalocephalus* all have elongate skulls with huge orbit extensions; they were apparently the largest predators in their habitats, recognized by their huge tusks. The correlation of tusk size and length of the keyhole-shaped orbit was highlighted by Beaumont (1977), who suggested that the anterior extension housed a powerful jaw-closing muscle. Unlike in colosteids and temnospondyls, the palate was entirely closed. The discovery of a basal baphetid, *Eucritta* from the Viséan of Scotland, has added substantial data on the postcranial skeleton and also sheds light on the early evolution of the group (Clack 2001). The relatively short-bodied *Eucritta* (~24 trunk vertebrae) was smaller than other baphetids, reaching less than 50 cm body length. The large squamosal embayment and wide, short-faced skull resemble the condition in small temnospondyls, but the detailed structures of the skull and limbs are quite different and more plesiomorphic, such as the halfmoon-shaped humerus or the two-headed ilium. A faint anterior extension of the orbit is interpreted as an incremental keyhole orbit (Milner *et al*. 2009). A further, most distinct baphetid is the Bashkirian genus *Spathicephalus*, which has a foreshortened skull table and a huge, parabolic snout not unlike extant giant salamanders (cryptobranchids). Its numerous teeth are equal-sized and chisel-shaped (Beaumont and Smithson 1998).

2.2 The amniote stem-group

By definition, amniote stem-group taxa are not amphibians in a phylogenetic sense, because they do not fall within lissamphibians or their stem group. After all, they are relatives of reptiles and mammals, but not salamanders and frogs. In an era dominated by character discussion, comparison of cladograms, and evolutionary scenarios, using the traditional concept of the Amphibia as an ecological rather than a phylogenetic group would be utterly confusing. Yet the more precise, analytical definition of amphibians as a natural group comes at a cost: the difficulties in recognizing and distinguishing stem-group taxa of amphibians and amniotes increase as one moves down the cladogram. Basal taxa on the tetrapod, amniote, and lissamphibian stems are often confused, especially when incompletely known. Only taxa with unambiguous amniote characters are described in the following section.

Stem-amniotes were superficially similar to extant salamanders in many ways. This is why they were traditionally described as amphibians, and lumped with stem-amphibians in poorly defined groups such as Stegocephalia or Labyrinthodontia (Romer 1947). However, on closer inspection, stem-amniotes carry some features that suggest a different physiology and mode of life of these aquatic to amphibious tetrapods. The most significant of these is the structure and size of the ribs: unlike in most tetrapodomorphs and stem-amphibians, the ribs of amniote ancestors are long and curved and could be moved inwards. This movement is practiced by all extant amniotes, permitting the rib basket to draw fresh air into the lungs as it is expanded and expel oxygen-depleted air during contraction. In the stem group of amniotes, such costal ventilation evolved as a key adaptation. It was a further major step in making amniotes more independent of the water, although many stem-amniotes still hatched from water-borne eggs.

Another, more obvious feature of these taxa is their elongate body: the number of vertebrae in the trunk skeleton is usually well beyond 30, contrasting with the situation in temnospondyls, the putative lissamphibian stem group, which mostly had 24. The rib basket of stem-amniotes was evidently more rigid and probably could assist the limbs in moving by lateral flexion. Many lineages on the amniote stem evolved eel- or snake-like body forms by successive increase in vertebral number and eventual reduction of the limbs – this happened in some anthracosaurs (which retained the limbs), four separate

lepospondyl clades (three of which lost limbs entirely), and some immediate amniote relatives, such as *Westlothiana*.

2.2.1 Anthracosauria

The anthracosaurs ("lizards of the coal") were a clade of Paleozoic tetrapods (Figure 2.11). Starting with small forms (50–80 cm) in the Early Carboniferous, they reached up to 3 m body length later in that period (Panchen 1980; Smithson 2000; Ruta and Clack 2006). Anthracosaurs had vertebrae in which both central elements (pleurocentrum and intercentrum) were well ossified and large. An anthracosaurian subclade is termed Embolomeri, in which both centra are disc-shaped (Panchen 1970). Anthracosaurs inhabited lakes and swamps within extensive coal forests of eastern North America and Europe, where they probably preyed on fishes and small tetrapods. They form common vertebrate finds in Pennsylvanian mudstones associated with coals of many sites in the British Isles, the Czech Republic, the Appalachians, and Ohio (see Chapter 3). Altogether, 19 genera and 24 species are known, ranging from the Mississippian (Viséan, 345 myr) to the Early Permian (Sakmarian, 285 myr).

In most phylogenetic hypotheses, anthracosaurs form the basalmost undisputed branch of the amniote stem group, which is indicated by two derived features of the group: the posterior skull table (sutural connection between parietal and tabular) and the elongated ribs (Figure 2.12). Despite their possession of amniote features, anthracosaurs appear to have been predominantly aquatic throughout their lives. Most species had well-established lateral line grooves on their skull bones and elongated skeletons with proportionally small limbs and very long swimming tails (Panchen 1970). Their elongated and narrow snouts bear large labyrinthodont teeth – this is why anthracosaurs were originally united with temnospondyls and other groups as labyrinthodonts (Romer 1947), which is today considered a polyphyletic assemblage.

A consistent feature of anthracosaurs is their massively ossified pelvic girdle and hindlimb, which is much larger than the forelimb. It is probable that they were able (and possibly often forced) to leave the water and undertake longer

Figure 2.11 Stratigraphic range of major crown tetrapod clades. After much initial diversity, only temnospondyls and chroniosuchians survived the Permo-Triassic boundary.

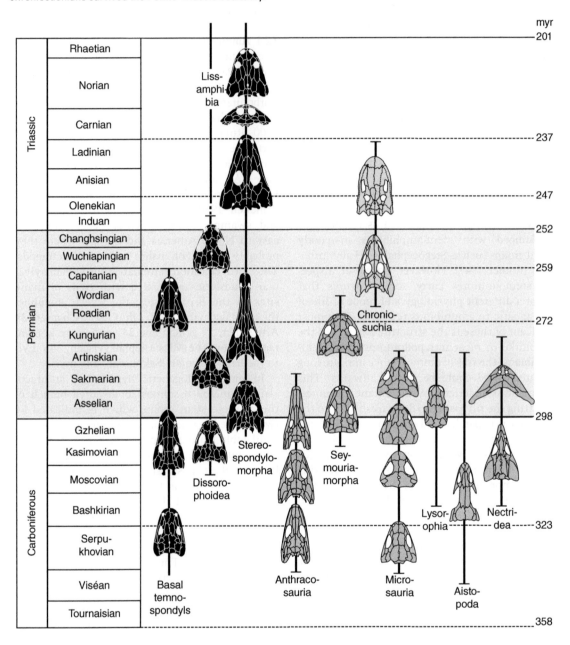

land excursions. Considering the coal-rich deposits in which they are found, this appears to make sense: similar habitats are today characterized by seasonal oxygen shortage. Turnover by wind or storms could have killed the lake fauna and force the surviving tetrapods to emigrate. The sediments yielding anthracosaurs range from sapropelic mudstones over coaly shales to ironstones, all of

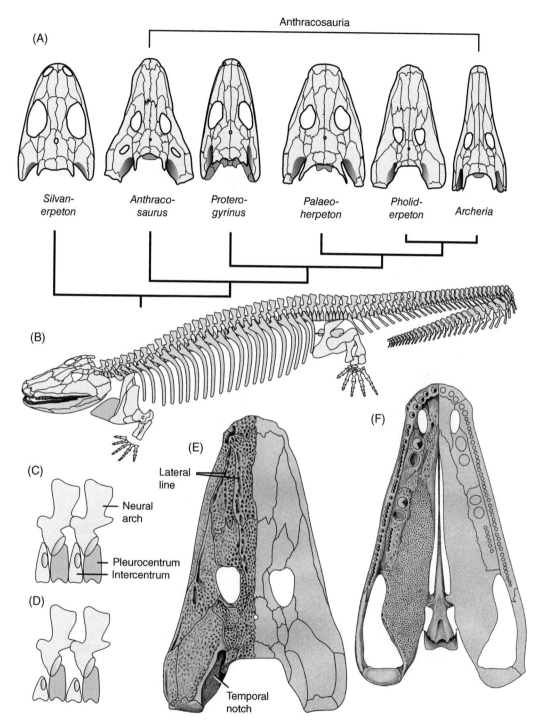

Figure 2.12 Anthracosaurs: (A) relationships; (B) the elongate skeleton of *Pholiderpeton* (Panchen 1972); (C, D) composition of trunk vertebrae (C, embolomerous; D, basal anthracosaur); (E) dorsal and (F) ventral side of skull in *Pholiderpeton* (adapted from Panchen 1972).

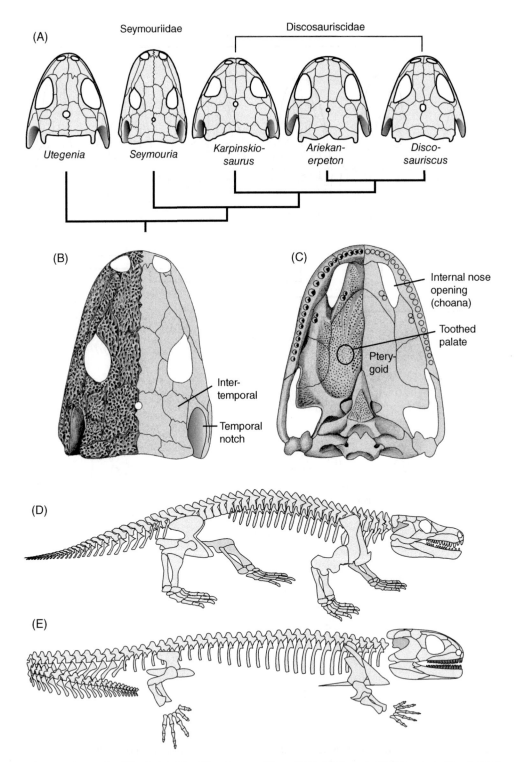

Figure 2.13 Seymouriamorphs: (A) relationships (Klembara and Ruta 2005); (B) dorsal and (C) ventral side of skull in *Seymouria* (adapted from White 1939); (D) the massive skeleton of terrestrial *Seymouria* (adapted from White 1939); (E) the feeble skeleton of larval *Discosauriscus* (adapted from Klembara and Ruta 2005).

which are interpreted as having formed under stagnant water conditions (Smithson 1985).

It is possible that anthracosaurs had abandoned gill breathing, as bony hyobranchial skeletons are absent in this group. A lamella on the dermal bones of the shoulder girdle usually associated with gill openings is also absent, indicating that the neck region was closed. The anthracosaur dermis housed a dense layer of thick bony scales, similar in arrangement and number to those of tetrapodomorph fishes. In contrast to amphibians, skin breathing is therefore not a probable means by which anthracosaurs respired. The remaining option is lung ventilation, which according to the rib morphology was practiced by means of costal aspiration. Anthracosaurs are therefore likely to have been primarily lung breathers who preferred to prey in lakes but were able to migrate between water bodies whenever the situation required it. This eventually set the stage for the evolution of amniotes.

The anthracosaurs are almost exclusively known from adult specimens, leaving their ontogenies unknown (a possible exception is the small skull of *Calligenethlon* from Nova Scotia, Canada). In contrast to amphibians, anthracosaurs had five fingers and also retained a phylogenetically ancient element in the skull table, the intertemporal. The cheek was moveable against the skull roof, a plesiomorphic feature shared with tetrapodomorph fishes. The known distribution of anthracosaurs suggests that they were tropical animals, living only within a few degrees latitude around the equator of Carboniferous times (Carroll 2009).

- **Mississippian anthracosaurs**. The basalmost anthracosaurs are known from the Early Carboniferous (late Viséan, ~330 myr) of Scotland. *Silvanerpeton* from East Kirkton was a small animal with large orbits and a short snout (Ruta and Clack 2006). The fact that it has well-ossified girdles and limbs suggests that it was an adult, contrasting with the rather immature skull morphology. *Eoherpeton* from Gilmerton near Edinburgh is larger with more robust tusks, lacks sensory grooves, and has vertebral centra that were not fully disc-shaped. The long hindlimbs indicate that the animals were capable of moving on land. *Proterogyrinus*

is from the Namurian A of Greer (Holmes 1984) and Cowdenbeath (Smithson 1986).
- **Predators in Late Carboniferous coal swamps**. *Anthracosaurus, Palaeoherpeton,* and *Pholiderpeton* were inhabitants of Pennsylvanian coal measures (Panchen 1972, 1977; Holmes 1984). Despite much morphological diversity, these taxa were all large predators with huge tusks, dentigerous palates, and powerful jaw musculature as inferred from attachments. *Palaeoherpeton* and *Pholiderpeton* had conspicuous lateral-line canals. *Anthracosaurus* had an akinetic skull and greatly enlarged tusks in the palate (Panchen 1977; Clack 1987).
- **Permian stream dwellers**. The Early Permian (~299–290 myr) genus *Archeria* was a gracile, long-snouted anthracosaur that inhabited rivers within coastal floodplains deposited in the Texas red beds (Holmes 1989).

2.2.2 Seymouriamorpha

Some substantial confusion between amphibians and amniotes arose from one particular group, the Permian Seymouriamorpha (Figure 2.13). Known from two different types of deposits, this relatively small clade of tetrapods falls into a terrestrial and an aquatic group. Only slowly was it realized that the larval forms from Europe (Discosauriscidae) and the terrestrial morphs from North America (Seymouriidae) belonged to the same group. Eventually, Klembara *et al.* (2006) were able to show that the two best-known genera, *Discosauriscus* and *Seymouria*, underwent very similar ontogenies but were still distinct taxa. Altogether, the seymouriamorphs include eight genera and 12 species, ranging throughout the Permian (~299–251 myr) (Laurin 2000; Klembara and Ruta 2005).

Seymouriidae. The terrestrial seymouriids were first recognized in Texan red-bed deposits, where they occur in stream and floodplain environments (Romer 1928, 1935). They are all included in the genus *Seymouria*, a 50–100 cm long animal with numerous adaptations for a terrestrial existence. *Seymouria* has a robust postcranial skeleton, with massive girdles, limb elements with fully formed joints, and large hand and foot skeletons. These and the ratio between upper and lower leg bones indicate an excellent capability to walk on land. This is consistent with finds of *Seymouria* in upland deposits at Tambach and Fort Sill. Composed of both pleurocentra and intercentra, the vertebral column is essentially similar to that of embolomeres. Here, however, the pleurocentrum is the main central element, bearing the unusually bulbous neural arch, and the intercentrum is reduced to a small wedge filling the gap between successive pleurocentra. The seymouriid skull is wide compared with that of anthracosaurs, not as deep, and has large orbits. As in many stem-amniotes, the intertemporal is retained, and the parietal broadly contacts the tabular. There are numerous rows of small teeth in the palate, and large recurved teeth along the jaw margins, indicating that seymouriids ranked among the larger terrestrial predators.

Discosauriscidae. In Europe, lakes and ponds were populated by aquatic tetrapods in great numbers during the Early Permian. In France and Germany, most water bodies were home to temnospondyls – the gilled branchiosaurids being the most abundant of these (Chapter 3). In east–central Europe, especially the Czech Republic, a different tetrapod clade was predominant, the Discosauriscidae. They are also known from some German and French localities, where they coexisted with temnospondyls. Superficially, their poorly ossified skeletons and broad skulls resemble those of branchiosaurids, and in the nineteenth century the two groups were often confused, sometimes even united in a single genus (e.g., *Melanerpeton*). However, numerous anatomical differences show that discosauriscids are stem-amniotes and closely related to seymouriids.

Discosauriscids were 10–20 cm long, salamander-like animals with external gills and long swimming tails (Klembara 1995). Although most elements in the braincase and girdles ossified earlier than in temnospondyls, larval seymouriamorphs were much less heavily built than their terrestrial adults (Klembara and Bartík 2000). The limbs developed slowly, and together with the sensory grooves and the fossilized gills they indicate an aquatic life. However, there was substantial variation across taxa. For instance, *Discosauriscus* retained sensory grooves until late in development, and attained sexual maturity in the water (Sanchez *et al.* 2008). Based on histological studies, the animals remained in the water for up to 10 years.

Klembara (2009) suggests that *Discosauriscus* left the water eventually, highlighting the loss of sensory grooves and more robust limbs bones in the largest specimens. *Seymouria*, on the other hand, attained its robust adult skeleton and adult skull morphology at earlier stages than *Discosauriscus*, apparently passing through the aquatic larval period more rapidly (Klembara *et al.* 2006). It is thus conceivable that discosauriscids and seymouriids modified the timing of developmental events in response to diverse habitats, as is also known from some temnospondyls (Schoch 2009b). However, it is also conceivable that *Discosauriscus* was an entirely aquatic form, as the data on sexual maturity in the larval state suggest.

The major difference between seymouriamorphs and extant amphibians is that seymouriamorphs did not undergo drastic morphological changes in a short period of time, but developed at a slow rate. Metamorphosis, as known from lissamphibians, was not an option for this group. From this perspective, seymouriamorphs were only similar to modern amphibians on a very gross scale. It is even more important, then, to highlight that this small clade managed to evolve both aquatic and terrestrial morphs, each of which pushed the adaptation to their particular habitat a bit further than other early tetrapods.

- *Seymouria*. Terrestrial forms from the Early Permian (~284–270 myr) of the United States (Texas, New Mexico, Oklahoma) and Germany (Thuringia) (White 1939; Berman *et al.* 2000; Klembara *et al.* 2006).
- *Discosauriscus*, *Makowskia*, and *Shpinarerpeton*. Aquatic taxa from the Early Permian of Moravia (Czech Republic), the classical discosauriscids (Klembara 1995, 1997).
- *Utegenia* and *Urumqia*. These basal seymouriamorphs are from the Early Permian of Kazakhstan and western China (Klembara and Ruta 2005).

2.2.3 Chroniosuchia

Chroniosuchians form a small but distinctive group of stem-amniotes with some affinities to anthracosaurs and seymouriamorphs. They were first recognized in Russian deposits of Late

Permian age (Vyushkov 1957), and are defined by a single row of complex dorsal osteoderms and a unique vertebral structure (Golubev 1998). The skull is also highly derived, but known only from a few taxa, while the girdles and limbs remain unknown with few exceptions. As a consequence, most of their taxonomy has been based on the osteoderms.

The most clear-cut derived characters of chroniosuchians are (1) osteoderms with two lateral wings and interconnected by joints and (2) pleurocentra deeply concave, articulating with spherical intercentra (Figure 2.14) (Golubev 1998). The ball-and-socket joint between pleurocentra and intercentra is unique among vertebrates and makes even fragmentary material readily identifiable. Functionally, this structure testifies to a high flexibility of the trunk, stabilized by the tight articulation between successive osteoderms. The chroniosuchian vertebrae are quite similar to those of seymouriamorphs in shape and size of pleurocentrum, intercentrum, and neural arch – only that the wedge-shaped intercentrum (e.g., *Seymouria*) has become ball-shaped in chroniosuchians. The pleurocentrum is fused to the neural arch, and the osteoderms were co-ossified with the neural spine, with one large osteoderm per vertebral segment.

As far as is currently understood, chroniosuchians fall into two separate groups differing in osteoderm morphology: the Chroniosuchidae (Late Permian of Russia and China and Middle/Late Triassic of Kyrgyzstan) and the Bystrowianidae (Late Permian–Middle Triassic of Russia and Germany). Chroniosuchids are by far the better-known group and form a well-supported clade. They have a peculiar skull morphology, with long paired openings in the elongate snout, not unlike the antorbital fenestrae of archosaurian reptiles. Although they retained anthracosaur features such as a broad tabular–parietal contact and an elongate tabular horn, the intertemporal was lost. In contrast to those of bystrowianids, the chroniosuchid osteoderms have additional joints.

Chroniosuchians occur in deposits yielding both aquatic and terrestrial taxa, and their skeletal features are somewhat ambiguous with respect to their mode of life. Whereas the lack of sensory grooves even in juveniles suggests that dependence on water was not as strong as in anthracosaurs and larval seymouriamorphs, the limbs and girdles are nowhere near as massive as in *Seymouria*, for instance (Clack and Klembara 2009). On the basis of bone density, Laurin *et al.* (2004) concluded that *Chroniosaurus* was terrestrial, whereas Golubev (2000) placed them in the aquatic community. The functional context of the large openings in the snout remains unsettled, but by analogy to archosaurs it is likely to have accommodated powerful jaw-closing muscles (pterygoideus portion). Golubev (2000) suggested that chroniosuchids were able to move the cheek against the skull table, as in anthracosaurs. At present, 11 genera of chroniosuchians are known, with six falling within the Chroniosuchidae and five in the Bystrowianidae.

- *Chroniosaurus* is a well-known chroniosuchid from the Late Permian of the Ural Forelands, Russia (Clack and Klembara 2009; Klembara *et al.* 2010). It has a slender and gracile skull with a large opening between the nares.
- *Madygenerpeton* from the Middle/Late Triassic of Kyrgyzstan has a highly derived skull morphology, closely resembling that of edopoid temnospondyls (Schoch *et al.* 2010). The 50 cm long predator has numerous tiny teeth and a very flattened skull with pustular ornamentation. The osteoderms are extremely broadened, and interlock tightly to give a rigid carapace (Buchwitz and Voigt 2010).
- *Bystrowiella* is a large (1.5 m) bystrowianid with tall neural spines from the Middle Triassic of southern Germany (Witzmann *et al.* 2008). The remains appear to have been washed into a lake that was not the habitat of *Bystrowiella*.

The relationships of chroniosuchians have just started to be studied, and currently there is no consensus. Based on palatal features shared with anthracosaurs, Clack and Klembara (2009) have argued for a close relationship with anthracosaurs. However, Schoch *et al.* (2010) found chroniosuchians to nest higher within the amniote stem, possibly between seymouriamorphs and crown amniotes.

Figure 2.14 Chroniosuchians: (A) relationships (Schoch *et al.* 2010, reproduced with permission of John Wiley & Sons); (B) two adjacent osteoderms in dorsal view (*Bystrowiella*, adapted from Witzmann *et al.* 2008); (C) vertebral segment in lateral view (*Bystrowiella*); (D) dorsal and (E) ventral view of skull in *Chroniosuchus* (adapted from Golubev 2000).

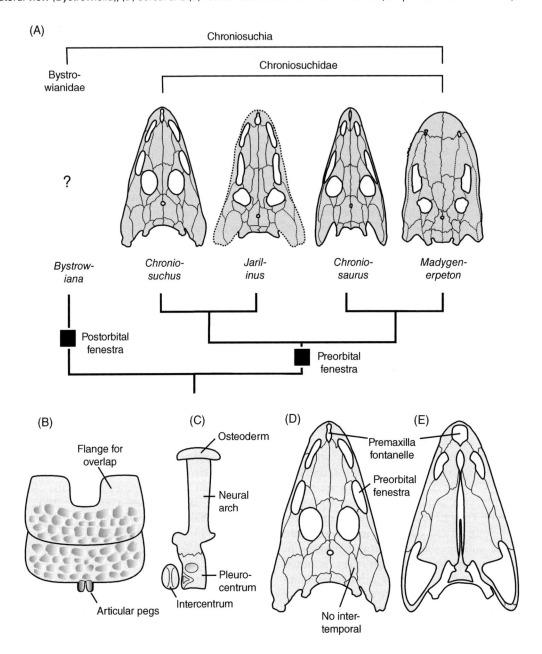

2.2.4 Lepospondyli

Lepospondyls are a diverse assemblage of early tetrapods, which has traditionally been viewed as monophyletic (Carroll and Chorn 1995). Most of them are small, not exceeding 5 cm skull length. Unlike the temnospondyls, lepospondyls had delicate skeletons (Figure 2.15), less readily preserved under many conditions. Therefore, they

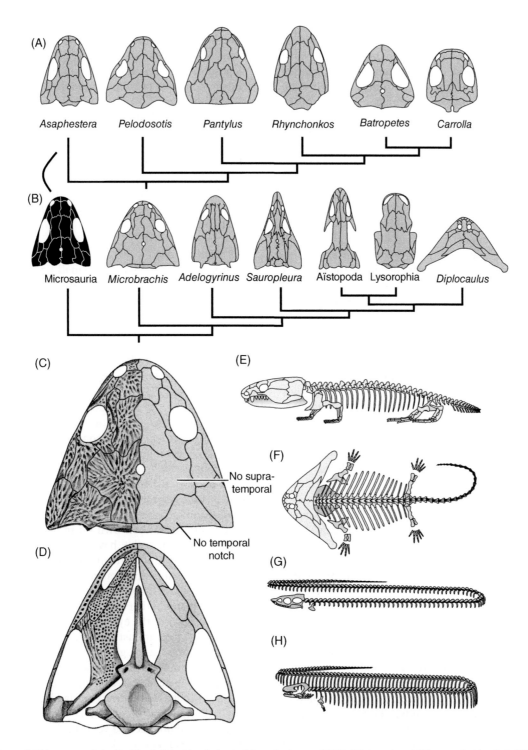

Figure 2.15 Lepospondyls: (A, B) relationships (adapted from Anderson 2001); (C) dorsal and (D) ventral side of skull in *Microbrachis* (adapted from Carroll and Gaskill 1978); (E) skeleton of *Pantylus* (adapted from Romer 1969); (F) skeleton of aquatic *Diplocaulus* (adapted from Milner 1996); (G) skeleton of aïstopod *Phlegethontia* (adapted from Anderson 2002); (H) lysorophian *Brachydectes* (adapted from Wellstead 1991).

are primarily known from coal and mudstone *Lagerstätten*. Nevertheless, they might have formed the most common land vertebrates, at least in terms of individual numbers, during the Late Carboniferous and Permian.

Currently, there are 62 lepospondyl genera and 84 species known. They were confined to North America, North Africa, Europe, and possibly Eurasia and China. Their stratigraphic range is from the Early Carboniferous (Viséan, ~340 myr) to the Early Permian (Artinskian, ~275 myr). They fall into six readily recognized groups: (1) the very diverse microsaurs, which include various caecilian-, salamander-, and lizard-like morphs, (2) the elongated lysorophians with rudimentary limbs, (3) the mostly aquatic nectrideans, (4) the limbless aïstopods with highly modified skulls, (5) the eel-like adelospondyls, and (6) the poorly known Acherontiscidae. Many of these lepospondyls were probably aquatic or amphibious, but terrestrial taxa are also known.

2.2.4.1 Lepospondyl characters

All the characters uniting lepospondyls are not exclusive to this assemblage. Some authors have therefore questioned their monophyly (Milner 1993; Ahlberg and Clack 1998), whereas most recent analyses have found them to be monophyletic (Carroll 1995; Anderson 2001; Ruta *et al.* 2003a; Vallin and Laurin 2004).

1. **Vertebral centra cylindrical**. The best-known feature of lepospondyls is their vertebral structure, in which the centrum forms a single unit of cylindrical shape. Often centrum and neural arch are fused. Only in some taxa do intercentra remain as ventral wedges between successive vertebrae, indicating that the centrum derives from the pleurocentrum originally. Cylindrical centra are also universally present in lissamphibians and albanerpetontids, as well as in some amphibamid temnospondyls.
2. **No squamosal embayment**. The deep notch between cheek and skull table is completely absent in all lepospondyls. However, this condition is also present in many other lineages.
3. **No palatal tusks**. There are no large, paired tusks on the vomer, palatine, and ectopterygoid in lepospondyls.

4. **Teeth not labyrinthodont**. The absence of dentine and enamel infolding is shared by lepospondyls but is also common among juveniles and larvae of temnospondyls, as well as lissamphibians.
5. **Odontoid peg and basioccipital**. Atlas with anterior projection (= odontoid peg) fitting into the concave basioccipital. An odontoid peg is also found in salamanders, the stem-gymnophionan *Eocaecilia*, the Albanerpetontidae, and some amphibamid temnospondyls, but in all these taxa it fits into a gap because the basioccipital is absent.

2.2.4.2 Microsauria

The name Microsauria (Greek: "small lizards") is quite fitting, because almost all of them are small and have often been confused with (or considered close relatives of) early amniotes – which indeed had a lizard-like appearance (Figure 2.15). There are three obvious reasons for a close resemblance between microsaurs and early amniotes: (1) lepospondyls hold a relatively high position within the amniote stem, (2) their robust limb skeletons have typical features of terrestrial animals, and (3) the miniature size results in structurally simplified bones, which makes convergences more difficult to identify than in larger taxa.

The group as a whole may well be paraphyletic with respect to other lepospondyl clades (e.g., nectrideans or lysorophians: see Anderson 2001 and Vallin and Laurin 2004). However, microsaurs are characterized by their own set of derived features: (1) the bony scales have numerous radial rods; (2) the number of digits in the hand is four or fewer; (3) the supratemporal and intertemporal elements are absent; (4) the snout is usually short; and (5) the distance between the eyes is wider than the diameter of the eye opening.

Microsaurs had robust skeletons, including fully ossified braincases, girdles, and limbs. This suggests that many taxa were capable of leaving the water, or were direct developers (without a larval stage) that lived entirely on land (Fröbisch *et al.* 2010). The latter option is employed by plethodontid salamanders, a very speciose, miniaturized group (Hanken 1983). That would explain why the smallest known juveniles of microsaurs already look like adults. If the analogy with salamanders holds true, then microsaurs were able to terminate growth at any body size without morphological differences – a strategy impossible for the slow-developing temnospondyls (Schoch 2009a). In microsaurs, skeletal development proceeded at a fast pace, without metamorphosis (Fröbisch *et al.* 2010).

Many microsaurs had strong, conical teeth and some had bulbous dentition indicating crushing bite habits. The diversity of tooth sizes and shapes is the highest among all early tetrapods, culminating in the huge battery of palatal teeth in the large (30 cm long) *Pantylus*. This indicates that the group may have occupied numerous niches held today by lizards, birds, and small mammals. A further feature correlating with feeding is the fenestration of the cheek, which is found in at least two separate microsaur lineages, based on Anderson's (2001) detailed phylogeny.

- **Basal microsaurs**. Two groups are recognized at the base of the microsaur radiation. The first comprises taxa with closed skulls (*Tuditanus*, *Asaphestera*, *Crinodon*), which reached a body size of 15–25 cm (Carroll and Gaskill 1978). These forms were similar to large land salamanders, differing in having stronger, fully mineralized teeth and by the retention of tooth patches in the almost closed palate. Their abbreviated posterior skull table is probably the primitive condition for microsaurs, with the tabular filling the gap left by the absent supratemporal and intertemporal bones. The second group (*Hapsidopareion*, *Llistrofus*) has a fenestrated cheek in which squamosal and jugal were widely separated and reduced to narrow struts. By analogy to salamanders and diapsid reptiles, the open cheek permitted jaw-closing musculature to expand dorsally and attach along the flank of the skull table and braincase. These basal microsaurs had large scales that completely covered their bodies.
- **A terrestrial giant dwarf**. A true giant – at least by microsaurian standards – was the 30 cm long Early Permian genus *Pantylus* (Romer 1969). Its huge skull and the stout body shape of this microsaur resemble those of the modern pinecone lizard (*Tiliqua*), a fruit-eating skink, to a surprising degree. The impressive

crushing dentition consists of numerous rows of strong teeth in the jaws and palate, and the openings for the jaw-closing muscles are huge, indicating the capability for powerful biting. Carboniferous relatives appear to be *Sparodus* from Nýřany and *Trachystegos* from Joggins.
- **Burrowing microsaurs**. Elongated bodies, small skulls, and tiny limbs were the typical features of the 20–40 cm long *Gymnarthridae*, *Ostodolepididae*, and *Goniorhynchidae*. Despite the small size, their skull bones are extremely thick, and the pointed triangular skull outline indicates burrowing behavior. The Early Permian genus *Rhynchonkos* shares features with the Jurassic *Eocaecilia* (Jenkins *et al.* 2007) such as the ventrally sloping snout, the wide parasphenoid, and the general arrangement of sutures. It has therefore been suggested to be a gymnophionan stem-group taxon (Anderson *et al.* 2008; Carroll 2009). The larger forms *Micraroter* and *Pelodosotis* have emarginated cheeks, probably for the attachment of external jaw-closing muscles.
- **Miniaturized forms**. The 5–7 cm long genera *Saxonerpeton*, *Batropetes*, and *Quasicaecilia* form a clade of miniaturized microsaurs (Carroll 1990). Like the plethodontid salamanders, these taxa surpassed a critical minimal size beyond which the skull morphology had to be rearranged. Large-scale fusion of the braincase, reduction of dermal bones to thin struts, and huge ear capsules make their skulls unique among tetrapods. The skeletons of these dwarfs retained the long, curved ribs and were fully ossified, suggesting a terrestrial mode of life.
- **Perennibranchiate microsaurs.** Only a few microsaurs appear to have been fully aquatic, such as *Microbrachis* (Carroll and Gaskill 1978; Vallin and Laurin 2004). These 15 cm long animals shared the elongate trunk (38 vertebrae) and small limbs with burrowing forms, but had a feeble skeleton in comparison. In *Microbrachis*, the dermal skull bones contain grooves for the lateral line and the ornament differs from that of all other lepospondyls. Unlike in larval branchiosaurids, the braincase and vertebrae were fully ossified and the ribs very long, whereas the shoulder and pelvic girdles remained even more cartilaginous than

in these temnospondyls. Although the largest known specimens of *Microbrachis* could be adult, they may still have been larvae of some unknown terrestrial adult. The presence of denticles in the gill region, similar to those of larval temnospondyls, indicates that these animals had a (cartilaginous) hyobranchial skeleton and open gill slits, but not necessarily functional gills.

2.2.4.3 Lysorophia

Lysorophians range from the Late Carboniferous (Westphalian A, ~315 myr) through the Early Permian (Artinskian, ~280 myr) (Milner 1987; Wellstead 1991). This small clade has strong affinities to microsaurs, especially the brachystelechids (Carroll 1995), but its resemblance to aïstopods has also been emphasized (Anderson 2001). It has been argued that lysorophians are nested within the Microsauria, making them a paraphyletic grade (Laurin and Reisz 1997). Lysorophians had extremely elongated trunks, containing more than 60 vertebrae, and tiny limbs which retained feeble hand and foot skeletons (Figure 2.15) (Wellstead 1991). The vertebrae have medially separated neural arches. The most diagnostic structure is the heavily ossified skull, which has broad medial bones (frontal, parietal, parasphenoid) and massive jaws, and lacks postfrontal, postorbital, and jugal (Wellstead 1991). The massive skull with its open sides recalls the condition in sirenid and proteid salamanders, the most neotenic caudates today. However, the heavy bones and common finds of lysorophians in fossilized burrows indicate that the animals led a burrowing life, possibly during aestivation. Indeed, the skull of the best-known lysorophian *Brachydectes* closely resembles that of amphisbaenians. By analogy to many squamates and lissamphibians, the open cheeks indicate the attachment of jaw-closing musculature along the braincase and medial skull bones.

Terrestrial locomotion appears to have been difficult for lysorophians, whose limbs were too small to lift the body off the ground for walking, and whose vertebrae lacked the specialized articulations required for snake-like creeping on the ground. Instead, they are more likely to have lived in water, swimming by lateral undulations – much like the extant salamander *Amphiuma*, which has similar rudimentary limbs and body proportions. This is consistent with their possession of a hyobranchial apparatus, a device probably used for inertial suction feeding under water (Wellstead 1991). This would imply the retention of open gill slits in adult lysorophians, but not necessarily functional gills. If this scenario holds true, the resemblance to larval or neotenic salamanders is based on convergent adaptations for locomotion and feeding rather than on common ancestry, as would be the resemblance to amphisbaenians in burrowing behavior. As evidenced by the genus *Brachydectes*, burrowing was obviously a response to seasonal drying of its aquatic habitat (Wellstead 1991).

2.2.4.4 Nectridea

The Nectridea were a primarily aquatic group, existing from the Pennsylvanian (Westphalian A, ~315 myr) to the Early Permian (Artinskian, ~280 myr) (Bossy and Milner 1998). They encompass three clades that differ significantly in skull morphology and body architecture: (1) the Urocordylidae were newt-like, gracile predators with very long tails, (2) the Diplocaulidae had short trunks and skulls with posterior projections (Figure 2.15), and (3) the Scincosauridae were a small terrestrial clade of elongate lizard-like appearance with a relatively tiny skull.

Two plesiomorphic features are interesting, indicating the primitive condition for the group. Nectrideans apparently retained five digits in the manus (A.R. Milner, personal communication 2012). Second, the supratemporal was retained in some nectrideans, attaining a very peculiar position in *Ptyonius*: instead of anterior, it lies lateral to the tabular, a situation otherwise only found in stem-amniotes (e.g., Gephyrostegidae). With the exception of the terrestrial *Scincosaurus*, nectrideans had well-established lateral-line grooves.

- **Urocordylidae.** These usually 15–20 cm long forms were able swimmers with tails longer than the rest of the body, composed of tall uniform vertebrae (Bossy and Milner 1998). The limbs were gracile, without bony carpals and tarsals, and propulsion was mainly generated by the laterally flattened tail. The skulls were narrow and high-sided, with a parabolic to triangular outline, some species bearing a pointed snout (rostrum) with numerous bulbous teeth with pointed tips, capable of impaling prey items (Bossy and Milner 1998). More lightly built than in other lepospondyls, the palate had substantial openings in some taxa. The basicranial joint was moveable, with a straight suture between cheek and skull table permitting some kinesis. Extensive sutures in the dermal skull bones indicate that the snout could be raised separately during feeding (Bossy and Milner 1998). These Pennsylvanian taxa probably spent their life in freshwater lakes and streams, most likely feeding on crustaceans and insect larvae.

- **Diplocaulidae.** These were larger forms (20–150 cm) with broad, flat skulls and prominent bony (tabular) horns in the cheek. Pennsylvanian taxa were small (10–15 cm range), with short horns and abbreviated trunks (Bossy and Milner 1998). The skull was completely akinetic, indicated by the firmly sutured palate bones, and the gape was short. In short-horned taxa (*Keraterpeton*, *Batrachiderpeton*), the tabular articulated with the cleithrum of the shoulder girdle to firmly anchor the skull with the girdle – this prevented lateral excursion of the head while swimming in undulations (Bossy and Milner 1998). In the Early Permian *Diplocaulus*, the cheek horns were extremely long, giving the outline of a "Napoleon hat" or boomerang. It has been argued that these extreme horns formed a hydrofoil. Wind-channel experiments revealed that this device maximizes lift at low current speed and low angle of attack (Cruickshank and Skews 1980). Imprints of animals with exactly this head shape were reported from a deposit in Thuringia (Germany), showing that *Diplocaulus* rested on the water bottom and had a region with soft folds behind the skull, presumably where the gills were located (Walter and Werneburg 1988). These imprints were found in red siltstones that formed in a river, with sedimentary structures indicating relatively fast-flowing water. This genus is known from Texas (Olson 1951) and Morocco (Germain 2009). In *Diplocaulus*, the skull and trunk were extremely flattened, with long straight ribs, whereas the tail was longer than head and trunk combined. Unlike in the short-horned taxa, the tabular and cleithrum were decoupled in *Diplocaulus*, which is consistent with the inferred different locomotory pattern. As the long horns prevented mouth opening while resting on the ground, the animals must have fed during swimming (Bossy and Milner 1998).

- **Scincosauridae.** Two terrestrial genera, *Scincosaurus* and *Sauravus*, were found in lake deposits at Nýřany (Czech Republic) as well as Blanzy and Autun (France). Unlike in other nectrideans, the limbs were robust and all elements were ossified. The teeth are very unusual: waisted crown with spatulate tip and two keels (Milner and Ruta 2009). Apparently, these 10–15 cm long lizard-like animals were feeding on small arthropods, which could be manipulated with their gripping dentition. The short limbs and small feet suggest that *Scincosaurus* was not a fast runner, but probably lived in the leaf-litter zone of forests (Bossy and Milner 1998).

2.2.4.5 Aïstopoda

The aïstopods, with their highly modified skulls, eel-like bodies, and no traces of limbs, are a most distinctive clade (Anderson 2003a, 2003b). This group is also the first among lepospondyls to appear in the fossil record, by mid-Viséan time (Early Carboniferous, ~340 myr). In fact, Germain (2008) recently suggested that the conquest of land was headed by aïstopods, with snake-like crawling predating four-legged walking. While this is in accordance with our present (very incomplete) stratigraphic knowledge of crown tetrapods, it is almost certainly wrong, as *Ichthyostega* is likely to have set foot on land by the Late Devonian (Clack 2012). In addition, *Ichthyostega*-like tetrapodomorph footprints of Middle Devonian age (mid-Eifelian, ~395 myr) have recently been discovered in Poland (Niedźwiedzki et al. 2010).

Aïstopods exemplify interesting evolutionary patterns not readily apparent in other tetrapods (Anderson et al. 2003). Starting with the opening of the cheek in basal aïstopods (*Lethiscus*), the dermal bones of the skull were successively reduced to thin platelets or rods, and some were eventually lost. As in other groups, fenestration of the cheek probably permitted the jaw-closing musculature to expand. The driving force behind this might have been miniaturization or a change in feeding. At any rate, the fenestration went much further than in other tetrapods, and the interesting aspect here is that the known aïstopods illustrate how this reduction progressed (Anderson *et al.* 2003): (1) a fontanelle remained open between the postorbital and squamosal (*Lethiscus*), (2) the postorbital was first reduced and then lost, permitting the jaw muscles to attach along a larger area (*Ophiderpeton* and *Oestocephalus*), and (3) the parietal was lost, giving yet more room for the muscles, as the underlying braincase was strengthened (*Phlegethontia*). At the same time, dermal bones were also lost in the palate (vomer, palatine) and the skull as a whole was largely simplified, with a heavily ossified braincase in the rear end and few remaining struts of dermal bone in the snout (Anderson *et al.* 2003). The developmental and evolutionary implications will be discussed in Chapter 8, section 8.6, but it should be stressed that the evolutionary pattern is unique in its detail and clarity. The described evolutionary trend correlates with a marked size reduction (*Phlegethontia* skulls measure only 3–10 mm), but also an enormous increase in the number of vertebrae (Anderson 2002).

The lifestyle of aïstopods remains an open question. The teeth range from small, recurved ones to robust ones with chisel-shaped crowns, much as in adelospondyls. Judging from the body architecture, both aquatic and terrestrial locomotion is conceivable. However, Anderson (2002) noted that despite their wide distribution, aïstopods are never common in any deposit. As most *Lagerstätten* formed in aquatic environments, aïstopods might well have been terrestrial, living along the shores of streams and ponds.

2.2.4.6 Adelospondyli

This small group shares only few features with other lepospondyls but has some interesting overlap with other early tetrapods. It is exclusively known from the late Viséan (Early Carboniferous, ~330 myr) of the Scottish Midland Valley (Andrews and Carroll 1991). Like lysorophians and aïstopods,

adelospondyls were extremely elongate, and they were also small in size (7 cm skull length). They retained dermal elements in the shoulder girdle, but lacked the pelvis and limbs (Andrews and Carroll 1991). Lateral lines and numerous hyobranchial bones indicate an aquatic mode of life. The teeth were small but chisel-shaped, a feature shared with some anthracosaurs and extant amphiumid salamanders (*Amphiuma* is known to feed on thick-shelled crustaceans). The enlarged posterior skull suggests the presence of voluminous jaw musculature, as in extant *Amphiuma*. In contrast to most early tetrapods, adelospondyls had an akinetic skull, with the squamosal and tabular apparently fused to a compound element (Andrews and Carroll 1991). The vertebrae are fully cylindrical, without rudiments of intercentra, but the ribs are remarkably short, as in temnospondyls.

2.2.4.7 Acherontiscidae

A further potential lepospondyl taxon is based on a single find of a tiny skeleton from the Late Carboniferous of Scotland, *Acherontiscus* (Carroll 1969). The skull measures just 1.5 cm in length, and the animal may well have been a larva or juvenile. Whereas the size and elongate trunk resemble aïstopods, adelospondyls, and lysorophians, the major difference is the retention of fully disc-shaped pleurocentra and intercentra. This is consistent with the condition in some anthracosaurs, although the poorly preserved skull is more similar to that of microsaurs. The snout is short, the posterior skull elongate, but the number of bones in the temporal region unclear (Carroll 1969). As in adelospondyls, there is an interclavicle and a clavicle, but no other girdle or limb bones. The ribs are moderately long, being consistent with those of juvenile microsaurs.

2.2.5 Gephyrostegida

Gephyrostegids are a small group of stem-amniotes from the Late Carboniferous of Europe and North America. They had a lizard-like body outline and fall into the 15–40 cm size range. Their skull resembles that of early amniotes in the configuration of the posterior skull table and general proportions (Figure 2.16). The conical teeth and the apparently moveable cheek suggest that they

Figure 2.16 Relationships of the amniote stem lineage (adapted from Ruta *et al.* 2003a and others).

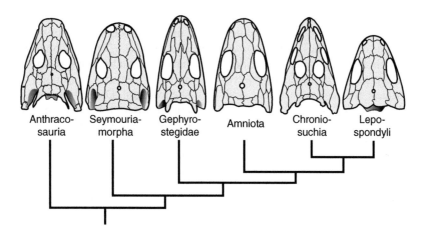

retained the feeding mechanism of anthracosaurs. The girdles and limbs are fully ossified and lateral lines are absent. In contrast to many stem-amniotes, gephyrostegids have short trunks, measuring only twice the length of the skull, a feature approaching the amniote condition.

2.2.6 Amniota

Extant amniotes have reached a tremendous diversity, and have evolved taxa as disjunct as turtles, snakes, birds, crocodiles, elephants, whales, and humans. Their success is based on their independence from the water, which involved major developmental and anatomic changes. Unfortunately, the most significant modifications occurred in the soft parts. Recognition of crown-group amniotes in the fossil record is therefore more difficult than, for instance, that of lissamphibians or other vertebrate groups with well-defined skeletal characters.

1. **No aquatic larva**. In contrast to lissamphibians, amniote young never undergo a larval phase. The complete loss of the aquatic larval stage implies the reduction of various additional features, most notably (1) the external larval gills, (2) the lateral-line system, (3) the tail fin, (4) the permeable skin, and (5) the larval hyobranchial skeleton. Whereas in most lissamphibians the bones form during the larval phase, amniotes go through a condensed embryonic period in which the skeleton is almost fully established.

2. **Large terrestrial eggs**. As fully land-dwelling animals, amniotes lay terrestrial eggs with a usually hard shell. Secondarily they may give birth to live young, omitting the production of egg shell. Unlike in many lissamphibians, hatchlings are more similar to adults. The embryo is nourished by a proportionally large quantity of yolk.

3. **Additional embryonic membranes**. In contrast to the relatively simple eggs of fishes and lissamphibians, amniotes have two additional embryonic membranes: amnion and chorion. These membranes protect the embryo and separate it from the egg shell, and each of them is derived from both ectodermal and mesodermal tissues. The amnion is the inner membrane, enclosing a water-filled cavity in which the embryo develops, while the chorion surrounds the amnion.

4. **Excretion**. The nitrogen-rich metabolic waste products leave the body as uric acid. Because it is almost insoluble, uric acid is much better suited for storage in the embryo. This is necessary because, unlike in lissamphibians, the shell of amniote eggs does not permit waste products to leave the egg.

5. **Epidermal scales**. The amniote skin is protected against water loss and damage by a

keratinized layer of the outer skin (epidermis). In most groups, the epidermis is parceled, with single units referred to as "scales." These scales are entirely epidermal, and thus differ from the bony scales of fishes and Paleozoic tetrapods, which develop in the deeper dermal layer of the skin. The borders between epidermal scales are less keratinized and more flexible. Epidermal scales are common to all reptiles, but also occur in extant mammals and are considered to be primitive for all amniotes. Fingernails and claws are other examples of regionally specialized keratinized epidermis.

6. **Penis.** In contrast to lissamphibians, amniotes have an unpaired intromittent organ that develops from the inner wall of the cloaca. (In lepidosaurian reptiles, this organ is lost, and in squamates paired penes have evolved: see Mickoleit 2004).

7. **Loss of cleithrum.** In the pectoral girdle, the dermal cleithrum was entirely reduced in the amniote stem-group and absent throughout the crown group.

8. **Pleurocentrum.** The intercentrum is either reduced to wedges or absent in extant amniotes. The pleurocentrum is always the main centrum, having a cylindrical structure and always bearing the neural arch, to which it is often fused.

9. **Transverse process.** The pterygoid has a deep-reaching, transversely aligned process that primitively bears a tooth row.

10. **Septomaxilla.** Where present, the small septomaxilla is a superficial dermal bone in early tetrapods and lissamphibians. In amniotes, it has lost contact with the skull roof and is located within the narial cavity.

2.2.6.1 Stem-amniotes and early crown amniotes

- **Westlothiana.** An elongate stem-amniote from East Kirkton (Scotland) which currently appears to be the earliest close amniote relative, having an Early Carboniferous age (~335 myr) (Smithson and Rolfe 1990).

- **Casineria.** A contemporary of *Westlothiana*, from Cheese Bay in Scotland, this taxon is based on an incomplete single skeleton without a skull (Paton *et al.* 1999).

- **Diadectidae.** This was one of the first fully terrestrial clades likely to have shared the key amniote features (amnion, chorion, hard-shelled eggs). Diadectids were herbivores and are known from Pennsylvanian–Early Permian deposits in Euramerica. Probably nesting below the mammal–reptile split, they fall outside the crown amniotes (Berman *et al.* 2004).

- **Hylonomus.** A crown-group amniote known from the tree-stump deposits of Joggins, Nova Scotia (Canada), dating ~315 myr. *Hylonomus* is a close relative of the Pennsylvanian *Petrolacosaurus*, which ranks among the oldest and most primitive diapsid reptiles (Müller and Reisz 2006).

2.3 The lissamphibian stem-group (Temnospondyli)

Temnospondyls were the most speciose clade of Paleozoic tetrapods (Figure 2.17), and survived well into the Mesozoic Era (Milner 1990; Holmes 2000). Comprising some 198 genera and 295 species to date, they are known from the late Viséan (Early Carboniferous, 330 myr) through the Aptian (Lower Cretaceous, 115 myr). In the case that lissamphibians really are temnospondyls, then the group survives to the present day, spanning the full 335 myr record – this would be the longest fossil record for any living tetrapod clade.

The name Temnospondyli (Greek *temneīn* = to cut; *spóndylos* = vertebra) refers to the compound structure of their vertebrae: their centra are divided into a wedge-shaped ventral element (intercentrum) and usually paired dorsal half-wedges (pleurocentra). In reality, this type of vertebra is common and more widespread among early tetrapods and even tetrapodomorphs. Truly diagnostic features are found elsewhere in the skeleton, but admittedly the group is less easy to define than others. Another feature often emphasized in temnospondyls is the complicated pattern by which dentine and enamel are folded in their

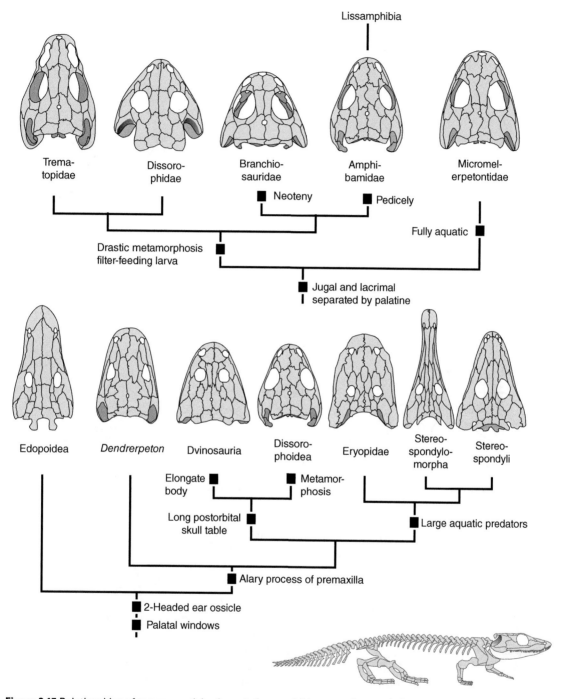

Figure 2.17 Relationships of temnospondyls, the putative amphibian stem lineage (adapted from Schoch 2012, reproduced by permission of John Wiley & Sons).

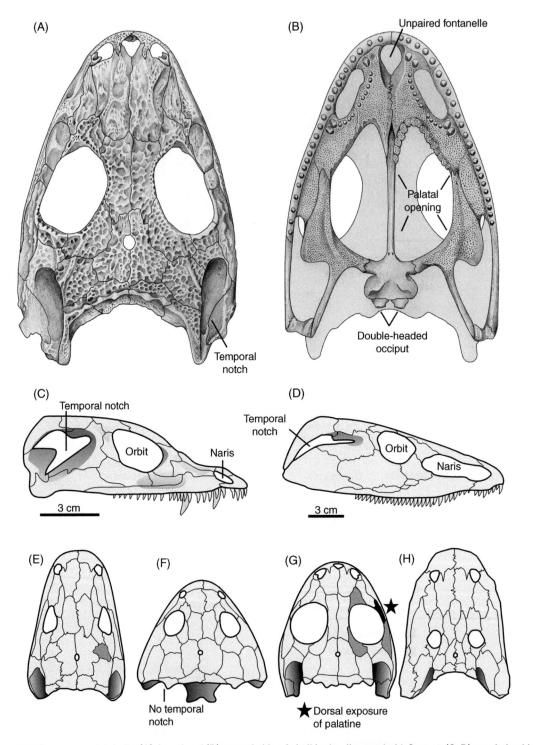

Figure 2.18 Temnospondyl skulls: (A) dorsal and (B) ventral side of skull in the dissorophoid *Cacops*; (C, D) crania in side view showing divergent ear structures in (C) dissorophid *Cacops* and (D) trematopid *Acheloma* (adapted from Polley and Reisz 2011, reproduced by permission of John Wiley & Sons); (E–H) skulls in dorsal view (E, *Dendrerpeton* adapted from Holmes *et al.* 1998, reproduced with permission of Taylor & Francis; F, *Dvinosaurus* adapted from Bystrow 1938, reproduced by permission of John Wiley & Sons; G, *Doleserpeton* adapted from Sigurdsen and Bolt 2009; H, *Eryops* adapted from Sawin 1941).

teeth. This is best seen in a cross-section of the large tusks. Resembling a maze, this structure had been coined "labyrinthodont" (labyrinth-toothed) by Richard Owen as early as 1841. The name "Labyrinthodontes" was an early synonym for Triassic temnospondyls, but has long been abandoned, as labyrinthodont teeth occur in many tetrapodomorphs and evolved separately in ichthyosaurs and monitor lizards.

Temnospondyl characters. Two of the crucial temnospondyl autapomorphies are also shared by the Lissamphibia. If the temnospondyl hypothesis of lissamphibian origin is preferred, then these characters indeed support temnospondyl monophyly. The other features are not exclusive to temnospondyls.

1. **Palatal openings**. All temnospondyls have wide (round or oval) openings in the palate (interpterygoid vacuities). These were covered by tissue bearing small, polygonal bony platelets, which were moveable against one another and permitted the eyeballs to be drawn into the mouth cavity during swallowing.
2. **Skull flat and braincase wide**. Another consequence of cranial flattening is the much wider braincase of temnospondyls and lissamphibians in comparison to all other tetrapods, already apparent in early embryonic stages of modern taxa (Goodrich 1930).
3. **Occiput**. A firm connection between the exoccipitals, postparietals, and tabulars is established by vertical and oblique, column-like processes, a feature only shared with colosteids.
4. **Wide vomers**. In the snout region, the vomers form wide plates, broadly separating the choanae. This contrasts with other tetrapods, but is also established in lissamphibians.
5. **Short ribs**. In most temnospondyls and all lissamphibians, the ribs are substantially shorter than in other tetrapods. The primitive condition for the group is indicated by *Dendrerpeton* and *Cochleosaurus*, whereas some later clades secondarily increased the length of the ribs, apparently to form heavy skeletons (e.g., *Mastodonsaurus*).
6. **Rod-like stapes**. Unlike in all basal tetrapods and stem-amniotes, the stapes is elongate and rod-like, having two proximal heads. Of these, the footplate pointed into the oval window, whereas the ventral process articulated with the parasphenoid (Bolt and Lombard 1985). The anuran stapes can be understood as a modification of the temnospondyl stapes in that the two heads are retained but the articulation with the palate was abandoned.

Most temnospondyls appear to be variations of one theme – a large, crocodile-like fish-eater inhabiting rivers, lakes, or even marine habitats (Figure 2.18, Figure 2.19). The skull is flat with a parabolic outline, and the numerous teeth are arranged in four rows (two in the upper jaw and palate, two in the mandible).

2.3.1 Edopoidea

Edopoids were 1–3 m long predators with body proportions similar to modern giant salamanders, but a skull superficially resembling that of alligators (Figure 2.19). They are found in lake deposits of the Pennsylvanian coal measures, river channels in Permian red beds, and the terrestrial tree-stump deposit of Florence. This pattern of occurrence suggests that at least *Cochleosaurus* was able to leave the water and crossed the tropical forest, either in search of prey or seeking another water body. The huge *Edops* must have been a top predator in rivers (Romer and Witter 1942), and it was eventually replaced by the similar-sized amphibious *Eryops* in the Texan floodplain environments. In the Nýřany peat lake, *Cochleosaurus* spent its youth in the water, as the large number of specimens and broad range of size classes indicate (Milner 1980b). *Cochleosaurus* and *Nigerpeton* further highlight that the lateral-line system is a feature that must be treated with caution: while the presence of lateral-line grooves is a good indicator of aquatic life in adults, their absence need not imply terrestrial habits. This became evident when lateral lines enclosed in ridges were discovered in *Nigerpeton*. This feature, along with the absence of bone in the carpals, tarsals, and pubic region, indicates that cochleosaurids were not fully terrestrial. These early temnospondyls are best characterized as amphibious generalists, which managed to cope with fluctuating habitats

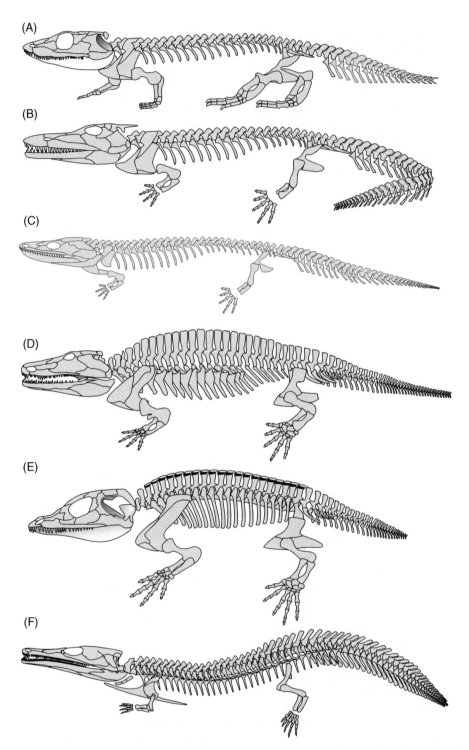

Figure 2.19 Temnospondyl skeletons: (A) *Dendrerpeton* (adapted from Holmes *et al.* 1998, reproduced with permission of Taylor & Francis.); (B) *Cochleosaurus* (adapted from Sequeira 2009); (C) *Trimerorhachis*, (D) *Eryops* (adapted from Romer 1966); (E) *Cacops* (adapted from Schoch 2012, reproduced with permission of John Wiley & Sons); (F) *Trematolestes* (adapted from Schoch 2009a).

in the tropical belt. A puzzling feature of all edopoids is their ear: unlike other temnospondyls, they had a huge and robust stapes pointing into a large, rounded squamosal embayment (Romer and Witter 1942).

2.3.2 Dendrerpeton and Balanerpeton

The early radiation of temnospondyls (Figure 2.20) also produced more fully terrestrial forms, but these were smaller than the edopoids. Two taxa are currently known: *Balanerpeton* from the

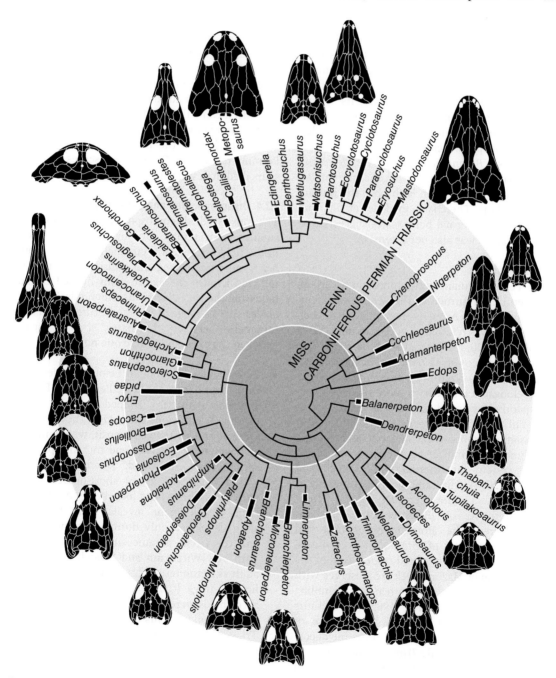

Figure 2.20 Diversification of temnospondyls, the largest Paleozoic tetrapod clade outside amniotes.

Viséan (330 myr) of East Kirkton, Scotland (Milner and Sequeira 1994) and *Dendrerpeton* from the Bashkirian (315 myr) of Joggins (Milner 1980a, 1996; Holmes *et al.* 1998). Both genera were in the 30–50 cm size range, had a full complement of bones in the limbs and girdles, were nevertheless lightly built, and occurred in terrestrial environments. *Dendrerpeton* is a common animal in the tree-stump fauna at Joggins, which demonstrates its ability to live in the densely vegetated forest habitat. *Balanerpeton* is exclusively known from the East Kirkton lake, in which its young apparently hatched and spent their youth, probably preying on ostracods. Both genera have slender, lightly built stapes which probably served as sound transmitters. *Balanerpeton* and *Dendrerpeton* were probably lung breathers but did not employ ribs in lung ventilation: in both taxa, the ribs are extremely short. There is no evidence of gills, and the retention of dermal bony scales suggests that skin respiration was also not extensive.

2.3.3 Dvinosauria

In the Pennsylvanian, one temnospondyl group, which entirely returned to the water, was abundant (Figure 2.20). After its monophyly had been confirmed, it was referred to as Dvinosauria (Yates and Warren 2000; Milner and Sequeira 2011). The best-known dvinosaur is *Trimerorhachis* from the Texas red beds, a 50 cm long animal with a short snout, elongated trunk with more than 30 vertebrae, poorly developed limbs, and a long swimming tail. Most notably, *Trimerorhachis* had a fully ossified hyobranchial skeleton, and concave ceratobranchials indicating the presence of an arterial system for internal gills (Schoch and Witzmann 2011). The same feature is present in *Dvinosaurus*, a Late Permian relative from northwestern Russia (Bystrow 1938; Shishkin 1973). Whereas the aforementioned taxa had apparently four pairs of *internal gills* as adults, juveniles of the closely related *Isodectes* from the Pennsylvanian of Mazon Creek are preserved with three pairs of *external gills*, a feature resembling larval salamanders (Milner 1982). Dvinosaurs appear to have lived under a broad range of conditions: *Isodectes* was found in marine deposits at Mazon Creek, while *Trimerorhachis* lived in small lakes on floodplains in Texas, then a lowland

setting under strong marine influence (Parrish 1978); it is therefore likely that it tolerated brackish conditions. *Dvinosaurus*, *Acroplous*, *Tupilakosaurus*, and *Thabanchuia* are known from aquatic deposits in Russia, North America, and South Africa, and *Erpetosaurus* from an oxbow lake at Linton, Ohio. The tiny limbs, elongated bodies, and lateral lines of all well-known dvinosaurs suggest that these animals lived in the water, and the gills indicate that they relied to a large extent on gills throughout their lives. This must have been a major advantage in many situations – but with it came the disadvantage of being trapped in the water body whatever happened to it. This is is exemplified by the bone beds in the Lower Permian red beds of Texas, where scores of dvinosaurs died in desiccating ponds (Parrish 1978).

2.3.4 Dissorophoidea and Zatracheidae

The most studied temnospondyl clade comprises small, terrestrial taxa. Although often overlooked in the field because of their small size, this group has a most fascinating and multifaceted story to tell. It includes the exotic spiny-headed zatracheids and the very diverse dissorophoids. The two clades share a range of unique features, such as a fontanelle between the nares, a wide space between the eyes, and a large otic notch.

- **Zatracheidae**. This small taxon includes only three genera from the Early Permian of Europe and North America. Named after the spike-bearing Texan genus *Zatrachys*, they are best known from a European deposit, where numerous skeletons of larvae and metamorphosing specimens were found: *Acanthostomatops* was a small (15–25 cm) short-bodied taxon with a very large head (Boy 1989). Analysis of larval development revealed that the wide skull developed only during metamorphosis, and the trunk became substantially shorter with age (Witzmann and Schoch 2006). The adult skeleton resembles the large carnivorous horned frogs (*Ceratophrys*), suggesting that zatracheids were sit-and-wait predators.
- **Dissorophoidea**. This is a vast Carboniferous–Triassic clade (five families, 45 genera, 72 species) (Figure 2.21). Ecologically, it encompasses three main groups: (1) heavily armored,

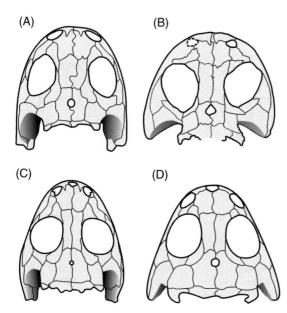

(A)

(B)

(C)

(D)

Figure 2.21 Four taxa considered as closely related to lissamphibians by various authors: (A) amphibamid *Amphibamus* (adapted from Milner 1982); (B) *Gerobatrachus*, a probable stem-batrachian (adapted from Anderson *et al.* 2008) (C) amphibamid *Doleserpeton* (adapted from Sigurdsen and Bolt 2009); (D) branchiosaurid *Apateon* (adapted from Schoch 1992).

terrestrial carnivores (Trematopidae, Dissorophidae), (2) tiny and unarmored terrestrial insectivores (Amphibamidae), and (3) aquatic perennibranchiates (Branchiosauridae, Micromelerpetidae). Dissorophoids evolved a drastic metamorphosis, by which a larva with long external gills transformed into a terrestrial adult. This is best exemplified by the amphibamids, which were all terrestrial, whereas branchiosaurids often delayed or abandoned metamorphosis to remain in the water. This gave rise to a cluster of species that were filter-feeders, surviving in habitats that were hostile to fishes (Boy and Sues 2000).

Dissorophoids have been considered as the stem-group of lissamphibians in some phylogenetic scenarios (Milner 1993; Ruta *et al.* 2003a; Zhang *et al.* 2005) (Figure 2.22), while most authors at least accept them as the stem-group of batrachians (frogs and salamanders). According to Laurin (1998), they

evolved batrachian characters in parallel. Irrespective of their relation to lissamphibians, dissorophoids underwent a remarkable evolutionary radiation during the Late Carboniferous and Early Permian (~307–270 myr). They were the only temnospondyl clade that evolved fully terrestrial as well as perennibranchiate taxa in the same genera, and evolved a constrained type of metamorphosis as a special version of developmental plasticity (Schoch 2009a). This enabled dissorophoids to conquer habitats otherwise inhabited by amniotes only (Reisz *et al.* 2009).

- **Micromelerpetidae.** The basal clade (or grade) of dissorophoids, containing only aquatic species (10–30 cm) with more or less larval appearance. Metamorphosis was slow and never completed in any of the known taxa (Boy 1995; Schoch 2009a).
- **Dissorophidae and Trematopidae.** Dissorophids, the most speciose dissorophoid clade (22 taxa) include 20–100 cm long highly terrestrial forms usually found in overbank or upland deposits (Schoch 2012). Large and robust limb bones, bony armor on the back, and a large ear region characterize these taxa (Reisz *et al.* 2009). The closely related trematopids were less heavily armored and had a different ear region (Polley and Reisz 2011). Both groups have huge fangs, indicating that they focused on larger prey than other dissorophoids. Larvae are so far only known from one trematopid, *Mordex*, and these appear to be remarkably similar to branchiosaurids (Milner 2007; Werneburg 2012).
- **Amphibamidae.** A range of miniaturized dissorophoids, probably monophyletic, in the 5–15 cm size range. They have broad skulls with short snouts and large eye openings, no armor, and a short tail in adults (Schoch and Rubidge 2005; Clack and Milner 2010). Some taxa (*Amphibamus*, *Doleserpeton*) have pedicellate and bicuspid teeth that closely resemble those of lissamphibians (Bolt 1969). As in batrachians, the palate bones are reduced to slender struts and the choana is transversely extended. The pleurocentrum has become the main element in the vertebra, approaching the lissamphibian condition in some genera (*Doleserpeton, Amphibamus*) (Sigurdsen and

Figure 2.22 Relationships of lissamphibians according to the temnospondyl hypothesis (adapted from Milner 1988; Ruta and Coates 2007). The extinct Albanerpetontidae form a separate branch within the lissamphibians.

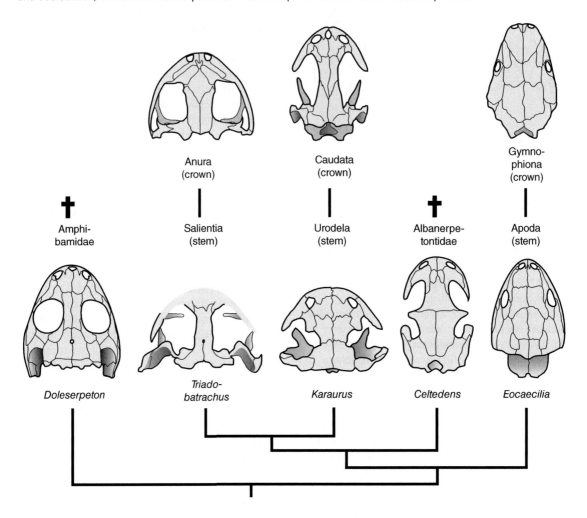

Bolt 2010), and some taxa lack bony scales (*Micropholis, Platyrhinops*).

- **Branchiosauridae**. A clade of small (5–12 cm) dissorophoids retaining larval features as adults, such as external gills, juvenile skull morphology, and the failure of many enchondral bones to ossify. Metamorphosed adults (known from *Apateon gracilis*) are very similar to amphibamids (Schoch and Fröbisch 2006). Branchiosaurids are the first non-lissamphibian taxon in which neoteny has been confirmed, after skeletochronology revealed that they attained sexual maturity while retaining a

larval morphology (Sanchez *et al.* 2010). Branchiosaurids are so abundant in some formations that some authors have argued over their biostratigraphic relevance (Werneburg and Schneider 2006).

2.3.5 Eryopoidea

During the Pennsylvanian, temnospondyls diversified in rivers and oxbow lakes, and probably also invaded coastal lagoons. The North American *Eryops* (with a length of up to 2.5 m) dwelled on floodplains and river shores, where it was one of the largest predators, possibly rivaling the equally large synapsid *Dimetrodon*. The robust limbs suggest that these heavy animals were capable of crossing dry land, but the long swimming tail and the typical fish-eater dentition indicate that they spent a lot of time in the water. The massive rib cage was floored by a dense sheet of bony scales, probably as a protection against damage to the belly while crawling over land. A close relative of *Eryops*, *Onchiodon*, is known from Europe (Boy 1990; Werneburg 2008). In southwestern Germany, the similar but slightly smaller *Sclerocephalus* (1.5–1.8 m) inhabited lakes of various sizes and ecological properties (Boy 1988). Preservation of larvae and juveniles in the same deposits revealed that *Sclerocephalus* was able to respond to different environmental conditions by modifying its larval development. Adult size, presence of lateral lines, length of swimming tail, and other features were adjusted to particular water conditions (Schoch 2009b). *Sclerocephalus* preyed on a particular genus of fish (*Paramblypterus*), which is always preserved in its gut contents. Although some large-growing populations of *Sclerocephalus* might have left the water occasionally, preserved tracks suggest that locomotion on land was strenuous for these sluggish animals. The largest Permian water body in Central Europe, the 80 km long Lake Humberg, was inhabited by *Archegosaurus*, a close relative of *Sclerocephalus* (Boy and Schindler 2000). This slender and gracile temnospondyl had a gharial-like elongated snout and evidently fed on acanthodian fishes. *Archegosaurus* was fully aquatic and thus less heavily ossified than either *Sclerocephalus* or *Eryops*, and is considered a basal relative of the dominant Triassic temnospondyls, the Stereospondyli (Witzmann 2006).

2.3.6 Stereospondyli

Despite the global impact of the end-Permian biotic crisis, temnospondyls managed to spread and diversify rapidly during the Early Triassic (Warren 2000). Most bones found in the rocks from that time window stem from these large amphibians, usually accumulated in pebbly or sandy river deposits. This, of course, is also due to size, as Triassic temnospondyls reached larger body size than ever before or since. Apparently, the relatives of Permian *Archegosaurus* continued

to exist in small lakes, rivers, and deltas before they evolved into more diverse niches. In South Africa, this diversification is well documented: the river- and lake-dwelling rhinesuchids (basal stereospondyls) were largely replaced by stereospondyls with terrestrial adaptations (*Lydekkerina*), and somewhat later also by fully aquatic forms (*Batrachosuchus*). In Europe, the river and pond faunas of the Buntsandstein were replaced by Middle Triassic swamps, deltas, and brackish marshes – all populated by diverse stereospondyl faunas. Most of these fully aquatic taxa differ conspicuously in head size, skull morphology, shape and arrangement of teeth, and body outline. They ranged from 1 to 6 m in length and were all predators, co-occurring with diverse fish faunas. Three clades are especially noteworthy.

- **Capitosauria**. These heavily built inhabitants of deltas, large rivers, and lakes are occasionally found in lagoonal and coastal marine deposits. The speciose clade was represented in all regions of Pangaea, with *Mastodonsaurus* in Europe and the Urals reaching 5–6 m in length. The crocodile-like body outline is well documented in the Australian genus *Paracyclotosaurus*. Capitosaurs had pachyostotic (extra-heavy) skeletons and probably lived on the bottom of large water bodies as ambush predators. Range: Early to Late Triassic (~250–200 myr).
- **Trematosauria**. These slender-bodied and long-snouted forms reached the widest distribution in the Early Triassic, probably due to their tolerance of (or even preference for) brackish and marine conditions. Marine trematosaurids are known from Svalbard, Madagascar, Pakistan, and Tasmania (Schoch and Milner 2000). They were able swimmers, with short trunks and very long tails, and probably captured fish by lateral sweeps of the head like the extant alligator gar *Lepisosteus*. The metoposaurids are a Late Triassic clade that evolved from trematosaurids and became aquatic bottom-dwellers similar to capitosaurs. Range: Early to early Late Triassic (~250–220 myr).
- **Plagiosauridae**. A clade of 1–3 m long bizarre flattened animals with extremely wide skulls, large eye openings, and rudimentary limbs

(Shishkin 1987; Hellrung 2003). Despite the small number of species this was an ecologically diverse clade, ranging from freshwater lakes and rivers over deltaic to brackish–marine habitats. The heavily armored *Gerrothorax* was not only one of the longest-lived genera of temnospondyls (40 myr), but also flexible with respect to its habitat: it was found in brackish lagoons, deltas and swamps, and large lakes, along the shores of hypersaline lakes, and in small rivers (Schoch and Witzmann 2012). Range: Middle to Late Triassic (~240–200 myr).

2.4 Albanerpetontidae

Amphibians are often described as falling into two entirely separate groups: the monophyletic lissamphibians and the Paleozoic–Mesozoic grade of early tetrapods. However, there is a third group – the Albanerpetontidae (Figure 2.23). They are mostly overlooked because the clade is small, articulated specimens are rare, and they are regarded as part of the problem rather than the solution. That is to say, they add little to clarify relationships between lissamphibians and

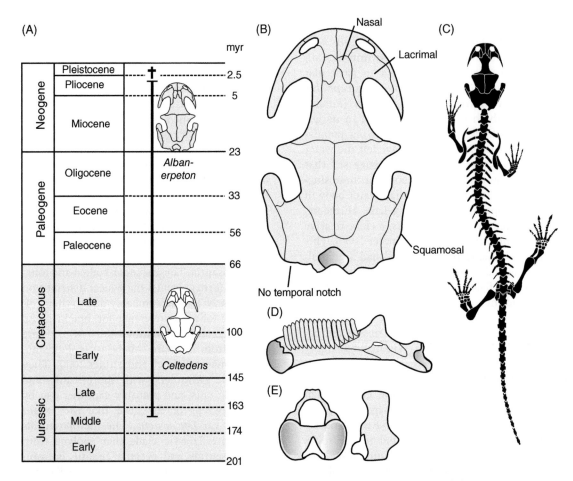

Figure 2.23 Albanerpetontidae: (A) stratigraphic range (adapted from Gardner and Böhme 2008); (B) skull roof of *Albanerpeton*; (C) skeleton restoration of *Celtedens*; (D) mandible of *Albanerpeton* in medial (inner) view; (E) atlas vertebra of *Albanerpeton* in anterior and lateral views. B, D, E adapted from Gardner 2001, reproduced with permission of John Wiley & Sons; C adapted from McGowan and Evans 1995, reproduced with permission of Nature Publishing Group.

Paleozoic groups, but pose additional problems, especially by revealing convergences between extant clades.

Deriving their name from their occurrence in fissure fills near Grive St Alban, France, albanerpetontids were first discovered in the Cretaceous of Italy by Costa (1864), who assigned them to salamanders. Superficially, they look like tiny land salamanders: tailed with well-developed limbs, four fingers, feeble skeletons, elongated vertebrae, and few elements in the skull. Most current workers therefore place them in Lissamphibia, arguing that they form the caudate sister taxon, a view that I follow here.

Albanerpetontids have a remarkable fossil record for two reasons: first, they span some 163 myr (Bathonian through Pliocene); and second, the clade became extinct only very recently (3 myr ago). One would almost expect to find a live one in a cave somewhere. The most numerous finds of the group are from Europe, with decreasing abundance in North America, Morocco, and Central Asia (Kazakhstan, Kyrgyzstan). The oldest finds are from western Europe, and only by the end of the Early Cretaceous (~112 myr) did they appear in North America, where they had already disappeared in the Paleogene (~55 myr). That said, the assignment of isolated material is often problematic: Curtis and Padian (1999) reported on vertebrae from the Eraly Jurassic Kayenta Formation which they assigned to salamanders, but Averianov et al. (2008) suggested they might equally well have come from albanerpetontids.

Ecologically, albanerpetontids are also an interesting group. They have been interpreted as fossorial, based on their cranial, mandibular, and vertebral structure (Estes and Hofstetter 1976). According to Wiechmann (2000), who studied a huge sample from the Middle Jurassic Guimarota mine (Portugal), they lived in humid soil, evidently in the vicinity of freshwater ponds. The robust skull could have been used as a ram or shovel in probing the soil, and the shape of the condyles and jaw joints would have permitted such movements. The albanerpetontid dentition possibly permitted a shearing bite (Gardner 2001), and they probably fed on arthropods with tough chitinous shells (Wiechmann 2000).

In some places, distinct species of *Albanerpeton* were reported, which differed in the morphology of skulls and body size. Apparently, these species fed on different prey and therefore evolved separate ecological niches (Gardner and Böhme 2008). Albanerpetontids occur in two main types of *Lagerstätten*: fissure fills and floodplain deposits. The best material stems from the Lower Cretaceous of Las Hoyas (Spain), where articulated specimens with skin preservation were found (McGowan and Evans 1995). The most important albanerpetontid characters are listed below.

1. **Fused frontals**. The frontals are the most characteristic elements, not only in their complete fusion, but also in the polygonal ornamentation.
2. **Interfingering joint in mandible**. The symphysis bears a ball-and-socket joint between left and right jaw. Nothing like this is found in any other tetrapod.
3. **Chisel-shaped teeth**. The marginal teeth are very long and robust, with chisel-shaped, tricuspid crowns. Unlike the teeth of most lissamphibians, they are not pedicellate.
4. **Modified cervical vertebrae**. The centra of atlas and axis form a tripartite facet.
5. **Bony scales**. The retention of thin ossified scales, similar to those of many Paleozoic tetrapods, is generally considered a plesiomorphy.

2.5 Lissamphibia

Modern amphibians (lissamphibians) are remarkably small compared to most early tetrapods. With few exceptions, they do not exceed 10 cm in length. They are ectotherms characterized by a skin rich in glands and a poorly ossified skeleton. The bony elements, especially the dermal bones of the skull and pectoral girdle, are substantially reduced compared to the basal tetrapod condition. The skull, for instance, contains less than half the number of bones found in that of a regular temnospondyl or lepospondyl. Most characteristic of many lissamphibians is their ontogeny, which involves an aquatic larval phase in which external larval gills are used as respiratory organs.

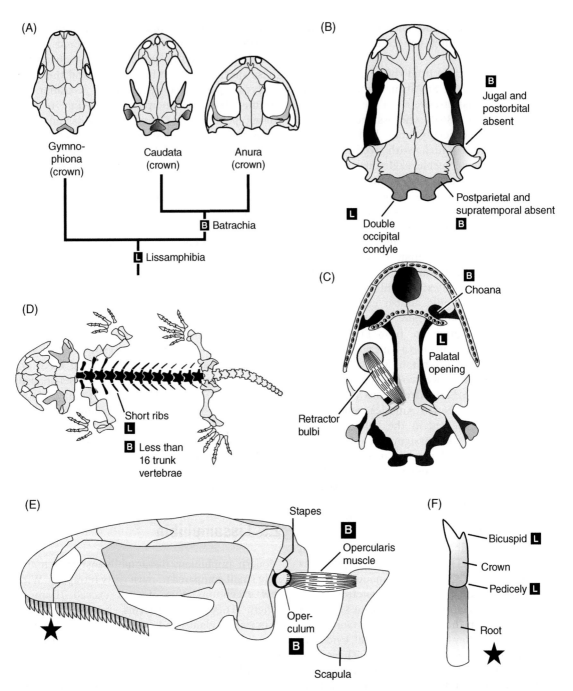

Figure 2.24 Lissamphibian characters: (A) Batrachia hypothesis (adapted from Milner 1988); (B) dorsal and (C) ventral views of skull of salamander *Dicamptodon*; (D) skeleton of *Karaurus* (adapted from Ivakhnenko 1978); (E) skull of *Dicamptodon* in lateral view; (F) pedicellate tooth of *Eocaecilia* (adapted from Jenkins *et al.* 2007).

This biphasic life cycle (larva–adult) is considered the primitive condition for lissamphibians, but many species in all three modern groups have modified this life cycle; this will be discussed in later chapters.

2.5.1 Lissamphibian characters

There are numerous features distinguishing modern amphibians from the bulk of Paleozoic tetrapods, but many of these are not exclusive to the Lissamphibia. A good example is the four-digited hand of salamanders and frogs. Whereas a herpetologist may be perfectly happy with this character, the paleontological perspective indicates problems. There are at least two lineages of Paleozoic tetrapods that share this character, and it originated almost certainly independently in both (temnospondyls and lepospondyls). This is indicated by numerous phylogenetic analyses, which place the temnospondyls at a very different node than the lepospondyls. A similar situation concerns the double occipital condyles, present in all lissamphibians. These problems are discussed in more detail in Chapter 9. Nevertheless, the Lissamphibia are firmly based on a range of anatomical autapomorphies (Figure 2.24) and therefore regarded as a well-established natural group (Parsons and Williams 1963; Milner 1988; Duellman and Trueb 1994; Mickoleit 2004). Molecular data strongly support Lissamphibia as the sister taxon of Amniota (Hedges and Maxson 1993; Feller and Hedges 1998; Zardoya and Meyer 2001; San Mauro *et al.* 2005; Zhang *et al.* 2005).

1. **Teeth pedicellate and bicuspid**. Most lissamphibians have small and not very solid teeth that are attached to the inner side of the jaws (pleurodonty). Adult teeth usually have a zone of weakness – formed by fibrous, poorly mineralized tissue – giving sufficient flexibility to permit the crown to bend inwards into the oral cavity. This condition results from a developmental peculiarity of lissamphibians: the base of the tooth (pedicel of dentine) and its enamel-covered crown mineralize from separate centers and fail to fuse during tooth formation (Smirnov and Vasil'eva 1995). This state is called pedicely, and in tetrapods has not been unequivocally proven outside the

Lissamphibia other than in the temnospondyls *Doleserpeton* and *Amphibamus* (Bolt 1969, 1979; Sigurdsen and Bolt 2010).

2. **Papilla amphibiorum**. Unlike amniotes, lissamphibians have two sense receptors in the inner ear. The first one is the papilla basiliaris, which is present in all tetrapods and supposed to have originated during the fish–tetrapod transition. The second receptor is the papilla amphibiorum, which is exclusive to caecilians, salamanders, and frogs. In lissamphibians, the basiliar papilla focuses on frequencies above 1000 Hz, whereas the amphibian papilla operates within the 600–1000 Hz range.

3. **Canalis perioticus**. A connecting channel between the perilymphatic sac and the perilymphatic cistern in the inner ear.

4. **Gonads with large fat bodies**. These develop ontogenetically from the genital fold and serve as an extra source of energy.

5. **Elbow joint**. In caudates and anurans, the radius and ulna articulate with a single, enlarged structure on the humerus (radial condyle). This condition is also present in the limbed stem-gymnophionan *Eocaecilia*, indicating that the feature is a derived character of Lissamphibia (Sigurdsen and Bolt 2009) that was subsequently lost in caecilians due to reduction of the limbs.

Other lissamphibian features are (6) the short ribs and (7) the palatal openings between pterygoid and parasphenoid (interpterygoid vacuities). These are not exclusive to Lissamphibia but occur in some Paleozoic taxa as well. A further character, the intermaxillary glands (8), are relevant for feeding: these glands are located in the anterior palate and produce sticky secretions that drop through a fontanelle onto the tongue. Similar fontanelles are present in dissorophoid and zatracheid temnospondyls.

Diversity. Currently more than 6300 present-day species of lissamphibians are known, a number that significantly exceeds that of living mammals and ranks about equal with that of squamates, the largest reptile clade alive (Haas 2010).

Distribution. All continents except Antarctica. Few species live north of the polar circle, and, in the southern hemisphere, they extend to Tierra del Fuego in southern Argentina. Lissamphibians are most speciose in the tropics and neotropics. A few anurans and salamanders manage to survive in deserts, where they rely on sporadic rainfalls.

2.5.2 Batrachia

Caecilians, salamanders, and frogs have a long evolutionary history, with each branch reaching back into the Late Paleozoic. Even though fossil evidence is poor, the presence of salientians by Early Triassic time indicates that the salamander and caecilian lineages must have separated at least in the Permian if not earlier. Molecular data suggest a still earlier branching (Zhang *et al.* 2005). It is not easy to spot skeletal synapomorphies between any two of the three extant clades, and many features have been proposed over the last two centuries that support one of the three different alternatives. By far the most preferred hypothesis has been that of the Batrachia (Anura and Caudata forming a clade). This hypothesis has the most robust support from skeletal and soft-tissue characters and has also been supported by recent molecular analyses (San Mauro *et al.* 2005; Zhang *et al.* 2005).

Batrachian characters
1. **Operculum and opercular muscle**. A separate cartilage or bone, located within the oval window of the ear capsule, serves as a second ear ossicle in addition to the stapes. This so-called operculum is not homologous to the gill-covering element in bony fishes. Instead, it has a cartilaginous precursor (which in some caudates never ossifies). The operculum and scapula are connected by means of a muscle (opercular muscle) that effectively connects the inner ear with the hand. Vibrations in the ground are thus transmitted via the forelimb and pectoral girdle to the inner ear.
2. **Scales absent**. Salamanders and frogs completely lack dermal bony scales, which are rarely present in caecilians and were fully retained in albanerpetontids.
3. **Choana**. The embryonic formation of the choana includes endoderm in addition to ectoderm. Morphologically, the choanae of metamorphosed salamanders and frogs are transversely elongate, which affects the outline of the vomer bone.
4. **Vertebral formation**. In batrachians, the vertebral centra originate from a single continuous cartilage cover of the notochord. They thus differ from all other vertebrates, where sclerotomes fall into distinct metameres.
5. **Macula neglecta absent**. The macula neglecta, the sensory cell cluster in the inner ear of bony fishes, is retained in gymnophionans and amniotes but absent in batrachians.
6. **Retina with green rods**. In addition to the "red" rods and cones, the batrachian retina contains also "green" rods (absorbing light in the 432 nm range).

A stem-batrachian? Anderson *et al.* (2008) have suggested that *Gerobatrachus*, a new taxon from the Early Permian (~270 myr) of Texas, is a stem-batrachian. This 10 cm long, broad-headed taxon has a palate similar to salientians and basal urodeles, and vertebrae composed of cylindrical pleurocentra that approach the lissamphibian condition. At the same time, the skull is essentially that of a dissorophoid temnospondyl – retaining all elements in the roof and palate – whereas the trunk is short with only 17 vertebrae. It is puzzling that the leg skeleton has a definitive caudate character (basale commune), and the atlas bears an odontoid peg (shared with caudates, albanerpetontids, *Eocaecilia*, and lepospondyls). This list shows how many unexpected combinations of characters a single new fossil can add. *Gerobatrachus* may indeed be a stem-batrachian, but it may equally plausibly turn out to be a stem-lissamphibian (more advanced than *Doleserpeton*) or a basal urodele, depending on the amount of reversals one is prepared to accept. At any rate, the discovery of *Gerobatrachus* has increased the plausibility of the temnospondyl hypothesis.

2.5.2.1 Anura (frogs and toads)

"A frog is a frog is a frog" – how else might we describe a group whose body plan has been modified so fundamentally that neither the baby nor the adult seems to have anything in common with other organisms? On closer inspection, of course, anurans share many features with salamanders, caecilians, and tetrapods in general. Frogs are extremely successful in terms of species number, range of habitats, and coping with harsh conditions. Perhaps the most remarkable feature is their metamorphosis, which transforms a highly specialized aquatic plankton-feeder into a leaping carnivore in a very short time. Sexual maturity is also reached during this brief phase. Frogs thus appear to have a highly constrained body plan – it is therefore surprising that anuran evolution has produced so many diverse adaptations on this common platform. Currently there are 5453 extant species of anurans (Haas 2010). Thus, frogs and toads are not only by far the largest lissamphibian clade, but even outnumber living mammals in species. The size range of modern anurans is more restricted than in salamanders, 1–30 cm, and the heaviest species weighs only 3.3 kg (*Conraua goliath*), compared with 40 kg in the Japanese giant salamander (*Andrias japonicus*).

The most characteristic feature of adult anurans is their ability to jump through the air and to cope with the impact of landing – numerous skeletal features shared by all anurans accomplish this (Figure 2.25). A functional complex unique to anurans is the urostyle (a rod composed of fused tail vertebrae), the forward-directed and elongated ilium, and the hinge joint between the sacrum and urostyle. In resting pose, the urostyle and trunk vertebrae are angled; when the frog jumps, the two move into one plane (Jenkins and Shubin 1998). As revealed by *Prosalirus*, the jumping ability was acquired by Early Jurassic times in the stem-group of anurans. This ability was therefore a property of the last common ancestor of anurans, even though it was lost in some lineages. Anurans inhabit all continents except Antarctica and the Arctic region. They have not managed to settle on the most remote Pacific islands and in extremely dry desert areas. In the tropics, they are most diverse.

2.5.2.1.1 Anuran characters

In contrast to salamanders, frogs are characterized by numerous definite autapomorphies. These features are unique among all vertebrates.

1. **Frontal and parietal fused**. Although not always fused in the midline, the frontals and parietals are co-ossified with each other in all anurans. This feature is already present in the early salientian *Triadobatrachus*.
2. **Parasphenoid T-shaped**. A condition that comes closest to the anuran morphology is

found in the Late Carboniferous temnospondyl *Amphibamus* (Milner 1982).
3. **Annulus tympanicus**. A cartilaginous ring spans the tympanum in adult frogs, which originates in the larva from an outgrowth of the quadrate.
4. **Urostyle**. The tail vertebrae are fused to a continuous rod, which articulates with the sacral vertebra by means of a hinge joint.
5. **Ribs fused to vertebrae**. The very short trunk ribs are co-ossified with the transverse processes (= flank projections) of the vertebrae.
6. **Radius and ulna, tibia and fibula, fused**. Resisting heavy stress during landing from a leap, the lower arm and leg bones are fused.
7. **Lower jaw without teeth**. Consistent throughout anurans (with one exception), this feature is functionally puzzling; there is no apparent adaptation known.

Other features are: (8) the number of trunk vertebrae is reduced to 10 or fewer, (9) the hindlimbs are much longer than the forelimbs, (10) the tibiale and fibulare are elongated, and (11) the intertarsal articulation.

In addition to the listed characters, anurans also share a few derived features with amniotes but not with caudates or gymnophionans. Most of these are located in the ear region: the stapes as an impedance-matching element, an air-filled middle ear cavity housing the stapes, and a eustachian tube connecting the middle ear with the buccal cavity. These characters form a functional complex. It is therefore generally held that they were acquired in the tetrapod stem-group and lost in salamanders and caecilians. Schmalhausen (1968) discussed some faint developmental evidence suggesting that some salamanders had a rudimentary middle ear cavity, but this remains an open question. It is also unknown whether albanerpetontids had a middle ear cavity, although structurally that region of their skulls resembles the caudate condition.

2.5.2.1.2 Mesozoic salientians (stem-group)

Salientia, the stem-group of anurans, probably dates back well into the Permian. This is concluded not from Permian fossils, but from the presence of the oldest salientians in Early Triassic

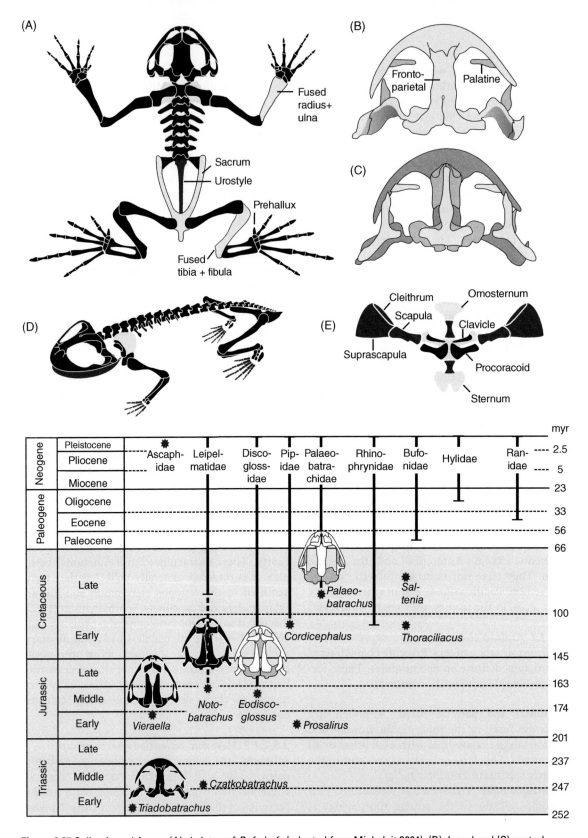

Figure 2.25 Salientia and Anura: (A) skeleton of *Bufo bufo* (adapted from Mickoleit 2004); (B) dorsal and (C) ventral views of salientian *Triadobatrachus*; (D) skeletal restoration of *Triadobatrachus*; (E) stratigraphic range of salientians (black) and anurans (white) (adapted from Sanchíz 1998). B–D adapted from Roček (2000).

rocks (Figure 2.25). These taxa already share a range of anuran characters, indicating that substantial evolution must have occurred during the Late Paleozoic. The fossil record improves in the Early Jurassic, where the first definitive jumping salientians are reported, and evidence mounts that some modern families (Leiopelmatidae, Discoglossidae) were already present by the Late Jurassic. The first anuran characters to form were those of the skull and pelvis, followed by a shortening of the trunk, the fusion of forearm and lower leg bones, and the formation of the urostyle (Jenkins and Shubin 1998). Most of these features suggest the skeleton was strengthened against forces produced by saltation, even though the modern anuran jumping apparatus was fully established only in Late Jurassic taxa. As in other cases, behavior probably paved the way, with structural changes in the skeleton following.

- *Triadobatrachus.* This most basal taxon is also the oldest (~250 myr). It is based on a single, nearly complete skeleton (10 cm) from a carbonate nodule of Early Triassic age in Madagascar (Rage and Roček 1989). *Triadobatrachus* has a moderately long trunk (14 vertebrae), retains a short tail with seven free vertebrae, and still has separate radius/ulna and tibia/fibula. The hindlimbs are only slightly longer than the forelimbs. Anuran characters are already well established: the parasphenoid is T-shaped, frontal and parietal are fused, and the elongate ilium pointed anteriorly. As in frogs, the mandible appears to lack teeth and the palatine forms an edentulous, transverse strut. In sum, *Triadobatrachus* still lacked the elaborate functional complex in the sacrum, but probably jumped in small leaps.
- *Czatkobatrachus.* Found in Early Triassic fissure fills in Poland (~245 myr), this small salientian (~5 cm) is known only from isolated bones (atlas, humerus, scapulocoracoid, ilium). The tail vertebrae were still separate but the ilium was slightly more frog-like than in *Triadobatrachus*, matching the slightly younger stratigraphic age (Evans and Borsuk-Białynicka 2009).
- *Prosalirus.* Based on several partially articulated specimens, this 5 cm long form is

considered the first salientian with some skeletal adaptations for anuran-like saltation (Jenkins and Shubin 1998). This taxon is from the Early Jurassic of northern Arizona (~189 myr). It has elongate hindlimbs, fused forearm and lower leg elements, and a still longer ilium. Most notably, the anuran sacral apparatus was fully established: (1) there was a urostyle, (2) the ilio-sacral joint was well in front of the ilio-femoral joint, and (3) the urostyle was connected to the sacrum by means of a hinge joint.

- *Viaerella.* A tiny form (3 cm) from the Middle Jurassic (~175 myr) of Patagonia, Argentina. It is further advanced towards the anuran condition in having only 10 trunk vertebrae and in that ribs 4–11 are fused to the vertebrae (Báez and Basso 1996).
- *Notobatrachus.* This large form (14 cm) has only nine trunk vertebrae and was found in Middle to Late Jurassic (~161 myr) deposits of Patagonia (Báez and Basso 1996).

2.5.2.1.3 Mesozoic and Cenozoic anurans (crown group)

During the Jurassic, definitive anurans occur in a range of deposits (UK, USA), followed by wider distribution during the Cretaceous (Argentina, Europe, Madagascar, Africa, Asia). By the Late Cretaceous (~88 myr), numerous anuran taxa were present across the continents (Sanchíz 1998; Roček 2000).

- *Eodiscoglossus.* The basal anuran family Discoglossidae was present by the late Mesozoic. A common genus is *Eodiscoglossus*, occurring in the Early Cretaceous (~141–145 myr) of Spain.
- *Shomronella.* Based on Early Cretaceous (~131 myr) tadpoles from Israel, this taxon testifies to the presence of fully established anuran larvae. Soft tissue preservation and the large size of the tadpoles reveal many anatomical details (Estes *et al.* 1978).
- *Beelzebufo.* A giant frog from the Late Cretaceous of Madagascar (65–70 myr), closely resembling the modern genus *Ceratophrys* from South America (Evans *et al.* 2008). This heavily built frog reached a length of 40 cm, exceeding the size of the extant goliath frog.

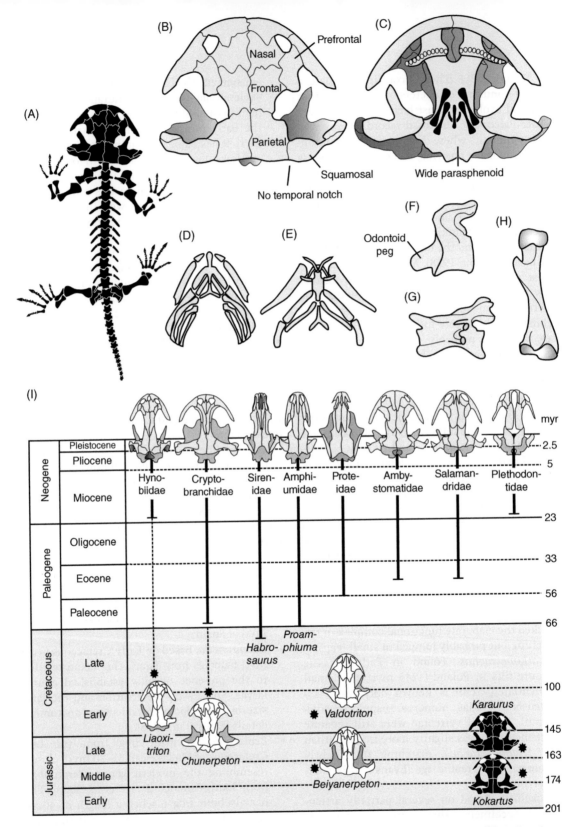

Figure 2.26 Urodela and Caudata. *Karaurus* (adapted from Ivakhnenko 1978): (A) skeleton restoration, (B) skull roof, (C) palate, with hyobranchial apparatus marked in black. (D) Hyobranchium of larval *Dicamptodon*, and (E) same of transformed specimen of same taxon (adapted from Rose 2003). (F) Atlas in lateral view; (G) trunk vertebra in lateral view; (H) femur in ventral view (adapted from Duellman and Trueb 1994). (I) Stratigraphic range of urodeles (black) and caudates (white), adapted from Milner (2000) and Gao and Shubin (2003).

- In the Early Cretaceous, only three families have been confirmed: Discoglossidae, Leiopelmatidae, and Pipidae. During the Late Cretaceous, several more clades made their first appearance (Leptodactylidae, Pelobatidae, and the two extinct families Palaeobatrachidae and Gobiatidae; see Roček 2000).

2.5.2.2 Caudata (salamanders)

The names caudate and salamander are here used interchangeably (Frost *et al.* 2006). They refer to the crown group of salamanders, and thus include all living species and numerous Mesozoic and Cenozoic taxa. All fossils outside the crown are referred to the more inclusive clade Urodela, which contains Caudata. (Note that Caudata and Urodela have been used with exactly the opposite meanings by former authors, and especially by paleontologists; here I follow the current convention as outlined by Frost *et al.* 2006.) Currently, 548 living species of caudates are known (Frost *et al.* 2006). They form only 10% of the known number of extant anuran species, but are more diverse in morphology, body size, and life span. There are 10 monophyletic families of caudates, but more than 50% of species belong to the Plethodontidae, the lungless salamanders.

2.5.2.2.1 Caudate characters

Despite numerous differences to Paleozoic tetrapods, caudates are not characterized by many autapomorphies (Figure 2.26). Most "typical" features are plesiomorphic, distinguishing the group from the more highly derived gymnophionans and anurans. There may be only one exclusive skeletal autapomorphy (namely the first in the following list).

1. **Palatine and palatoquadrate remodeling**. The palatoquadrate region is partially resorbed during metamorphosis, giving the eye more space and extending the attachment for eye-moving musculature.
2. **Parasphenoid process wide and flat**. Although this character is found in a range of temnospondyls (dvinosaurians, brachyopoids) and lepospondyls (microsaurs, lysorophians), it is unique to the Caudata among extant tetrapods.
3. **Odontoid peg**. Atlas with projection at anterior margin that fits into the space between the occipital condyles, forming a hinge joint. This character is not only present in extant caudates, but also occurs in various other extinct taxa: (1) the apodan *Eocaecilia*, (2) the Albanerpetontidae, (3) some amphibamid temnospondyls, and (4) the Lepospondyli.

Further characters that distinguish salamanders from other lissamphibians are not unique to the group: (4) the scapula and coracoid ossify as a single unit (shared with many temnospondyls), (5) the stapes is short and stout and directed towards the quadrate (shared with many lepospondyls and a few temnospondyls), and (6) the trunk ribs have two heads (shared with most Paleozoic tetrapods).

2.5.2.2.2 Mesozoic urodeles (stem-group)

The monophyletic group that includes caudates and all their stem taxa is referred to as Urodela (Frost *et al.* 2006). Despite their likely origin in the Late Permian, concluded from the presence of salientians in the Early Triassic, definitive urodeles make their first appearance in the Middle Jurassic (Milner 1994) (Figure 2.26). A poorly preserved skeleton from the Middle or Late Triassic of Kyrgyzstan, *Triassurus*, was suggested as a basal urodele (Ivakhnenko 1978), but this specimen is in need of reinvestigation.

- *Marmorerpeton*. The earliest unambiguous evidence of urodeles stems from Bathonian (~167 myr old) microvertebrate localities in England (Evans and Milner 1994). These finds already comprise a whole fauna of stem-salamanders, as evidenced by the diversity of atlas vertebrae (Milner 2000). The best-represented of these taxa is *Marmorerpeton*, a form similar to *Karaurus* based on cranial and vertebral material. Other, disarticulated material of similar age was found in western Siberia, representing a 20 cm long urodele named *Urupia* (Skutchas and Krasnolutskii 2011).
- *Karaurus*. This is the best-preserved Mesozoic urodele, known from a complete, articulated skeleton found in the Late Jurassic (~161 myr) Karatau lake deposit of Kazakhstan (Ivakhnenko 1978). Together with its close relative *Kokartus* from the Middle Jurassic (~165 myr) of Kyrgyzstan (Skutchas and Martin 2011), it

forms a basal urodele clade. Both genera are large (20 cm), heavily built, with broad-parabolic skulls and dermal bones ornamented as in many temnospondyls and anurans. Their skulls resemble those of extant *Dicamptodon* and *Ambystoma*.

- *Sinerpeton* and *Laccotriton*. From Late Jurassic (~151 myr) deposits of Hebei, north China (Gao and Shubin 2001). Up to 500 specimens of these stem-salamanders were found in a small deposit that formed during a pyroclastic eruption.

- *Pangerpeton* and *Jeholotriton*. Closer to the crown group are two urodeles from the Jurassic–Cretaceous boundary (~145 myr) of Liaoning, northeast China. *Pangerpeton* has only 14 trunk vertebrae, approaching the caudate condition (Wang and Evans 2006).

2.5.2.2.3 Mesozoic caudates (crown group)

- *Valdotriton*. Small land salamanders are known from good material from the Lower Cretaceous of Spain, with 8 cm long *Valdotriton* based on a few complete skeletons (Evans and Milner 1996). The presence of a salamandriform character (fused prearticular and angular in the mandible) suggests that *Valdotriton* is a caudate nesting above the cryptobranchoids. This is confirmed by the presence of a cryptobranchoid (*Chunerpeton*) in the coeval Jehol Biota, indicating that salamandriforms must have already existed.

- *Chunerpeton*. A late Middle Jurassic (~164 myr), well-preserved salamander from Inner Mongolia, northeast China, may be the earliest record of the giant cryptobranchoid salamanders (Gao and Shubin 2003). The moderately large (16 cm) form has one-headed ribs, only three rib-bearing caudal vertebrae, and no lacrimal bone. Both larvae and adults have been reported, the latter with branchial denticles resembling those of branchiosaurids (Gao and Shubin 2003).

- *Beiyanerpeton*. This is a Late Jurassic caudate with features of salamandroids (e.g., separated nasals) from Liaoning Province, north China (Gao and Shubin 2012). This larval or neotenic form is one of the few lissamphibians to preserve branchial denticles similar to thoe of branchiosaurids.

- *Batrachosauroides*. These salamanders belong to an extinct non-metamorphosing clade (Batracho-sauroididae) showing some affinities to the Proteidae, although these are all related to neoteny (Milner 2000). They range from the Late Cretaceous through the Pliocene (~99–4 myr).

- *Scapherpeton*. This second family of extinct caudates is recognized by their vertebrae, which resemble those of cryptobranchids (Milner 2000). They range from the Late Cretaceous through the Eocene (~99–50 myr).

The other extant caudate families can mostly be traced back in the fossil record as follows (Evans *et al.* 1996; Milner 2000): the Sirenidae to the Late Cretaceous (North America and Africa), the Hynobiidae to the Miocene (Europe), the Cryptobranchidae to the Paleocene (Eurasia), the Proteidae to the Paleocene (North America), the Plethodontidae to the Miocene (western North America), the Ambystomatidae to the Miocene (North America), the Dicamptodontidae to the Paleocene (North America), and the Salamandridae to the Late Cretaceous (~70 myr, Spain).

2.5.2.3 Gymnophiona (caecilians)

Gymnophionans form the smallest lissamphibian clade, comprising only some 175 species today (Figure 2.27). They are not well known outside herpetology, as they are confined to the tropics and most species lead a burrowing life in the soil. At first sight, they may be confused with large earthworms, but their powerful jaws and teeth are undisputed vertebrate features on closer inspection. All gymnophionans are limbless; they have an elongated and segmented trunk and an abbreviated tail. They reach lengths ranging from 11 to 150 cm. Caecilians are nocturnal and feed on earthworms and arthropods, especially beetles and termites. Larger species also prey on lizards, snakes, and small birds. They live in the top soil layers, but may appear on the surface after heavy rainfall. They inhabit a full range of environments, from decaying plant material, humus, wet mud of river banks, to fully aquatic habitats. The genus *Ichthyophis* is known to build burrow systems and uses mucous secretions to ease digging and stabilize the burrow walls (Haas 2010).

Figure 2.27 Apoda and Gymnophiona: (A) stratigraphic range of apodans (adapted from Carroll 2009). (B, D) apodan *Eocaecilia*; (C, E) gymnophionan *Ichthyophis* (adapted from Jenkins *et al.* 2007).

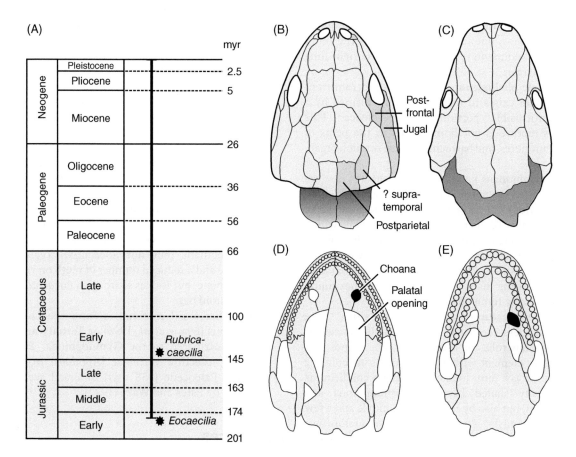

2.5.2.3.1 *Gymnophionan characters*

1. **Trunk greatly elongated**. The number of presacral vertebrae is greatly increased, ranging from 95 to 285 vertebrae within the group.
2. **Limbs and girdles completely absent**. All extant gymnophionans lack limbs and girdles. Their presence in the stem taxon *Eocaecilia* might reveal interesting details of the reduction process.
3. **Tail skeleton short**. Despite their elongate bodies, caecilians have short tail skeletons or have entirely lost them.
4. **Skull massive with compound bones**. The skull may be fenestrate or entirely closed, but is always very solid by the fusion of bones: maxilla and palatine, pterygoid and quadrate,

and os basale (braincase, parasphenoid). The strengthening of the skull meets the demands of extensive probing in the soil.
5. **Skin segmented in trunk**. Numerous rings (annuli) segment the presacral body. Primary rings are continuous and correlate with myosepta, the bordering sheets between trunk muscles. Secondary rings are located in between primary ones.
6. **Eyes largely reduced**. Connected with their existence in the dark, caecilians have rudimentary, small eyes. In some species, they are even covered by dermal bones.
7. **Tentacular organ**. A chemosensory organ located between the eye and nose, formed by an outgrowth of the narial passage. A former

eye muscle (retractor bulbi) serves as retractor of the tentacle, when the organ is to be withdrawn inside the skull.

8. **Asymmetric lungs**. As in snakes and amphisbaenians, one lung is enlarged and the other one rudimentary; in caecilians, the right lung is larger.

9. **Male phallodeum**. An unpaired, intromittent organ exists in male caecilians, formed by a protrusible portion of the cloaca. This structure is not homologous to the penis of amniotes, and batrachians have no such organ.

Gymnophionans have numerous features that are considered symplesiomorphies, and by which they differ from Batrachia. For instance, they have bony scales (Zylberberg and Wake 1990), which are considered a plesiomorphic retention of the dermal scales of early tetrapods. However, the heavily ossified skull is not a primitive character state, bearing many apomorphic traits not found in Paleozoic tetrapods.

Other features are more puzzling. Like salamanders, caecilians have no middle ear cavity, eustachian tube, or tympanum. The stapes is a massive element with a large process articulating with the jaw joint (quadrate). Such a condition is not only shared between Gymnophiona and Caudata, but also by many lepospondyls and a few temnospondyls. The ear region is therefore not necessarily derived in caecilians and salamanders, but may be plesiomorphic. Another possibility, more subtle though, is that it is a pedomorphic condition that evolved several times in parallel. Studies of salamander larvae support this (personal observations).

2.5.2.3.2 Mesozoic gymnophionans and stem taxa

The fossil record of gymnophionans and their stem-group (Apoda) was very poor until recently. The discovery of basal apodans from the Mesozoic has profoundly changed the situation (Jenkins et al. 2007; Evans and Sigogneau-Russell 2001).

- *Eocaecilia*. Based on numerous articulated specimens, this Early Jurassic (~189 myr) taxon has many, but by no means all characters of the Gymnophiona. It thus falls into the Apoda (stem-group). *Eocaecilia* has apparently a closed (stegokrotaphic) skull table, retaining several bones not present in caecilians (postfrontal, jugal, postparietal). The existence of a tabular is unclear; the homology of a small round element in the skull table is equivocal. The palate has oval openings, and a palatine–vomerine tooth row lateral to the choana, a gymnophionan feature. The braincase was described on the basis of CT scans (Maddin et al. 2012). The atlas has an odontoid peg unlike that in caecilians but similar to that in salamanders. The tiny limbs are already rudimentary, but the femur has a pronounced trochanter as in caudates.

- *Rubricacaecilia*. Evans and Sigogneau-Russell (2001) reported an apodan from the Early Cretaceous (~145 myr) of Morocco. It shares with caecilians (but not *Eocaecilia*) keeled vertebrae and a reduced number of teeth on the splenial bone, but it does share with *Eocaecilia* the odontoid peg.

- Most other fossil material is from the Late Cretaceous (Evans et al. 1996) or Tertiary and can be attributed to the Gymnophiona. For instance, caeciliid vertebrae (*Apodops*) from the late Paleocene (~58 myr) of Brazil were reported by Estes and Wake (1972).

References

Ahlberg, P.E. & Clack, J.A. (1998) Lower jaws, lower tetrapods ± a review based on the Devonian genus Acanthostega. *Transactions of the Royal Society of Edinburgh: Earth Sciences* **89**, 11–46.

Ahlberg, P.E. & Clack, J.A. (2005) The axial skeleton of the Devonian tetrapod Ichthyostega. *Nature* **437**, 137–140.

Ahlberg, P.E. & Johannson, Z. (1998) Osteolepiforms and the ancestry of tetrapods. *Nature* **395**, 792–794.

Ahlberg, P.E. & Milner, A.R. (1994) The origin and early diversification of tetrapods. *Nature* **368**, 507–514.

Ahlberg, P.E., Lukševics, E., & Lebedev, O. (1994) The first tetrapod finds from the Devonian (Upper Famennian) of Latvia. *Philosophical*

Transactions of the Royal Society of London B **343**, 303–328.

Ahlberg, P.E., Lukševics, E., & Mark-Kurik, E. (2000) A near-tetrapod from the Baltic Middle Devonian. *Palaeontology* **43**, 533–548.

Ahlberg, P.E., Clack, J.A., Luksevics, E., Blom, H., & Zupins, I. (2008) *Ventastega curonica* and the origin of tetrapod morphology. *Nature* **453**, 1199–1204.

Anderson, J.S. (2001) The phylogenetic trunk: maximal inclusion of taxa with missing data in an analysis of the Lepospondyli (Vertebrata, Tetrapoda). *Systematic Biology* **50**, 170–193.

Anderson, J.S. (2002) Revision of the aïstopod genus *Phlegethontia* (Tetrapoda: Lepospondyli). *Journal of Paleontology* **76**, 1029–1046.

Anderson, J.S. (2003a) A new aïstopod (Tetrapoda: Lepospondyli) from Mazon Creek, Illinois. *Journal of Vertebrate Paleontology* **23**, 79–88.

Anderson, J.S. (2003b) Cranial anatomy of *Coloraderpeton brilli*, postcranial anatomy of *Oestocephalus amphiuminus*, and reconsideration of Ophiderpetontidae (Tetrapoda: Lepospondyli: Aïstopoda). *Journal of Vertebrate Paleontology* **23**, 532–543.

Anderson, J.S., Carroll, R.L., & Rowe, T.B. (2003) New information on *Lethiscus stocki* (Tetrapoda: Lepospondyli) from high-resolution computed tomography and a phylogenetic analysis of Aïstopoda. *Canadian Journal of Earth Sciences* **40**, 1071–1083.

Anderson, J.S., Reisz, R.R., Scott, D., Fröbisch, N.B., & Sumida, S.S. (2008) A stem batrachian from the Early Permian of Texas and the origin of frogs and salamanders. *Nature* **453**, 515–518.

Andrews, S.M. & Carroll, R.L. (1991) The order Adelospondyli: Carboniferous lepospondyl amphibians. *Transactions of the Royal Society of Edinburgh: Earth Sciences* **82**, 239–275.

Andrews, S. M. & Westoll, T.S. (1970) The postcranial skeleton of *Eusthenopteron foordi* Whiteaves. *Transactions of the Royal Society of Edinburgh* **68**, 207–329.

Averianov, A.O., Martin, T., Skutschas, P.P., Rezvyi, A.S., & Bakirov, A.A. (2008) Amphibians from the Middle Jurassic Balabansai Svita in the Fergana Depression, Kyrgyzstan (Central Asia). *Palaeontology* **51**, 471–485.

Báez, A.M. & Basso, N.G. (1996) The earliest known frogs of the Jurassic of South America: review and cladistic appraisal of their relationships. *Münchner geowissenschaftliche Abhandlungen A* **30**, 131–158.

Beaumont, E.H. (1977) Cranial morphology of the Loxommatidae (Amphibia: Labyrinthodontia). *Philosophical Transactions of the Royal Society of London B* **280**, 29–101.

Beaumont, E.H. & Smithson, T. (1998) The cranial morphology and relationships of the aberrant Carboniferous amphibian *Spathicephalus mirus* Watson. *Zoological Journal of the Linnean Society* **122**, 187–209.

Berman, D.S., Henrici, A.C., Sumida, S.S., & Martens, T. (2000) Redescription of *Seymouria sanjuanensis* (Seymouriamorpha) from the Lower Permian of Germany based on complete, mature specimens with a discussion of paleoecology of the Bromacker locality assemblage. *Journal of Vertebrate Paleontology* **20**, 253–268.

Berman, D.S., Henrici, A., Kissel, R.A., Sumida, S., & Martens, T. (2004) A new diadectid (Diadectomorpha), *Orobates pabsti*, from the Early Permian of central Germany. *Bulletin of Carnegie Museum of Natural History* **35**, 1–36.

Boisvert, C., Mark-Kurik, E., & Ahlberg, P. (2008) The pectoral fin of *Panderichthys* and the origin of digits. *Nature* **456**, 636–638.

Bolt, J.R. (1969) Lissamphibian origins: possible protolissamphibian from the Lower Permian of Oklahoma. *Science* **166**, 888–891.

Bolt, J.R. (1979) *Amphibamus grandiceps* as a juvenile dissorophid: evidence and implication. In: M.H. Nitecki (ed.), *Mazon Creek Fossils*. New York: Academic Press, pp. 529–563.

Bolt, J.R. & Lombard, R.E. (1985) Evolution of the amphibian tympanic ear and the origin of frogs. *Biological Journal of the Linnean Society* **24**, 83–99.

Bolt, J.R. & Lombard, R.E. (2000) Palaeobiology of *Whatcheeria deltae*, a primitive Mississippian tetrapod. In: H. Heatwole & R.L. Carroll (eds.), *Amphibian Biology. Volume 4. Palaeontology.* Chipping Norton, NSW: Surrey Beatty, pp. 1044–1052.

Bossy, K.A. & Milner, A.C. (1998) Order Nectridea. In: P. Wellnhofer (ed.), *Handbuch der Paläontologie.* Munich: Pfeil, Vol. 1, pp. 73–131.

Boy, J.A. (1988) Über einige Vertreter der Eryopoidea (Amphibia: Temnospondyli) aus dem europäischen Rotliegend (? höchstes Karbon – Perm). 1. *Sclerocephalus*. *Paläontologische Zeitschrift* **62**, 107–132.

Boy, J. A. (1989) Über einige Vertreter der Eryopoidea (Amphibia: Temnospondyli) aus dem europäischen Rotliegend (? höchstes Karbon – Perm). 2. Acanthostomatops. *Paläontologische Zeitschrift* **63**, 133–151.

Boy, J.A. (1990) Über einige Vertreter der Eryopoidea (Amphibia: Temnospondyli) aus dem europäischen Rotliegend (? höchstes Karbon – Perm). 3. *Onchiodon*. Paläontologische Zeitschrift **64**, 287–312.

Boy, J.A. (1995) Über die Micromelerpetontidae (Amphibia: Temnospondyli). 1. Morphologie und Paläoökologie des *Micromelerpeton credneri* (Unter-Perm; SW-Deutschland). *Paläontologische Zeitschrift* **69**, 429–457.

Boy, J.A. & Schindler, T. (2000) Ökostratigraphische Bioevents im Grenzbereich Stefanium/Autunium (höchstes Karbon) des Saar-Nahe-Beckens (SW-Deutschland) und benachbarter Gebiete. *Neues Jahrbuch für Geologie und Palaontologie, Abhandlungen* **216**, 89–152.

Boy, J.A. & Sues, H.-D. (2000) Branchiosaurs: larvae, metamorphosis and heterochrony in temnospondyls and seymouriamorphs. In: H. Heatwole & R.L. Carroll (eds.), *Amphibian Biology. Volume 4. Palaeontology*. Chipping Norton, NSW: Surrey Beatty, pp. 973–1496.

Buchwitz, M. & Voigt, S. (2010) Peculiar carapace structure of a Triassic chroniosuchian implies evolutionary shift in trunk flexibility. *Journal of Vertebrate Paleontology* **30**, 1697–1708.

Bystrow, A.P. (1938) *Dvinosaurus* als neotenische Form der Stegocephalen. *Acta Zoologica* **19**, 209–295.

Carroll, R.L. (1969) A new family of Carboniferous amphibians. *Palaeontology* **12**, 537–548.

Carroll, R.L. (1990) A tiny microsaur: size constraints in Palaeozoic tetrapods. *Palaeontology* **33**, 1–17.

Carroll, R.L. (1995) Problems of the phylogenetic analysis of Paleozoic choanates. *Bulletin du Muséum National d'Histoire Naturelle Paris C* **17**, 380–445.

Carroll, R.L. (2009) *The Rise of Amphibians*. Baltimore: Johns Hopkins University Press.

Carroll, R.L. & Chorn, J. (1995) Vertebral development of the oldest microsaur and the problem of

"lepospondyl" relationships. *Journal of Vertebrate Paleontology* **15**, 37–56.

Carroll, R.L. & Gaskill, P. (1978) The order Microsauria. *Memoirs of the American Philospohical Society* **126**, 1–211.

Castanet, J., Francillon-Vielleiot, H., de Ricqlès, A., & Zylberberg, L. (2003) The skeletal histology of the Amphibia. In: H. Heatwole & R.L. Carroll (eds.), *Amphibian Biology. Volume 5. Osteology*. Chipping Norton, NSW: Surrey Beatty, pp. 1597–1683.

Clack, J.A. (1987) Two new specimens of *Anthracosaurus* (Amphibia: Anthracosauria) from the Northumberland Coal Measures. *Palaeontology* **30**, 15–26.

Clack, J.A. (1994) *Acanthostega gunnari*, a Devonian tetrapod from East Greenland; the snout, palate and ventral parts of the braincase, with a discussion of their significance. *Meddelelser om Grønland Geoscience* **31**, 1–24.

Clack, J.A. (1998a) The neurocranium of *Acanthostega gunnari* Jarvik and the evolution of the otic region in tetrapods. *Zoological Journal of the Linnean Society* **122**, 61–97.

Clack, J.A. (1998b) The Scottish Carboniferous tetrapod *Crassigyrinus scoticus* (Lydekker) - cranial anatomy and relationships. *Transactions of the Royal Society of Edinburgh: Earth Sciences* **88**, 127–142.

Clack, J.A. (2001) *Eucritta melanolimnetes* from the Early Carboniferous of Scotland: a stem tetrapod showing a mosaic of characteristics. *Transactions of the Royal Society of Edinburgh Earth Sciences* **92**, 75–95.

Clack, J. A. (2009) The fin to limb transition: new data, intepretations, and hypotheses from paleontology and developmental biology. *Annual Reviews of Earth and Planetary Sciences* **37**, 163–179.

Clack, J.A. (2012) *Gaining Ground: the Origin and Evolution of Tetrapods*, 2nd edition. Bloomington: Indiana University Press.

Clack, J.A. & Carroll, R.L. (2000). Early Carboniferous tetrapods. In: H. Heatwole & R.L. Carroll (eds.), *Amphibian Biology. Volume 4. Palaeontology*. Chipping Norton, NSW: Surrey Beatty, pp. 1030–1043.

Clack, J.A. & Coates, M.I. (1995) *Acanthostega gunnari*, a primitive, aquatic tetrapod? *Bulletin* du Muséum national d'Histoire naturelle Paris **17**, 359–372.

Clack, J. A. & Finney, S. M. (2005) *Pederpes finneyae*, an articulated tetrapod from the Tournaisian of Western Scotland. *Journal of Systematic Palaeontology* **2**, 311–346.

Clack, J.A. & Klembara, J. (2009) An articulated specimen of *Chroniosaurus dongusensis*, and the morphology and relationships of the chroniosuchids. *Special Papers in Palaeontology* **81**, 15–42.

Clack, J.A. & Milner, A.R. (2010) *Platyrhinops* from the Upper Carboniferous of Linton and Nýřany and the family Amphibamidae (Amphibia: Temnospondyli). *Transactions of the Royal Society of Edinburgh: Earth and Environmental Sciences* **100**, 275–295.

Coates, M.I. (1996) The Devonian tetrapod *Acanthostega gunnari* Jarvik: postcranial anatomy, basal tetrapod interrelationships and patterns of skeletal evolution. *Transactions of the Royal Society of Edinburgh: Earth Sciences* **87**, 363–421.

Coates, M.I. & Clack, J.A. (1990) Polydactyly in the earliest known tetrapod limbs. *Nature* **347**, 66–69.

Coates, M.I. & Clack, J.A. (1991) Fish-like gills and breathing in the earliest known tetrapod. *Nature* **352**, 234–236.

Costa, O. (1864) Paleontologia del regno di Napoli contenente la desrizione e figure di tutti gli avanzi organici fossili racciusi nel suolo di questo regno. *Atti Academia Pontaniana* **8**, 1–198.

Cruickshank, A.R.I. & Skews, B.W. (1980) The functional significance of nectridean tabular horns. *Proceedings of the Royal Society London B* **209**, 513–537.

Curtis, K. & Padian, K. (1999) An early Jurassic microvertebrate fauna from the Kayenta Formation of northeastern Arizona: microfaunal change across the Triassic-Jurassic boundary. *PaleoBios* **19**, 19–37.

Daeschler, E.B., Shubin, N.H., & Jenkins, F.A. (2006) A Devonian tetrapod-like fish and the evolution of the tetrapod body plan. *Nature* **440**, 757–763.

Daeschler, E.B., Clack, J.A., & Shubin, N.H. (2009). Late Devonian tetrapod remains from Red Hill, Pennsylvania, USA: how much diversity? *Acta Zoologica* **90** (Supplement 1), 306–317.

Dilkes D.W. & Brown, L.E. (2007) Biomechanics of the vertebrae and associated osteoderms of the Early Permian amphibian *Cacops aspidephorus*. *Journal of Zoology* **271**, 396–407.

Duellman, W.E. & Trueb, L. (1994) *Biology of Amphibians*. Baltimore: Johns Hopkins University Press.

Estes, R. & Hofstetter, R. (1976) Les Urodèles du Miocène de La Grive-Saint-Alban (Isère, France). *Bulletin du Muséum National d'Histoire Naturelle, 3 serie, 398, Sciences de la Terre* **57**, 297–343.

Estes, R. & Wake, M.H. (1972) The first fossil record of caecilian amphibians. *Nature* **239**, 228–231.

Estes, R., Spinar, Z.V., & Nevo, E. (1978) Early Cretaceous pipid tadpoles from Israel (Amphibia: Anura). *Herpetologica* **34**, 374–393.

Evans, S.E. & Borsuk-Białynicka, M. (2009) The Early Triassic stem-frog *Czatkobatrachus* from Poland. *Palaeontologica Polonica* **65**, 79–105.

Evans, S.E. & Milner, A.R. (1994) Middle Jurassic microvertebrate assemblages from the British Isles. In: N.C. Fraser & H.-D. Sues (eds.), *In the Shadow of the Dinosaurs: Early Mesozoic Tetrapods*. New York: Cambridge University Press, pp. 303–321.

Evans, S.E. & Milner, A.R. (1996) A metamorphosed salamander from the Early Cretaceous of Las Hoyas, Spain. *Philosophical Transactions of the Royal Society B* **351**, 627–646.

Evans, S.E. & Sigogneau-Russell, D. (2001) A stem-group caecilian (Amphibia: Lissamphibia) from the Lower Cretaceous of Morocco. *Palaeontology* **44**, 259–273

Evans, S.E., Milner, A.R., & Werner, C. (1996) Sirenid salamanders and a gymnophionan amphibian from the Cretaceous of the Sudan. *Palaeontology* **39**, 77–95.

Evans, S.E., Jones, M.E.H., & Krause, D.W. (2008) A giant frog with South American affinities from the Late Cretaceous of Madagascar. *Proceedings of the National Academy of Sciences* **105**, 2951–2956.

Feller, A.E. & Hedges, S.B. (1998) Molecular evidence for the early history of living amphibians.

Molecular Phylogenetics and Evolution **9**, 509–516.

Fröbisch, N.B., Olori, J., Schoch, R.R, & Witzmann, F. (2010) Amphibian development in the fossil record. *Seminars in Cell and Developmental Biology* **21**, 424–431.

Frost, D.R., Grant, T., Faivovich, J., *et al.* (2006) The amphibian tree of life. *Bulletin of the American Museum of Natural History* **297**, 1–370.

Gao, K.Q. & Shubin, N.H. (2001) Late Jurassic salamanders from northern China. *Nature* **410**, 574–576.

Gao, K.Q. & Shubin, N.H. (2003) Earliest known crown-group salamanders. *Nature* **422**, 424–428.

Gao, K.Q. & Shubin, N.H. (2012) Late Jurassic salamandroid from western Liaoning, China. *Proceedings of the National Academy of Sciences* **109**, 5767–5772.

Gardner, J.D. (2001) Monophyly and affinities of albanerpetontid amphibians (Temnospondyli; Lissamphibia). *Zoological Journal of the Linnean Society* **131**, 309–352.

Gardner, J.D. & Böhme, M. (2008) Review of the Albanerpetontidae (Lissamphibia), with comments on the paleoecological preferences of European Tertiary albanerpetontids. In: J.T. Sankey & S. Baszio (eds.), *Vertebrate Microfossil Assemblages: Their Role in Paleoecology and Paleobiogeography*. Bloomington: Indiana University Press, p. 178–218.

Germain, D. (2008) A new phlegethontiid specimen (Lepospondyli, Aïstopoda) from the Late Carboniferous of Montceau-les-Mines (Saone-et-Loire, France). *Geodiversitas* **30**, 669–680.

Germain, D. (2009) The Moroccan diplocaulid: the last lepospondyl, the single one on Gondwana. *Historical Biology* **22**, 4–39.

Godfrey, S.J. (1988) Isolated tetrapod remains from the Carboniferous of West Virginia. *Kirtlandia* **43**, 27–36.

Godfrey, S.J. (1989) Ontogenetic changes in the skull of the Carboniferous tetrapod *Greererpeton burkemorani* Romer 1969. *Philosophical Transactions of the Royal Society London B* **323**, 135–153.

Golubev, V.K. (1998) Revision of the Late Permian chroniosuchians (Amphibia, Anthracosauromorpha) from Eastern Europe. *Paleontological Journal* **32**, 390–401.

Golubev, V.K. (2000) [Permian and Triassic chroniosuchians and biostratigraphy of the Upper Tatarian series in Eastern Europe]. *Trudy Paleontologiceskogo Instituta RAS* **276**, 1–172. (In Russian.)

Goodrich, E.S. (1930). *Studies on the Structure and Development of Vertebrates*. London: Macmillan.

Gould, S.J. & Vrba, E. (1982) Exaptation: a missing term in the science of form. *Paleobiology* **8**, 4–15.

Haas, A. (2010) Lissamphibia. In: W. Westheide & R. Rieger (eds.), *Spezielle Zoologie. Wirbel- oder Schädeltiere*. Heidelberg: Spektrum Akademischer Verlag, pp. 330–359.

Hanken, J. (1983) Miniaturization and its effects on cranial morphology in plethodontid salamanders, genus *Thorius* (Amphibia: Plethodontidae). II. The fate of the brain and sense organs and their role in skull morphogenesis and evolution. *Journal of Morphology* **177**, 155–268.

Hedges, S. B. & Maxson, L.R. (1993) A molecular perspective on lissamphibian phylogeny. *Herpetological Monographs* **7**, 27–42.

Hellrung, H. (2003) *Gerrothorax pustuloglomeratus*, ein Temnospondyle (Amphibia) mit knöcherner Branchialkammer aus dem Unteren Keuper von Kupferzell (Süddeutschland). *Stuttgarter Beiträge zur Naturkunde* **330**, 1–130.

Hennig, W. (1966) *Phylogenetic Systematics*. Urbana: University of Illinois Press.

Holmes, R. (1984) The Carboniferous amphibian *Proterogyrinus scheelei* Romer, and the early evolution of tetrapods. *Philosophical Transactions of the Royal Society London B* **306**, 431–524.

Holmes, R. (1989) The skull and axial skeleton of the Lower Permian anthracosaur amphibian *Archeria crassidisca* Cope. *Palaeontographica A* **207**, 161–206.

Holmes, R. (2000) Palaeozoic temnospondyls. In: H. Heatwole & R.L. Carroll (eds.), *Amphibian Biology. Volume 4. Palaeontology*. Chipping Norton, NSW: Surrey Beatty, pp. 1081–1120.

Holmes, R.B., Carroll, R.L., & Reisz, R.R. (1998) The first articulated skeleton of *Dendrerpeton acadianum* (Temnospondyli, Dendrerpetontidae) from the Late Carboniferous locality of Joggins, Nova Scotia, and a review of its relationships. *Journal of Vertebrate Paleontology* **18**, 64–79.

Hook, R.W. (1983) *Colosteus scutellatus* (Newberry), a primitive temnospondyl amphibian from the middle Late Carboniferous of Linton, Ohio. *American Museum Novitates* **2770**, 1–41.

Ivakhnenko, M.F. (1978) Urodeles from the Triassic and Jurassic of Soviet Central Asia. *Paleontological Journal* **12**, 362–368. (In Russian).

Jacob, F. (1977) Evolution and tinkering. *Science* **196**, 1161–1166.

Janvier, P. (1996) *Early Vertebrates*. Oxford Monographs on Geology and Geophysics 33. Oxford: Oxford Univesity Press.

Jarvik, E. (1954) On the visceral skeleton of *Eusthenopteron* with a discussion of the parasphenoid and palatoquadrate in fishes. *Kunglik Svenska Vetenskapsakademien Handlingar* **5**, 1–104.

Jarvik, E. (1980) *Basic Structure and Evolution of Vertebrates*. Vols. **1–2**. London and New York: Academic Press.

Jarvik, E. (1996) *The Devonian Tetrapod* Ichthyostega. Fossils and Strata 40. Oslo: Scandinavian University Press.

Jenkins, F. Jr. & Shubin, N. (1998) *Prosalirus bitis* and the anuran caudopelvic mechanism. *Journal of Vertebrate Paleontology* **18**, 495–510.

Jenkins, F.A., Jr., Walsh, D.M., & Carroll, R.L. (2007) Anatomy of *Eocaecilia micropodia*, a limbed caecilian of the Early Jurassic. *Bulletin of the Museum of Comparative Zoology Harvard* **158**, 285–366.

Klembara, J. (1995) The external gills and ornamentation of skull-roof bones of the Lower Permian Discosauriscus (Kuhn 1933) with remarks to its ontogeny. *Paläontologische Zeitschrift* **69**, 265–281.

Klembara, J. (1997) The cranial anatomy of *Discosauriscus* Kuhn, a seymouriamorph tetrapod from the Lower Permian of Boscovice Furrow (Czech Republic). *Philosophical Transactions of the Royal Society London B* **352**, 257–302.

Klembara, J. (2009) New cranial and dental features of *Discosauriscus austriacus* (Seymouriamorpha, Discosauriscidae) and the ontogenetic conditions of *Discosauriscus*. *Special Papers in Palaeontology* **81**, 61–69.

Klembara, J. & Bartík, I. (2000) The postcranial skeleton of *Discosauriscus* Kuhn, a seymouriamorph tetrapod from the Lower Permian of the Boskovice Furrow (Czech Republic). *Transactions of the Royal Society of Edinburgh: Earth Sciences* **90**, 287–316

Klembara, J. & Ruta, M. (2005) The seymouriamorph tetrapod Utegenia shpinari from the ?Upper Carboniferous-Lower Permian of Kazakhstan. Part I: Cranial anatomy and ontogeny. *Transactions of the Royal Society of Edinburgh: Earth Sciences* **94**, 45–74.

Klembara, J., Berman, D.S., Henrici, A.C., Cernansky, A., & Werneburg, R. (2006) Comparison of cranial anatomy and proportions of similarly sized Seymouria sanjuanensis and Discosauriscus austriacus. *Annals of the Carnegie Museum* **75**, 37–49.

Klembara, J., Clack, J.A., & Cernansky, A. (2010) The anatomy of the palate of *Chroniosaurus dongusensis* (Chroniosuchia, Chroniosuchidae) from the Upper Permian of Russia. *Palaeontology* **53**, 1147–1153.

Laurin, M. (1998) The importance of global parsimony and historical bias in understanding tetrapod evolution. Part I. Systematics, middle ear evolution, and jaw suspension. *Annales des Sciences naturelles* **19**, 1–42.

Laurin, M. (2000) Seymouriamorphs. In: H. Heatwole & R.L. Carroll (eds.), *Amphibian Biology. Volume 4. Palaeontology*. Chipping Norton, NSW: Surrey Beatty, pp. 1064–1080.

Laurin, M. & Anderson, J.S. (2004) Meaning of the name Tetrapoda in the scientific literature: an exchange. *Systematic Biology* **53**, 68–80.

Laurin, M. & Reisz, R.R. (1997) A new perspective on tetrapod phylogeny. In: S.S. Sumida & K.L.M. Martin (eds.), *Amniote Origins: Completing The Transition To Land*. London: Academic Press, pp. 9–59.

Lebedev, O.A. & Coates, M.I. (1995) The postcranial skeleton of the Devonian tetrapod *Tulerpeton curtum* Lebedev. *Zoological Journal of the Linnean Society* **114**, 307–348.

Lombard, R.E. & Bolt, J.R. (1995) A new primitive tetrapod, *Whatcheeria deltae*, from the Lower Carboniferous of Iowa. *Palaeontology* **38**, 471–494.

Maddin, H., Jenkins, F.A., & Anderson, J.S. (2012) The braincase of *Eocaecilia micropodia* (Lissamphibia, Gymnophiona) and the origin of caecilians. *PloS ONE* **7**, e50743.

McGowan, G. & Evans, S.E. (1995) Albanerpetontid amphibians from the Cretaceous of Spain. *Nature* **373**, 143–145.

Mickoleit, G. (2004) *Phylogenetische Systematik der Wirbeltiere*. Munich: Pfeil.

Milner, A.C. & Ruta, M. (2009) A revision of *Scincosaurus* (Tetrapoda, Nectridea) from the Moscovian of Nýřany, Czech Republic, and the phylogeny and interrelationships of nectrideans. *Special Papers in Palaeontology* **81**, 71–89.

Milner, A.C., Milner, A.R., & Walsh, S.A. (2009) A new specimen of *Baphetes* from Nýřany, Czech Republic and the intrinsic relationships of the Baphetidae. *Acta Zoologica* **90**, 318–334.

Milner, A.R. (1980a) The temnospondyl amphibian *Dendrerpeton* from the Upper Carboniferous of Ireland. *Palaeontology* **23**, 125–141.

Milner, A.R. (1980b) The tetrapod assemblage from Nýřany, Czechoslovakia. In: A.L. Panchen (ed.), *The Terrestrial Environment and the Origin of Land Vertebrates*. London and New York: Academic Press, pp. 439–496.

Milner, A.R. (1982) Small temnospondyl amphibians from the Middle Late Carboniferous of Illinois. *Palaeontology* **25**, 635–664.

Milner, A.R. (1987) The Westphalian tetrapod fauna; some aspects of its geography and ecology. *Journal of the Geological Society of London* **144**, 495–506.

Milner, A.R. (1988) The relationships and origin of living amphibians. In: M.J. Benton (ed.), *The Phylogeny and Classification of the Tetrapods*. Oxford: Clarendon Press, Vol. 1, pp. 59–102.

Milner, A.R. (1990) The radiations of temnospondyl amphibians. In: P.D. Taylor & G.P. Larwood (eds.), *Major Evolutionary Radiations*. Oxford: Clarendon Press, pp. 321–349.

Milner, A.R. (1993) The Paleozoic relatives of lissamphibians. In: D. Cannatella & D. Hillis (eds.), *Amphibian relationships: phylogenetic analysis of morphology and molecules*. *Herpetological Monographs* **7**, 8–27.

Milner, A.R. (1994) Late Triassic and Jurassic amphibians. In: N.C. Fraser & H.-D. Sues (eds.), *In the Shadow of the Dinosaurs: Early Mesozoic Tetrapods*. New York: Cambridge University Press, pp. 5–22.

Milner, A.R. (1996) A revision of the temnospondyl amphibians from the Upper Carboniferous of Joggins, Nova Scotia. *Special Papers in Palaeontology* **52**, 81–103.

Milner, A.R. (2000) Mesozoic and Tertriary Caudata and Albanerpetontidae. In: H. Heatwole & R.L. Carroll (eds.), *Amphibian Biology. Volume 4. Palaeontology*. Chipping Norton, NSW: Surrey Beatty, pp. 1412–1444.

Milner, A.R. (2007) *Mordex laticeps* and the base of the Trematopidae. *Journal of Vertebrate Paleontology* **27**, 118A.

Milner, A.R. & Sequeira, S.E.K. (1994) The temnospondyl amphibians from the Viséan of East Kirkton, West Lothian, Scotland. *Transactions of the Royal Society of Edinburgh: Earth Sciences* **84**, 331–361.

Milner, A.R. & Sequeira, S.E.K. (2011) The amphibian *Erpetosaurus radiatus* (Temnospondyli, Dvinosauria) from the Middle Pennsylvanian of Linton, Ohio: morphology and systematic position. *Special Papers in Palaeontology* **86**, 57–73.

Müller, J. & Reisz, R.R. (2006) The phylogeny of early eureptiles: comparing parsimony and Bayesian approaches in the investigation of a basal fossil clade. *Systematic Biology* **55**, 503–511.

Niedźwiedzki, G., Szrek, P., Narkiewicz, K., Narkiewicz, M., & Ahlberg, P.E. (2010) Tetrapod trackways from the Middle Devonian Period of Poland. *Nature* **463**, 43–48.

Olson, E.C. (1951) *Diplocaulus*, a study in growth and variation. *Fieldiana: Geology* **11**, 57–149.

Panchen, A.L. (1970) Anthracosauria. In: O. Kuhn (ed.), *Encyclopedia of Paleoherpetology*, Vol. 5A. Stuttgart: Fischer.

Panchen, A.L. (1972) The skull and skeleton of *Eogyrinus attheyi* Watson (Amphibia: Labyrinthodontia). *Philosophical Transactions of the Royal Society London B* **263**, 279–326.

Panchen, A.L. (1973) On *Crassigyrinus scoticus* Watson, a primitive amphibian from the Lower Carboniferous of Scotland. *Palaeontology* **16**, 179–193.

Panchen, A.L. (1977) On *Anthracosaurus russelli* Huxley (Amphibia: Labyrinthodontia) and the family Anthracosauridae. *Philosophical Transactions of the Royal Society London B* **279**, 447–512.

Panchen, A.L. (1980) The origin and relationships of the anthracosaur Amphibia from the late

Palaeozoic. In: A.L. Panchen (ed.), *The Terrestrial Environment and the Origin of Land Vertebrates.* London and New York: Academic Press, pp. 319–350.

Panchen, A.L. (1985) On the amphibian *Crassigyrinus scoticus* Watson from the Carboniferous of Scotland. *Philosophical Transactions of the Royal Society London B* **309**, 505–568.

Parrish, W.C. (1978) Paleoenvironmental analysis of a Lower Permian bonebed and adjacent sediments, Wichita County, Texas. *Palaeogeography, Palaeoclimatology, Palaeoecology* **24**, 209–237.

Parsons, T.S. & Williams, E.E. (1963) The relationships of the modern Amphibia: a re-examination. *Quarterly Review of Biology* **38**, 26–53.

Paton, R.L., Smithson, T.R., & Clack, J.A. (1999) An amniote-like skeleton from the Early Carboniferous of Scotland. *Nature* **398**, 508–513.

Pierce, S.E., Clack, J.A., & Hutchinson, J.R. (2012) Three-dimensional limb joint mobility in the early tetrapod *Ichthyostega*. *Nature* **486**, 523–526.

Polley, B.P. & Reisz, R.R. (2011) A new Lower Permian trematopid (Temnospondyli: Dissorophoidea) from Richards Spur, Oklahoma. *Zoological Journal of the Linnean Society* **161**, 789–815.

Rage, J.-C. & Roček, Z. (1989) Redescription of *Triadobatrachus massinoti* (Piveteau, 1936) an anuran amphibian from the Early Triassic. *Palaeontographica A* **206**, 1–16.

Reisz, R.R., Schoch, R.R., & Anderson, J.S. (2009) The armoured dissorophid *Cacops* from the Early Permian of Oklahoma and the exploitation of the terrestrial realm by amphibians. *Naturwissenschaften* **96**, 789–796.

Roček, Z. (2000) Mesozoic anurans. In: H. Heatwole & R.L. Carroll (eds.), *Amphibian Biology. Volume 4. Palaeontology.* Chipping Norton, NSW: Surrey Beatty, pp. 1295–1331.

Romer, A.S. (1928) Vertebrate faunal horizons in the Texas Permo-Carboniferous red beds. *University of Texas Bulletin* **2801**, 67–108.

Romer, A.S. (1935) Early history of Texas redbeds vertebrates. *Bulletin of the Geological Society of America* **46**, 1597–1658.

Romer, A.S. (1947) Review of the Labyrinthodontia. *Bulletin of the Museum of Comparative Zoology Harvard College* **99**, 1–368.

Romer, A.S. (1966) *Vertebrate Paleontology.* Chicago: University of Chicago Press.

Romer, A.S. (1969) The cranial anatomy of the Permian amphibian *Pantylus*. *Breviora* **314**, 1–37.

Romer, A.S. & Witter, R.V. (1942) *Edops*, a primitive rhachitomous amphibian from the Texas red beds. *Journal of Geology* **50**, 925–960.

Rose, C.S. (2003) The developmental morphology of salamander skulls. In: H. Heatwole & R.L. Carroll (eds.), *Amphibian Biology. Volume 5. Osteology.* Chipping Norton, NSW: Surrey Beatty, pp. 1684–1781.

Ruta, M. & Clack, J.A. (2006) A review of *Silvanerpeton miripedes*, a stem amniote from the Lower Carboniferous of East Kirkton, West Lothian, Scotland. *Transactions of the Royal Society of Edinburgh: Earth Sciences* **97**, 31–63.

Ruta, M. & Coates, M.I. (2007) Dates, nodes and character conflict: addressing the lissamphibian origin problem. *Journal of Systematic Palaeontology* **5**, 69–122.

Ruta, M., Coates, M.I., & Quicke, D.L.J. (2003a) Early tetrapod relationships revisited. *Biological Reviews* **78**, 251–345.

Ruta, M., Jeffrey, E., & Coates, M.I. (2003b) A supertree of early tetrapods. *Proceedings of the Royal Society of London B* **270**, 2507–2516.

Sanchez, S., Klembara, J., Castanet, J., & Steyer, J.S. (2008) Salamander-like development in a seymouriamorph revealed by palaeohistology. *Biology Letters* **4**, 411–414.

Sanchez, S., de Ricqlès, A., Schoch, R.R., & Steyer, J.S. (2010) Developmental plasticity of limb bone microstructural organization in *Apateon*: histological evidence of paedomorphic conditions in branchiosurs. *Evolution and Development* **12**, 315–328.

Sanchíz, B. (1998) Salientia. In: P. Wellnhofer (ed.), *Handbuch der Paläontologie.* Munich: Pfeil, Vol. 4, pp. 1–275.

San Mauro, D., Vences, M., Alcobendas, M., Zardoya, R., Meyer, A. (2005) Initial diversification of living amphibians predated the breakup of Pangea. *American Naturalist* **165**, 590–599.

Säve-Söderbergh, G. (1932) Preliminary note on Devonian stegocephalians from East Greenland. *Meddelelser om Grønland* **94**, 1–107.

Sawin, H. J. (1941) The cranial anatomy of *Eryops megacephalus*. *Bulletin of the Museum of Comparative Zoology Harvard* **89**, 407–463.

Schmalhausen, I.I. (1968) *The Origin of Terrestrial Vertebrates*. London and New York: Academic Press.

Schoch, R.R. (1992) Comparative ontogeny of Early Permian branchiosaurid amphibians from southwestern Germany. Developmental stages. *Palaeontographica A* **222**, 43–83.

Schoch, R.R. (2009a) The evolution of life cycles in early amphibians. *Annual Review of Earth and Planetary Sciences* **37**, 135–162.

Schoch, R.R. (2009b) Developmental evolution as a response to diverse lake habitats in Paleozoic amphibians. *Evolution* **63**, 2738–2749.

Schoch, R.R. (2010) Heterochrony: the interplay between development and ecology in an extinct amphibian clade. *Paleobiology* **36**, 318–334.

Schoch, R.R. (2012) Character distribution and phylogeny of the dissorophid amphibians. *Fossil Record* **15**, 121–137.

Schoch, R. R. & Fröbisch, N.B. (2006) Metamorphosis and neoteny: alternative developmental pathways in an extinct amphibian clade. *Evolution* **60**, 1467–1475.

Schoch, R.R. & Milner, A.R. (2000) Stereospondyli. In: P. Wellnhofer (ed.), *Handbuch der Paläontologie*. Munich: Pfeil, Vol. 3B.

Schoch, R.R. & Rubidge, B.S. (2005) The amphibamid *Micropholis* from the Lystrosaurus Assemblage Zone of South Africa. *Journal of Vertebrate Paleontology* **25**, 502–522.

Schoch, R.R. & Witzmann, F. (2011) Bystrow's paradox: gills, forssils, and the fish-to-tetrapod transition. *Acta Zoologica* **92**, 251–265.

Schoch, R.R. & Witzmann, F. (2012) Cranial morphology of the plagiosaurid *Gerrothorax pulcherrimus* as an extreme example of evolutionary stasis. *Acta Zoologica* **92**, 251–265.

Schoch, R.R., Voigt, S., & Buchwitz, M. (2010) A chroniosuchid from the Triassic of Kyrgyzstan and analysis of chroniosuchian relationships. *Zoological Journal of the Linnean Society* **160**, 515–530.

Schultze, H.-P. (1997) Umweltbedingungen beim Übergang von Fisch zu Tetrapode. *Sitzungsberichte der Gesellschaft Naturforschender Freunde zu Berlin Neue Folge* **36**, 59–77.

Schultze, H.-P. & Arsenault, M. (1985) The panderichthyid fish *Elpistostege*: a close relative of tetrapods? *Palaeontology* **28**, 293–310.

Sequeira, S.E.K. (2009) The postcranium of *Cochleosaurus bohemicus* Frič, a primitive Upper Carboniferous temnospondyl from the Czech Republic. *Special Papers in Palaeontology* **81**, 137–153.

Shishkin, M.A. (1973) The morphology of the early Amphibia and some problems of lower tetrapod evolution. *Trudy Paleontologiceskogo Instituta Akademij Nauk SSSR* **137**, 1–257. (In Russian.)

Shishkin, M.A. (1987) The evolution of early amphibians (Plagiosauroidea). *Trudy Paleontologiceskogo Instituta Akademiya Nauk SSSR* **225**, 1–143. (In Russian.)

Shubin, N. H. (2008) *Your Inner Fish: a Journey into the 3.5-Billion-Year History of the Human Body*. New York: Pantheon Press.

Shubin, N.H., Daeschler, E.B., & Jenkins, F.A., Jr. (2006) The pectoral fin of *Tiktaalik roseae* and the origin of the tetrapod limb. *Nature* **440**, 764–771.

Sigurdsen, T. & Bolt, J.R. (2009) The lissamphibian humerus and elbow joint, and the origins of modern amphibians. *Journal of Morphology* **270**, 1443–1453.

Sigurdsen, T. & Bolt, J.R. (2010) The Lower Permian amphibamid *Doleserpeton* (Temnospondyli: Dissorophoidea), the interrelationships of amphibamids, and the origin of modern amphibians. *Journal of Vertebrate Paleontology* **30**, 1360–1377.

Skutchas, P.P. & Krasnoluzkii, S.A. (2011) A new genus and species of basal salamanders from the Middle Jurassic of Western Siberia, Russia. *Proceedings of the Zoological Institute RAS* **315**, 167–175.

Skutchas, P.P. & Martin, T. (2011) Cranial anatomy of the stem salamander *Kokartus honorarius* (Amphibia: Caudata) from the Middle Jurassic of Kyrgyzstan. *Zoological Journal of the Linnean Society* **161**, 816–831.

Smirnov, S.V. & Vasil'eva, A.B. (1995) Anuran dentition: development and evolution. *Russian Journal of Herpetology* **2**, 120–128.

Smithson, T.R. (1982) The cranial morphology of *Greererpeton burkemorani* Romer (Amphibia:

Temnospondyli). *Zoological Journal of the Linnean Society* **76**, 29–90.

Smithson, T.R. (1985) The morphology and relationships of the Carboniferous amphibian Eoherpeton watsoni Panchen. *Zoological Journal of the Linnean Society* **85**, 317–410.

Smithson, T. R. (1986) A new anthracosaur amphibian from the Carboniferous of Scotland. *Palaeontology* **29**, 603–628.

Smithson, T. R. (2000) Anthracosaurs. In: H. Heatwole & R.L. Carroll (eds.), *Amphibian Biology. Volume 4. Palaeontology.* Chipping Norton, NSW: Surrey Beatty, pp. 1053–1063.

Smithson, T.R. & Rolfe, W.D.I. (1990) *Westlothiana* gen. nov.: naming the earliest known reptile. *Scottish Journal of Geology* **26**, 137–138.

Vallin, G. & Laurin, M. (2004) Cranial morphology and affinities of *Microbrachis*, and a reappraisal of the phylogeny and lifestyle of the first amphibians. *Journal of Vertebrate Paleontology* **24**, 56–72.

Vorobyeva, E. & Schultze, H.P. (1991) Description and systematics of panderichthyid fishes with comments on their relationship to tetrapods. In: H.-P. Schultze & L. Trueb (eds.) *Origins of the Higher Groups of Tetrapods: Controversy and Consensus.* Ithaca: Cornell University Press, pp. 68–109.

Vyushkov, B.P. (1957) [New unusual animals from the Tatarian deposits of the European part of the USSR]. *Doklady Akademij Nauk SSSR* **113**, 183–186. (In Russian.)

Walter, H. & Werneburg, R. (1988) Über Liegespuren (Cubichnia) aquatischer Tetrapoden (?Diplocauliden, Nectridea) aus den Rotteroder Schichten (Rotliegendes, Thüringer Wald/DDR). *Freiberger Forschungshefte* **419**: 96–106.

Wang, Y. & Evans, S.E. (2006) A new short-bodied salamander from the Upper Jurassic/Lower Cretaceous of China. *Acta Palaeontologica Polonica* **51**, 127–130.

Warren, A.A. (2000) Secondarily aquatic temnospondyls of the Upper Permian and Mesozoic. In: H. Heatwole & R.L. Carroll (eds.), *Amphibian Biology. Volume 4. Palaeontology.* Chipping Norton, NSW: Surrey Beatty, pp. 1121–1149.

Warren, A.A. (2007) New data on *Ossinodus pueri*, a stem-tetrapod from the Early Carboniferous of Australia. *Journal of Vertebrate Paleontology* **27**, 850–862.

Wellstead, C.F. (1991) Taxonomic revision of the Lysorophia, Permo-Carboniferous lepospondyl amphibians. *Bulletin of the American Museum of Natural History* **209**, 1–90.

Werneburg, R. (2008) Der "Manebacher Saurier" – ein neuer großer Eryopide (*Onchiodon*) aus dem Rotliegend (Unter-Perm) des Thüringer Waldes. *Veröffentlichungen des Naturhistorischen Museums Schleusingen* **22**, 3–40.

Werneburg, R. (2012) Dissorophoide Amphibien aus dem Westphalian D (Ober-Karbon) von Nýřany in Böhmen (Tschechische Republik) – der Schlüssel zum Verständnis der frühen "Branchiosaurier". *Veröffentlichungen des Naturhistorischen Museums Schleusingen* **27**, 3–50.

Werneburg, R. & Schneider, J. (2006) Amphibian biostratigraphy of the European Permo-Carboniferous. *Geological Society Special Publications* **265**, 201–215.

White, T.E. (1939) Osteology of *Seymouria baylorensis* Broili. *Bulletin of the Museum of Comparative Zoology, Harvard College* **85**, 325–409.

Wiechmann, M.P. (2000) The albanerpetontids from the Guimarota mine. In: T. Martin & B. Krebs (eds.), *Guimarota: a Jurassic Ecosystem.* Munich: Pfeil, pp. 51–54.

Witzmann F. (2006). Morphology and palaeobiology of the Permo-Carboniferous temnospondyl amphibian *Archegosaurus decheni* Goldfuss, 1847 from the Saar-Nahe Basin, Germany. *Transactions of the Royal Society of Edinburgh: Earth Sciences* **96**, 131–162.

Witzmann, F. (2007) The evolution of the scalation pattern in temnospondyl amphibians. *Zoological Journal of the Linnean Society* **150**, 815–834.

Witzmann, F. & Schoch, R.R. (2006) Skeletal development of *Acanthostomatops vorax* from the Döhlen Basin of Saxony. *Transactions of the Royal Society of Edinburgh: Earth Sciences* **96**, 365–385.

Witzmann, F., Schoch, R.R., & Maisch, M.W. (2008) A relic basal tetrapod from the Middle Triassic of Germany. *Naturwissenschaften* **95**, 67–72.

Yates, A.M. & Warren, A.A. (2000). The phylogeny of the 'higher' temnospondyls (Vertebrata: Choanata) and its implications for the monophyly and origins of the Stereospondyli. *Zoological Journal of the Linnean Society* **128**, 77–121.

Zardoya, R. & Meyer, A. (2001) On the origin of and phylogenetic relationships among living amphibians. *Proceedings of the National Academy of Sciences* **98**, 7380–7383.

Zhang, P., Zhou, H., Chen, Y.Q., Liu, Y.F., & Qu, L.H. (2005) Mitogenomic perspectives on the origin and phylogeny of living amphibians. *Systematic Biology* **54**, 391–400.

Zylberberg, L. & Wake, M.H. (1990) Structure of the scales of *Dermophis* and *Microcaecilia* (Amphibia: Gymnophiona), and a comparison to dermal ossifications of other vertebrates. *Journal of Morphology* **206**, 25–43.

3 Amphibian Life Through Time

The fossil record is extremely heterogeneous and full of major gaps.
Some geological intervals and geographic regions are rich in fossils,
whereas others are almost empty or have not been studied extensively to
date. Early tetrapods are known from a few rich deposits, whose study sheds
light on exotic habitats and long-extinct vertebrate communities that often
differ radically from those of modern ecosystems. During most of amphibian
evolution, Earth was a very different planet from what it is today:
supercontinents, huge oceans, and mountain ranges that have long since
disappeared formed the setting in which rainforests, glaciers, deserts, lakes,
and huge deltas replaced one another repeatedly. Early tetrapods and ancient
amphibians inhabited many different regions and manifold zones from dry
uplands to the sea. A walk through the fossil record reveals that some periods
are best documented from equatorial regions (Devonian–Carboniferous),
whereas others are known from almost globally distributed deposits (Triassic,
Cenozoic). The faunal assemblages of which the early tetrapods were part also
form the raw material for studies of paleoecology, evolution, and extinction.
In this chapter, the changes in early tetrapod and amphibian faunas are
illustrated by brief descriptions of exceptional fossil deposits (Figure 3.1).
It is not only interesting to report the fossils occurring in these deposits, but
also worth taking a look at the conditions under which they formed and the
geographical and climatic setting in which the faunas existed.

Amphibian Evolution: The Life of Early Land Vertebrates, First Edition. Rainer R. Schoch.
© 2014 Rainer R. Schoch. Published 2014 by John Wiley & Sons, Ltd.

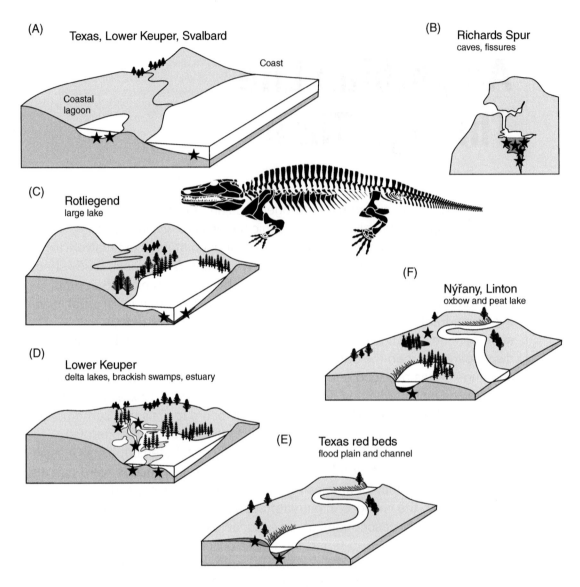

Figure 3.1 Typical deposits in which early amphibians occur. (A) Coastal lagoons provide calm sedimentation, while marine deposits usually preserve less complete skeletal material. (B) Caves and fissures are exceptional but offer exquisite preservation. (C) Large lakes contribute disproportionately to our knowledge of Paleozoic tetrapods, especially where the bottom was free of oxygen (black shales). (D) Delta settings with estuaries and swamps often preserve a wide range of faunas, including brackish and freshwater forms. (E) Floodplains and channels accumulate skeletons and isolated bones in various ways, but the occurrence of fossils is less predictable. (F) Small oxbow and peat lakes preserve skeletons in coal or coaly mudstones, and they are often complete but not always well preserved. Adapted from Boy (1977) and Milner (1987).

3.1 Aquatic predators prepare for land

The Late Devonian world. In the last phase of the Devonian period (374–359 myr), global geography, climate, and environments were very different from today (Figure 3.2). The continental crust was divided into three major units: (1) Laurussia or Euramerica (encompassing North America, Greenland, and northern Europe), (2) Siberia and Kazakhstan (as closely located but tectonically distinct units), and (3) the vast Paleozoic super-continent Gondwana (South America, Africa, India, Australia, and Antarctica). Although forming a single large continent, Euramerica was partially flooded by epicontinental seas along its margins. The Euramerican landscape was partially subdivided by a large mountain range (the Caledonides) crossing East Greenland, Scotland, and Norway. The southern margins of the continent were bordered by small continental plates, which separated Euramerica from the huge Gondwanan landmass to the south.

The world climate underwent major changes during the Devonian and was regionally differentiated. In the Frasnian (385–374 myr), temperatures were warm and the level of atmospheric carbon dioxide was high, whereas oxygen levels were low (Berner 2006; Clack 2007). During the Famennian (374–359 myr), temperatures dropped in the southern hemisphere, leading to a glaciation in South America, whereas the north apparently lacked a polar ice cap (Streel *et al.* 2000). Two major extinction events mark the beginning and the end of the Famennian, probably related to major changes in sea level (Pujol *et al.* 2006).

The Greenland deposits. The richest stem-tetrapod deposits so far discovered are all located on the Euramerican mainland: Ellesmere Island in Arctic Canada (*Tiktaalik*), Greenland (*Ichthyostega*, *Acanthostega*), and the Baltic region (*Tulerpeton*, *Ventastega*). Among these, the fluvial sandstone deposits of eastern Greenland form the richest *Lagerstätte*, having yielded as many as 500 tetrapodomorph specimens to date (Blom *et al.* 2007). They formed in an environment of river channels, small deltas, and larger lakes (Marshall and Stephenson 1997). The setting forms part of the so-called East Greenland Basin formed in a tectonically active fault zone within the Caledonian mountain range. During the Middle and Late Devonian, the basin was flooded by a large and deep lake, followed by rivers flowing in a southerly direction. The sequence of sediments in the basin starts with meandering rivers and coastal floodplains (390 myr, Givetian age), followed by river deposits that grade into aeolian sediments (wind-dominated "sand seas"). Finally, the red and grey river deposits of the latest Devonian (374–359 myr, Famennian age) include the beds in which the famous stem-tetrapods were collected (Blom *et al.* 2007). The Greenland localities occupied a latitude of approximately 15° south, situated in the arid belt of the large Euramerican continent, but also within the range of the summer monsoon (Olsen 1993). The formations that yielded the numerous specimens of *Ichthyostega* and *Acanthostega* were deposited in a large alluvial fan. In this environment, mud-rich floodplains contained clay-dominated soils (vertisols) formed during humid phases and carbonate-rich soils (aridisols) characterizing the drier periods. The formation of such soils especially the aridisols, requires a long time, often thousands of years. This indicates fluctuating levels of humidity (Retallack 1997). The fossil-bearing sequence spans several hundred meters of red, green, and purple siltstones and sandstones. The rivers in which the sandstones were deposited often existed for short time intervals only, interrupted by long phases of aridity or larger flooding events (sheet floods) during wetter seasons. Both *Ichthyostega* and *Acanthostega* appear to have lived in more permanent aquatic habitats, such as deeper water holes, larger channels, or lakes. The accumulation of complete *Acanthostega* skeletons at one site was explained by the animals seeking refuge in a deeper water pit during a dry period.

3.2 Hot springs, scorpions, and little creepers

The Early Carboniferous world. The Carboniferous Period received its name from the abundance of coal in many Late Paleozoic strata. During the Industrial Revolution, numerous coal deposits in

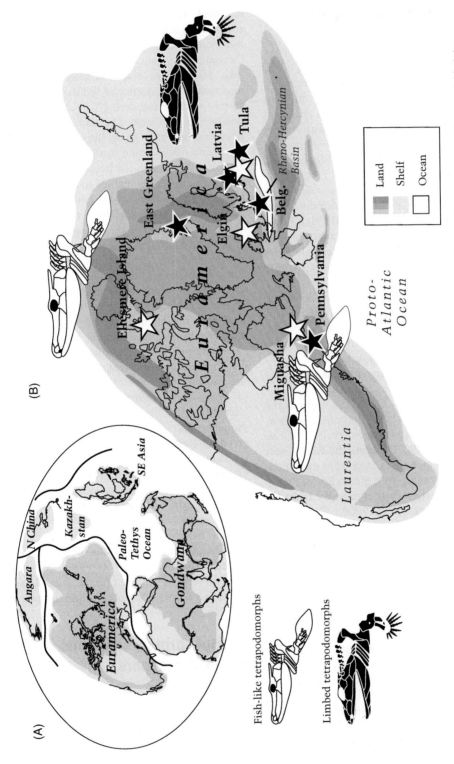

Figure 3.2 The earliest tetrapodomorphs with digits lived in the Late Devonian of Euramerica, among which the Greenland deposits have yielded the largest quantities of finds (*Acanthostega*, *Ichthyostega*). Recently, Ellesmere Island produced the important finds of the tetrapodomorph fish *Tiktaalik*. Adapted from Paleogeography based on www.scotese.com and Ziegler (1989). White stars, fish-like taxa; black stars, limbed tetrapodomorphs.

the British Isles, Central Europe, and the Appalachian Mountains of North America were exploited, which led to the discovery of scores of Carboniferous fossils. The coal formed when extensive rainforests occupied Euramerica (Figure 3.3), supported by an increasingly warm and humid climate (Falcon-Lang 1999). Britain and other coal-rich regions lay close to the equator in Early Carboniferous times and were subject to a seasonal climate, as indicated by growth rings in trees. Monsoonal circulation played an important role, triggered by the large Paleotethys Ocean south of Euramerica. The Early Carboniferous is referred to as the Mississippian in North America.

The East Kirkton deposit. The richest Early Carboniferous locality is a former limestone quarry at Bathgate in the Scottish Midland Valley. Intensive excavation produced numerous skeletons of tetrapods, both stem-amphibians and stem-amniotes, of various taxa (Rolfe *et al.* 1994). The dark limestone is of late Viséan age (330 myr) and was deposited in a hilly landscape dominated by volcanoes. In phases of volcanic activity, hot ash and lava set the dense forests on fire, as evidenced by finds of charcoal (Clack 2012). The limestone itself formed in a freshwater lake, and silica originated from volcanic ash that was washed into the lake. The fossil-rich layers preserve the skeletons of numerous land-living invertebrates and tetrapods but no fishes, along with some definitive aquatic taxa. This lake was populated by ostracods and juveniles of *Balanerpeton*. Remains of terrestrial tetrapods were washed in from nearby water bodies, probably small creeks. The setting resembles that of today's Yellowstone National Park in the United States, with hot springs, boiling creeks, and carbon dioxide pockets killing unlucky animals in an instant. Careful excavation at East Kirkton revealed that the lake became confluent with other water bodies in the last phase of its existence and was eventually inhabited by fishes. The entire sequence of lacustrine strata was probably laid down in several hundreds or a few thousand years (Clack 2012). The East Kirkton lake is peculiar in preserving almost exclusively allochthonous organisms. These include conifers, large scorpions, myriapods, harvestmen, and amphibious euryp-terids (large filter-feeding relatives of horseshoe crabs). The vertebrate fauna comprises small tetrapods, such as the temnospondyl *Balaner-peton*, the baphetid *Eucritta*, the anthracosaur *Silvanerpeton*, the limbless aïstopod *Ophider-peton*, and the enigmatic stem-amniotes *Eldeceeon* and *Westlothiana* (Smithson 1994). All tetrapods appear to have been either fully terres-trial or capable of longer excursions on land. The presence of juvenile specimens of *Balanerpeton* is more puzzling, because they were almost cer-tainly aquatic. This suggests either that the lake was habitable in some phases but poor in species (*Balanerpeton* and its putative but unpreserved prey), or that *Balanerpeton* laid eggs in a nearby water body and juveniles were regularly washed into the East Kirkton lake. Whichever processes were responsible for the accumulation of skele-tons at East Kirkton, they assembled the oldest and richest tetrapod fauna known so far. Other Early Carboniferous tetrapod localities are rare and have produced few specimens, such as from the Tournaisian (359–345 myr) of Dunbarton in Scotland (*Pederpes*) and the middle Viséan (340 myr) of What Cheer in Iowa (*Whatcheeria*) (Clack 2012).

3.3 Life in the tropical coal forest

The Late Carboniferous world. During the Late Carboniferous, the three main continental units had approached one another closely. Euramerica and the Siberia–Kazakhstan continent were near-ing collision. The southern rim of the Euramerican land mass had already started to collide with Gondwana, resulting in a mountain range that was several thousand kilometers long. This range includes the Appalachians in the eastern United States, the North African Atlas Mountains, and the so-called Variscan belt extending across most of Central Europe. The huge mountain range harbored numerous basins and valleys in which dense forests and coal swamps existed, such as those preserved in the coal measures of Britain and the Czech Republic (Figure 3.4). In North America, coal forests formed in coastal areas, with famous tetrapod localities in Illinois, Ohio, and Nova Scotia. Whereas these regions were located

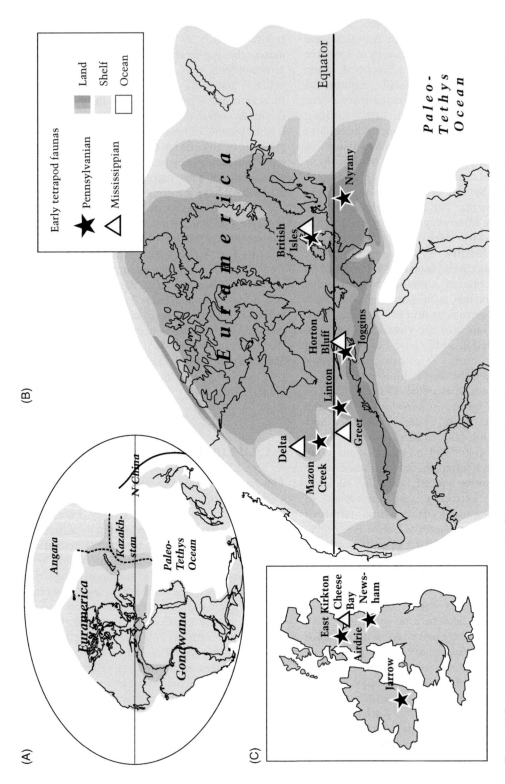

Figure 3.3 Early tetrapods were widespread in the Carboniferous, albeit fossil deposits are restricted to then tropical regions. The largest number of finds was made in Scotland and northern England (classical "coal measures"). Adapted from Paleogeography based on www.scotese.com and Ziegler (1989).

Figure 3.4 Most finds dating around the Carboniferous–Permian boundary come from a belt just north of the equator. In North America, coastal lagoons and other lowland areas housed diverse tetrapod faunas, whereas the Variscan highlands of Europe preserved tetrapods in lake deposits of Spain, France, Germany, and the Czech Republic. Adapted from Paleogeography based on www.scotese.com and Ziegler (1989).

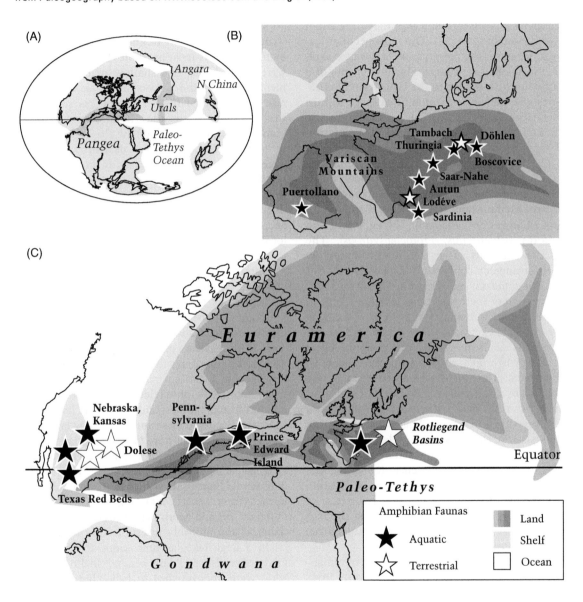

in the tropical belt, much of the southern hemisphere (Africa, India, Australia) was still covered by a large ice cap. In this section, four localities representing different environmental settings are used to illustrate the tropical forest ecosystems of the Late Carboniferous: the coastal deposit at Joggins (Nova Scotia, Canada), the shale beds at Mazon Creek (Illinois, USA), the abandoned river-channel site at Linton (Ohio, USA), and the small peat lake at Nýřany near Plzeň (Czech Republic).

Joggins. The coastal cliffs of Nova Scotia are rich in Paleozoic fossil localities, exemplified by the Joggins site that yielded tetrapods in very good states of preservation. Weathering plays the main role in the discovery of fossils at this site – specimens are usually found after they have fallen down the eroding sea cliff. The most peculiar feature at Joggins is the preservation of tetrapod skeletons in fossil tree stumps. Flooding events of a nearby brackish sea led to the deposition of mudstones and sandstones, which filled these stumps (Falcon-Lang *et al*. 2006). The sediments formed in a tropical forest densely covered by large lycopsid trees (*Sigillaria*) that reached diameters of 90–120 cm. Apart from these large trees, the flora consisted of calamiteans, ferns, pteridosperms, and cordaitaleans. The tetrapods include small terrestrial temnospondyls (*Dendrerpeton*), microsaurs (*Asaphestera, Ricnodon, Hylerpeton*), and amniotes (*Hylonomus, Protoclepsydrops*). The preserved animals either fell into the hollow rotten stumps or deliberately explored them for food, but could not get out again. This favored the preservation of their skeletons, which are mostly disarticulated, probably as a result of scavenging (Boy 1977). The tetrapods are found together with land snails, giant myriapods (including the up to 2 m long *Arthropleura*), eurypterids, arachnids, and insects. The formation of the deposit was triggered by the frequent flooding of the forest, which led to the decay of tree stumps that could then serve as animal traps. This also explains the small size of the preserved skeletons: animals longer than a meter have not been found. A similar tree-stump deposit at Florence (Nova Scotia) yielded a very different tetrapod fauna, suggesting an ecological differentiation of the two sites, or alternatively faunal changes in time, as the two deposits stem from slightly different time slices within the Pennsylvanian (Boy 1977).

Mazon Creek. In northern Illinois, Pennsylvanian-aged shales are quarried for coal, yielding ironstone nodules that are rich in marine and freshwater fossils. Although exceptional among the fauna, tetrapod skeletons were reported that shed light on the larval development of some groups. The temnospondyls *Amphibamus* and *Isodectes* were found, along with the small amniote *Cephalerpeton*. The preservation includes the outline of the body, external gills, and sediment fillings of intestine (Milner 1982). Interestingly, the tetrapods are found in the marine sequence of the rocks, co-occurring with medusae, marine bivalves, invertebrate tracks, crustaceans, up to 30 species of fish, and the enigmatic *Tullimonstrum*, a metazoan of unknown affinities (Baird *et al*. 1986). It is most probable that the tetrapods were washed into this shallow marine environment from the shore. Even the larvae are far too rare to have lived in that habitat, as shown by the contrast with the hundreds of larval branchiosaurids in lakes of the *Rotliegend* facies in Europe. The freshwater deposits at Mazon Creek are rich in plants, myriapods, and arthropods but lack tetrapods (Baird *et al*. 1986). The coal deposits of Mazon Creek formed in a deltaic setting, densely vegetated by large lycopsid trees and horsetails. As at the somewhat older Joggins locality, these forests were repeatedly flooded by the sea, here indicated by thick tidal sediments.

Linton and Nýřany. Two coal mines have produced extremely informative samples of Late Carboniferous tetrapods. The Diamond Mine at Linton in Ohio bears cannel coal that formed in an oxbow lake, an abandoned channel of a meandering river. In Pennsylvanian times, Ohio formed part of a coastal plain along the western margin of the Appalachian mountain range. Linton has yielded many hundreds of skeletons of fishes, stem-amniotes, and temnospondyls, which lived in a short-term ecosystem within the oxbow lake. The "gas coal" (*Gaskohle*) from Nýřany, a small mining town in the western Czech Republic, is famous for a diverse fauna that preserves animals from several different habitats (Milner 1980). These habitats formed in a basin containing water bodies of different size and depth, which were rich in plant material and probably intermittently overgrown. (1) In a large lake, anthracosaurs and baphetids predated on fish. (2) Poorly aerated, shallow swamp lakes were impoverished in fishes but populated by small aquatic tetrapods: the lepospondyls *Oestocephalus, Scincosaurus, Sauropleura*, and *Microbrachis* and the temnospondyls *Limnogyrinus* and *Cochleosaurus*, which are represented by numerous size classes. Larvae of terrestrial temnospondyls were also present, but size distribution indicates that larger individuals left the lake after metamorphosis (Milner 2007). The animals were possibly killed by poisoning due to seasonal turnovers in the lakes. (3) The densely vegetated lowland floodplain was inhabited by microsaurs (*Hyloplesion*, *Sparodus*, *Ricnodon*, *Crinodon*), aïstopods (*Phlegethontia*), terrestrial temnospondyls (*Amphibamus*, *Mordex*), the stem-amniotes *Gephyrostegus* and *Solenodonsaurus*, and true amniotes (*Archaeothyris*), some of which might have come from uplands in the vicinity (Milner 1980).

3.4 Neotenes explore unfavorable waters

The world around the Carboniferous–Permian boundary. By 300 myr ago, the supercontinent Pangaea had finally formed. It united most of the major continental units in a single structure, although shallow seas covered large areas, so that some regions were separated by seaways. For instance, Euramerica and Siberia were separated by an epicontinental sea, although the continental crusts below that sea had already collided. The collision of Siberia–Kazakhstan with Euramerica had resulted in the formation of the Ural mountain range, and the Variscan mountains traversed an extensive part of central Europe. The southern hemisphere was still covered by huge glaciers, as shallow seas covered large parts of the American southwest as well as South America and northern Africa. In the east, a gigantic ocean, called the Tethys, had formed. The Appalachian–Variscan mountains reached a breadth of more than 1000 km in some regions. Numerous basins formed within this mountain range, preserving sediment fillings many hundreds of meters thick. In the internal belt of the mountain range (France, Germany, Poland, and the Czech Republic), the basins frequently housed lakes, some of which reached a length of 50–80 km. The aquatic fauna was usually poor in species, confined to a few bony fishes and stem-amphibians; amniotes are rare and have left only tracks in phases when the basins were dominated by river deposits. The impoverished vertebrate faunas – compared with the tropical faunas of the Late Carboniferous coal measures – have been suggested to indicate a high altitude of the basins within the mountain belt (Boy and Schindler

2000), but this is disputed by others (Schultze and Soler-Gijon 2004). The chemical properties of the water bodies have also been a matter of debate, with interpretation of lake faunas ranging from freshwater (Boy and Sues 2000) to saline (Schultze 2009). The sediment-fills of the European Variscan basins have been referred to as *Rotliegend* ("red beds": Figure 3.5), although the color of sediments is variable and includes conglomerates and sandstones of various shades (channel-fills) intercalated with grey mudstones and yellow dolomites (lake sediments) in numerous successive cycles.

Odernheim. Vineyards at Odernheim, a small town in southwest Germany, expose thin layers of hard limestone that are extremely rich in larval temnospondyls (branchiosaurids). Some beds are so full of skeletons that an area of $1\,m^2$ contains 20 specimens or more. The black bones stand out against the light-colored limestones, and often parts of the skin are preserved as brown shadows revealing the body contours. The small branchiosaurid larvae are famous for the preservation of long external gills. Private collectors found thousands of specimens at the site, which were later distributed to museums all over the world. The collection of the specimens is made difficult by the very hard limestone. However, splitting the limestone along single layers may be enhanced by freezing and then heating the rocks. The fauna includes a single species of actinopterygian fish (*Paramblypterus*), a common neotenic branchiosaurid (*Apateon*), and two rare, larger temnospondyls (*Micromelerpeton*, *Sclerocephalus*). The ecological properties of Lake Odernheim are discussed in detail in Chapter 7 (Paleoecology).

Niederhäslich. In the nineteenth century, a Lower Permian limestone was mined in the small town of Niederhäslich near Dresden in Saxony, Germany. The deposit extends over a few hundred meters only within the small Döhlen basin, a 20 km long depression filled with Permian river and lake sediments (Schneider 1993). This basin was located at the northeastern margin of the Variscan mountains, which during the time of deposition (Sakmarian, 290 myr) must have been largely eroded. The pale brown beds are only a few tens of centimeters thick but contain a rich tetrapod fauna in certain bedding planes. Dissolution of the white bone is the rule in these rocks, leaving only imprints of skeletons. In casting the

Figure 3.5 Environmental setting in central Europe around the Carboniferous–Permian boundary. Located within the Variscan mountain belt, lakes of various sizes harbored tetrapod faunas poor in taxa but very rich in individuals (*Rotliegend* faunas). Skeletons are common in lake mudstones, and tracks of terrestrial amphibians and early amniotes occur in floodplain deposits near river channels.

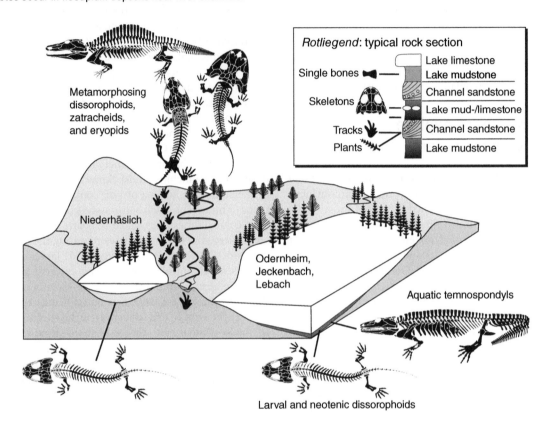

impression with silicone or plaster, researchers have obtained exquisite positives of the skulls, teeth, and limb bones. The limestone formed in a small water body that was rich in larvae of terrestrial temnospondyls (*Apateon, Acanthostomatops, Onchiodon*) and the neotenic dissorophoid *Branchierpeton*. A few larval specimens of the *Seymouria*-like *Discosauricus* were also found. Obviously the lake, which must have been small and shallow, was almost exclusively inhabited by larvae and *Branchierpeton*. It probably could not support larger aquatic vertebrates, as fishes are absent and adult temnospondyls are all metamorphosed and rare, indicating only seasonal visits during breeding (Boy 1990). Land-living amniotes and microsaurs are much more common than in

other Permian deposits of Europe. The terrestrial fauna comprises the pelycosaurs *Haptodus* and *Edaphosaurus*, the diapsid *Kadaliosaurus*, the diadectid *Phanerosaurus*, and the microsaurs *Batropetes* and *Saxonerpeton*. The unique and very local deposit of Niederhäslich thus combines faunal elements of uplands (Tambach, New Mexico) with those of typical *Rotliegend* lakes.

3.5 Lowlands, uplands, and a cave

The Early Permian world. During the first 20 million years of the Permian, large epicontinental seas withdrew from northern Africa and South America. The vast Appalachian–Variscan mountain belt had already passed its maximal extent. The glaciers in southern Gondwana were slowly decreasing in area, and a huge water-filled basin in Brazil and South Africa appeared, the Irati Sea. Black shales and limestones deposited in this isolated water body yield the oldest marine amniotes, the mesosaurs. In the southwestern United States, an epicontinental gulf covered large parts of Texas, Oklahoma, and Kansas. There, the coastline was subject to frequent changes, as evidenced by intercalated marine limestones and fluvial mudstone/sandstone sequences. The terrestrial sediments are predominantly colored mudstones, so-called red beds. Similar rocks are known from Central Europe, where they form part of the higher *Rotliegend* sequence. In Europe, conditions became more arid, as the Variscan mountains were successively eroded. The plants indicate major climatic differences between three principal regions or "provinces" of Pangaea: (1) the Euramerican flora consisted mainly of pteridosperms, ferns, and gingkophytes existing in seasonally dry areas; (2) the Cathaysian province was a tropical rainforest flora, restricted to the isolated Chinese continental units in the Tethys, and was dominated by lycopsids, ferns, and sphenophytes; and (3) the Gondwanan flora in the southern hemisphere was a cool-temperate plant association with *Glossopteris*, cordaites, and pinales (Rees *et al.* 2002). In some regions such as Oman (then northeast Gondwana), the three floras intermingled (Fluteau *et al.* 2001). Paleobotanical data also shed light on the distribution of climates across the supercontinent (Rees *et al.* 2002): the tropical permanently wet belt included most of the Variscan mountains, South China, and North China; the tropical summer-wet zones extended from Texas over Britain into the Urals in the north, and across Venezuela, Algeria, and the Middle East in the south; the arid belts were large and covered most of North America and northern Europe, as well as Brazil and West Africa within the southern belt; finally, the cool-temperate to cold regions extended over most of Siberia in the north (the so-called Angara region, housing a Late Permian endemic flora), and at least half of the Gondwanan area in the south. During the Permian, the deserts expanded at the expense of the temperate and cool climate belts (Rees *et al.* 2002). The existence of large glaciers in the Late Carboniferous and Early Permian is indicated by different lines of geological evidence: striated rock surfaces suggest glacier movements, while dropped stones (tillites) were left by icebergs (Link 2009). Isotope ratios of carbon, oxygen, and strontium permit conclusions to be drawn concerning atmospheric levels of CO_2 and O_2. The 90 myr long Late Paleozoic phase of cold climates in Siberia and Gondwana is referred to as an "icehouse" condition. Evidence for this glaciation was used by Alfred Wegener as an argument in his pioneering studies of continental drift, because evidence of glaciation was found on all former Gondwana continents. The icehouse conditions weakened during the Permian, paving the way for a "greenhouse" climate to dominate the Mesozoic Era (Link 2009). In contrast to icehouse conditions, greenhouse climates are characterized by nutrient upwelling, marine transgression, and carbonate production in the sea (Fischer 1986). The oxygen content of the atmosphere (Berner 1990) was high during the Late Carboniferous and Early Permian (> 30%), but dropped drastically during the Permian to reach a low near the Permian–Triassic boundary (< 15%), in contrast with a present-day value of 21%.

Texas red beds. This is one of the classic regions for Paleozoic tetrapod fossils, where continued collecting since the nineteenth century has accumulated great numbers of specimens (Romer 1935). The Early Permian red beds cover large areas in the American southwest, especially parts of Texas, Oklahoma, and New Mexico, and amounting to some 1.5 km in thickness. The mudstones, siltstones, and sandstones are typical fluvial deposits, formed in river channels, floodplains, and small lakes (Hentz 1988) (Figure 3.6). The plants grew mostly along river banks and pond sides and include ferns, pteridosperms, and conifers. The river and pond faunas consist of sharks, bony fishes, temnospondyls, nectrideans, and anthracosaurs (Romer 1935). Most common are the fully aquatic *Diplocaulus*, *Trimerorhachis*, and *Archeria*, accompanied by the amphibious top predator *Eryops*. The same deposits may contain larger numbers of terrestrial tetrapods as well. The classical pelycosaurs (stem-mammals) *Dimetrodon*, *Sphenacodon*, and *Edaphosaurus* and the

Figure 3.6 Environmental setting during the Early Permian in Texas and New Mexico. The classic red-bed deposits formed in lowland areas under warm and humid conditions. The tetrapod faunas were much richer than in mountainous Europe, with terrestrial temnospondyls and seymouriamorphs forming abundant components.

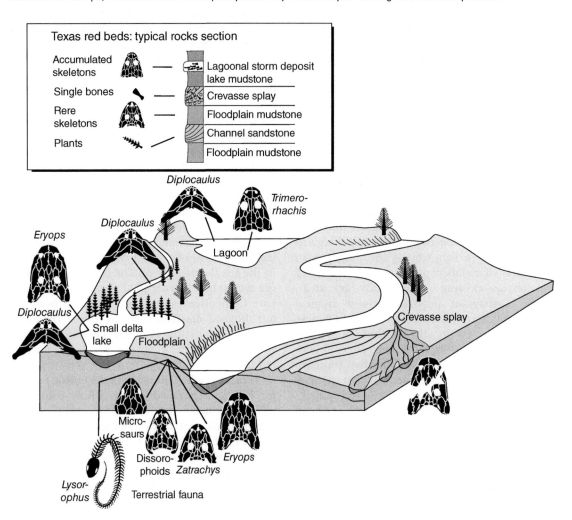

stem-amniotes *Seymouria* and *Diadectes* are prominent examples. The relatively smaller microsaurs *Euryodus*, *Gymnarthrus*, and *Pantylus* are common, as well as the fully terrestrial temnospondyls *Dissorophus*, *Cacops*, *Platyhistrix*, *Trematops*, various tiny amphibamids, and the spiny-skulled temnospondyl *Zatrachys*. A bone bed at Thrift, Texas, has been analyzed in detail by Parrish (1978), who found evidence for an inland incursion of storm tides, destroying small lakes on a floodplain. The inhabitants of the lakes (*Xenacanthus*, *Trimerorhachis*) were killed and their skeletons covered many hundred square meters. The setting was evidently close to the sea and under strong marine influence, as indicated by salt marshes. The ponds sometimes dried, which resulted in the accumulation of temnospondyl skeletons (Case 1935) – especially of the gill-breathing *Trimerorhachis*, which was unable to leave the water. It is expected that some of the aquatic tetrapods tolerated brackish or even marine conditions, but geochemical evidence is required to test that for each taxon and habitat separately. The Geraldine bone bed was studied by Sander (1989), who found charcoal evidence for forest fires that might have killed animals that were later deposited in a lake. Based on large samples of limb bones, Bakker (1982) suggested that young *Eryops* inhabited swamps, whereas the 2 m long adults preferred floodplains and streams, where they probably preyed on fishes and smaller tetrapods. This is probably the setting in which more amphibious forms such as *Eryops* migrated between water bodies. When speaking of the Texas red beds, it must be borne in mind that they include numerous very different localities, most of which are poorly understood; the faunas appear very diverse, but this may boil down on closer inspection to a few taxa per locality and particular horizon (Romer 1928). Finally, Olson (1958) reported the exceptionally well-documented case of a Permian floodplain pond that received its water from a small stream. This locality permitted the identification of three neighboring habitats: (1) the pond itself was populated by the large nectridean *Diplocaulus magnicornis*, which probably fed on aquatic invertebrates; (2) in the stream, the shark *Xenacanthus* and a small relative of the tetrapod in the pond, *Diplocaulus brevirostris*, dominated; and (3) the surrounding floodplain was home to the pelycosaurian top predator *Dimetrodon* and the herbivore *Diadectes* (Olson 1958). Remains of the large temnospondyl *Eryops* were washed into the pond, but its habitat seems to have been elsewhere.

Tambach (Bromacker). A small Permian basin in Thuringia (central Germany) has yielded lake deposits similar to those of other Variscan regions (Werneburg 2001). The tetrapod faunas of Europe and North America were always perceived as distinct, based on the different facies – upland lake deposits in the Variscan mountains here, lowland floodplain deposits in the American Southwest there. An exceptional locality that bridges the gap is a sandstone quarry at the Bromacker locality of Tambach (Thuringia), where red beds similar to those of New Mexico and Texas had long been famous for their richness in vertebrate tracks. Later, Martens (1989) reported tetrapod skeletons co-occurring with tracks in the same deposit, which is an exceptional occurrence. Usually in the fossil record, vertebrate tracks and their producers are not found in the same strata. Continued excavation by Thomas Martens and American colleagues unearthed a bonanza of early Permian tetrapods at Tambach, including the plant-eating stem-amniotes *Diadectes* and *Orobates*, the carnivorous pelycosaur *Dimetrodon*, the carnivorous stem-amniote *Seymouria*, and the terrestrial dissorophoids *Tambachia* and *Georgenthalia* (Eberth *et al.* 2000). A surprising find was an apparently bipedal small amniote, the parareptile *Eudibamus*, which probably fed on plants (Berman *et al.* 2000). Aquatic and amphibious tetrapods are entirely absent in this upland deposit, which formed in a small valley in the northern foothills of the Variscan mountains. Juveniles and larvae of *Seymouria* and temnospondyls are absent, but their preservation in somewhat older horizons of the same general area (Klembara 1995; Werneburg 2001) shows that reproduction still relied on water in both groups. Eberth *et al.* (2000) analyzed the depositional history of the Tambach site, concluding that the red-brown sandstones and siltstones formed in a relatively dry basin that was seasonally flooded. Ponds and streams did not persist for long, and the animals are believed to have been killed by floods. The savanna-like climate was hot year-round, with conifers and seed ferns predominating. The abundance of high-fiber plant-eating diadectids and the absence of amphibious tetrapods highlight the aridity of the habitat. The Tambach site is further exceptional in that it first permitted fossil tracks to be matched with the skeletons of the track-makers: the researchers were able to show that two track species of *Ichniotherium* precisely matched the limb skeletons of *Diadectes* and *Orobates*. Martens (2005) reported burrows from the same site that must have been produced by animals with diadectid body proportions. This highlights the habit of some Permian herbivores to dig long helical (coiled) burrows, probably as a means of coping with hot and dry conditions.

Fort Sill. The Dolese limestone quarry near Fort Sill (Oklahoma) has yielded countless tetrapod bones over a period of seven decades (Olson 1967). Continued excavation by private collectors has produced ever more material, amounting to thousands of bones and teeth and more recently also articulated skeletons. To date as many as 36 tetrapod taxa have been reported from the fissure fills, which according to Reisz (2007) formed in a large Early Permian cave system. Such deposits are extreme-concentration *Lagerstätten*, in which vertebrate remains accumulate over many thousands of years. Preservation at Fort Sill is often very good, and some of the best-preserved three-dimensional temnospondyl and microsaur skeletons have been described from here. This site had already produced the classic material of *Doleserpeton*, a small dissorophoid with pedicellate teeth (Bolt 1969). Recent finds include the trematopid *Acheloma*, the dissorophid *Cacops*, and the amphibamids *Tersomius* and *Pasawioops* (Fröbisch and Reisz

2008; Reisz *et al.* 2009; Polley and Reisz 2011). With the entirely terrestrial and probably arid conditions at Fort Sill, the Paleozoic amphibians reached an extreme point of a broadened range of successfully colonized environments.

3.6 Hide and protect: extreme life in the hothouse

The world across the Permo-Triassic boundary. The Late Permian was a time of substantial changes (Figure 3.7): sea levels dropped to a low, cold polar regions transformed into temperate ones, continent-wide ice sheets disappeared, carbon dioxide levels were on the rise, and the oxygen

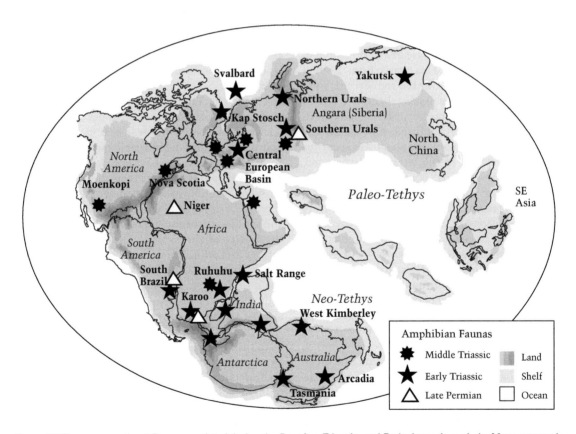

Figure 3.7 The supercontinent Pangaea existed during the Permian, Triassic, and Early Jurassic periods. Many tetrapods reached a global distribution, and by the Early Triassic rich tetrapod faunas were established – ranging from fully marine to arid mid-continental regions. Adapted from Paleogeography based on www.scotese.com and http://jan.ucc.nau.edu/rcb7/globaltext2.html.

content of the atmosphere declined (Erwin 2001). With so many severe changes in a geologically short interval (~5 myr), it is no wonder that the Permo-Triassic (P–T) boundary marks the most severe extinction event of Phanerozoic history: as many as 90% of marine species appear to have become extinct (Benton 2003). The climate was clearly entering a new phase, starting with a steady increase in atmospheric carbon dioxide (Saunders and Reichow 2009). Evidence for substantially higher CO_2 levels is consistent with the worldwide expansion of hot and dry conditions: fossil soils indicate that hot climate belts were expanding (at the expense of the wet equatorial tropical belt) and vegetation underwent substantial changes (Ward *et al.* 2000). Even areas close to the South Pole (at 80–85°) were temperate, as indicated by a deciduous forest in Antarctica with fast-growing trees (Taylor *et al.* 1992) – the cold climate belts had obviously disappeared.

The South African Karoo Basin best exemplifies the changes across the P–T boundary. Here a nearly continuous Permo-Triassic rock record preserved continental deposits so rich in tetrapod remains that they are used as stratigraphic index fossils (Rubidge 1995). These involve a reddening of sediments, a major reduction in the area of water-loving plants, and a drying of floodplain habitats (Smith 1995). The river systems changed from meandering into braided – that is, they covered much larger areas because the protection of river banks by rooted plants had vanished. This has been interpreted as a consequence of a catastrophic die-off of vegetation, a trend also confirmed in other parts of the world (Ward *et al.* 2000). In the oceans, evidence of anoxic and alkaline conditions was found (Woods 2005). What seems certain is the evidence for a worldwide change to a very dry "hothouse" climate within the last 5 myr of the Permian. The radiometric dating of extensive basalt fields in the Siberian highlands revealed a "smoking gun" for the origin of the hothouse: eruptions of vast flood basalts occurred just before the beginning of the Triassic, and they are believed to have had a substantial impact on the climate-relevant gases in the atmosphere (Saunders and Reichow 2009).

During the Early Triassic, marine and terrestrial ecosystems slowly recovered (Erwin 2001). The disappearance of metazoan reefs was followed by a reorganization of marine ecosystems, after which microbes were the only reef builders for a 5–6 myr period (Pruss and Bottjer 2005). Extinction of land plants has only recently been reported for this period, with the opportunistic pioneer lycopsid *Pleuromeia* dominating Early Triassic landscapes in both Euramerica and Gondwana (Grauvogel-Stamm and Ash 2005). Vertebrates show a mixed picture: during the late Early Triassic, reptiles experienced a rapid evolutionary diversification in the sea, with sauropterygians, ichthyosaurs, and smaller clades distributed globally, whereas the typically Permian land vertebrate communities, dominated by synapsids, were decimated (Smith and Botha 2005). The terrestrial fauna includes generally small taxa, such as small terrestrial temnospondyls, lizard-like parareptiles, smaller plant-eating dicynodonts, and insectivorous cynodonts.

In the long run, the tiny and opportunistic lissamphibians were among the winners in this crisis, as were the huge aquatic temnospondyls, the top predators in freshwater ecosystems around the world. Among the amniotes, diapsid reptiles became the dominant predators in both terrestrial and marine realms, whereas the synapsids slowly became extinct, with the exception of the single lineage that gave rise to mammals at the end of the Triassic. Ultimately, the evolutionary success of lissamphibians, mammals, and birds was only possible after the P–T extinction had destroyed the synapsid-dominated faunas of the Permian.

Karoo Basin. Covering almost two-thirds the area of South Africa, the Karoo Basin contains 12 000 m of terrestrial sediments (Carboniferous–Jurassic). Its Permo-Triassic strata have become famous for their richness in tetrapods, exemplified by the abundant dicynodont *Lystrosaurus*. This 1 m long plant-eating synapsid inhabited huge areas across Pangaea (Africa, India, Antarctica, China, and Russia) – its distribution provided powerful evidence for Alfred Wegener in the 1920s, as he argued for the existence of an early Mesozoic supercontinent. Stratigraphy across and beyond the Karoo Basin is also based on vertebrate index fossils, and part of the Early Triassic is known as the *Lystrosaurus* Assemblage Zone, for instance. The amphibian record of the Karoo Basin changed essentially through stratigraphy: contrasting the general trend across the P–T boundary, diversity was low in the Permian but increased successively during the earliest Triassic. In the Late Permian, 2–3 m long aquatic fish-eaters inhabited riverbanks and oxbow lakes on vast floodplains (temnospondyl *Rhinesuchus*). This landscape, which provided rich habitats for the large plant-eater *Dicynodon*, was replaced by a dry alluvial plain with sparse vegetation and hostile playa lakes (Smith and Botha 2005). This was the habitat of *Lystrosaurus*, which fed on lush ferns and clubmosses and probably lived in burrows. *Lystrosaurus* is often preserved in mass accumulations, where the animals probably died in the periphery of shrinking water holes. In this setting, large crocodile-like temnospondyls (*Uranocentrodon*) persisted in the few larger rivers, but were accompanied by a range of small temnospondyls that were more terrestrial (*Micropholis*, *Lydekkerina*, *Broomistega*) (Smith and Botha 2005). The small eel-like temnospondyl *Thabanchuia* might have survived dry seasons in burrows. In contrast, most of the small amphibians had robust limb skeletons, suggesting that they were able to cross longer distances between water bodies – clearly an advantage in a seasonal and unpredictable climate. In the early Middle Triassic, humid conditions returned, the diversity in land plants and herbivores increased, and huge archosauriform predators evolved, such as 5 m long *Erythrosuchus*. Freshwater bodies were again populated by large amphibians, among them the 1–3 m long stereospondyls *Parotosuchus* and *Batrachosuchus*.

Czatkowice cave. The limestone deposits at Czatkowice, southern Poland, include horizontal funnels and corridors within Paleozoic carbonates (Paszkowski 2009). They formed in fissures filled by sands and silts during the early Triassic (Olenekian, 249–245 myr) which probably formed in a small cave. In these, vertebrate bones accumulated in a breccia. The walls and roof of the cave were covered by crystalline, pink flowstones formed earlier in the Permian under hydrothermal conditions. The Triassic bone breccia suggests that the vertebrate deposit formed in a collapsed doline, in which the fine sediments preserve traces of the roof collapse. The bone material is exquisitely and three-dimensionally preserved but disarticulated and usually broken

into pieces (Paszkowski 2009). The dating of the bone breccia is based on lungfish teeth – as usual for vertebrate index taxa, such correlations operate on a rather coarse scale. Analysis of surface structure and geochemistry revealed that the bones were probably reworked from ephemeral pond deposits in the vicinity (Borsuk-Białynicka and Evans 2009). The Czatkowice deposit is famous for its proto-frog *Czatkobatrachus*, the second-oldest salientian after Madagascan *Triadobatrachus*. The vertebrate fauna includes a rich assemblage of juvenile temnospondyls (capitosaurs, brachyopids), parareptiles, small diapsids, and archosauromorphs (Borsuk-Białynicka and Evans 2009). The problem with this deposit, which, like Fort Sill, preserves large quantities of excellent bones, is the complete lack of articulated specimens. This makes referral of any two elements to the same taxon hypothetical and restricts ecological analysis.

Marine deposits of Spitsbergen. The most surprising finds of Triassic amphibians were made in fully marine rocks at Spitsbergen (Svalbard) in the Arctic Sea (Harland 1997). Occasional finds of temnospondyls in marine rocks are common in the Triassic. However, in central Spitsbergen, rich remains of eight different species were found in *Posidonomya* beds of Early Triassic age. In siltstones and carbonate nodules regularly yielding two unequivocal marine groups (ammonites and ichthyosaurs), abundant remains of stereospondyl amphibians occur (Wiman 1914). While the fragmentary nature of many finds suggests transport before deposition, coprolites indicate the presence of the amphibians at the localities (Lindemann 1991). The Early Triassic appears to have been the peak time for marine temnospondyls – similar habitats were also reported from Greenland (Wordie Creek), Western Australia (Blina Shale), western Pakistan (Mianwali), and Madagascar (Middle Sakamena Beds). However, the story may be more complicated: Lindemann (1991) analyzed the strontium isotope ratio of temnospondyl bones, their coprolites, and undoubtedly marine taxa from the Spitsbergen sample. He found that temnospondyl bones and coprolites had ratios quite different from the marine controls, suggesting migration between freshwater and sea in a coastal environment, as performed by eels and salmon today.

3.7 Predators in deltas, lakes, and brackish swamps

The Middle–Late Triassic world. During the entire Triassic Period, Pangaea existed as a single supercontinent, but the Tethys Ocean experienced rapid expansion in the north (Neo-Tethys). In East Asia, the northern and southern Chinese blocks had collided and together with Xinjiang were fused to Pangaea. Iran and Tibet were isolated microcontinents approaching the Eurasian landmass, eventually colliding with it in the Late Triassic. During that time, an extensive rift valley formed inside the old Appalachian–Moroccan fold belt, later connecting to rifts in the Arctic between Greenland and Norway. This paved the way for the Jurassic breakup of Pangaea and initiated the birth of the North Atlantic Ocean.

The climate was still warm-temperate well into high latitudes, with the eastern United States, the Urals, and parts of China falling in the tropical belt. Although significantly shrunken since the latest Permian, semi-arid zones still covered huge areas of Pangaea: most of Brazil and north-central Africa in the southern hemisphere, the southwestern United States, eastern Canada, and central Europe in the north. The Tethys covered vast shelf areas in Europe, Arabia, India, and China, which were populated by crinoid and bivalve reefs, a diverse fish fauna, and marine reptiles. The Late Triassic was also a crucial time for the evolution of tetrapods: sphenodontians, turtles, crocodile-like archosaurs, pterosaurs, dinosaurs, and the immediate precursors of mammals first appeared during this 35 myr epoch. During the Triassic, only two major amphibian clades survived: the large stereospondyls and the tiny lissamphibians.

The Lower Keuper delta. In western and central Europe conditions were still semi-arid, but the Central European Basin now housed large streams that were nourished by subtropical rainfall in the Baltic region. This led to the deposition of delta sandstones and playa lakes, interrupted by marine transgression of the Tethys (Figure 3.8). Middle Triassic limestones and coal have been quarried since the late seventeenth century across central Europe, yielding rich finds of vertebrate fossils. The coal deposits formed in a large delta that spanned most of Germany. Some marine incursions left a landscape with numerous lakes, swamps, and marshes. Lined by horsetail stands, these water bodies ranged between a few hundred meters and tens of kilometers in length and changed at a geologically fast pace ($\sim 10^2$–10^4 years). The frequent marine influence, both by increasing sea levels and tropical storms, led to a broad range of brackish to hypersaline lakes. The favorable ones among these were inhabited by marine reptiles (nothosaurs, placodonts, pachypleurosaurs), but bivalves indicate that most of these lakes offered rather harsh conditions. In contrast, the freshwater lakes – oxbows, coal seams, and ponds – harbored rich fish and amphibian faunas (see Chapter 7, Paleoecology).

Two localities have yielded rich material in the last few decades: a roadcut near Kupferzell and a limestone quarry at Vellberg. In both sites, grey mudstones bear mass accumulations of skulls and bones, and in places pockets with articulated skeletons have been found. The 5–6 m long temnospondyl *Mastodonsaurus* is almost universally present, accompanied by smaller aquatic predators in the 1–3 m range (*Kupferzellia*, *Trematolestes*, *Callistomordax*, *Plagiosuchus*, and *Gerrothorax*). These taxa occur with variable frequency in different lake deposits. A typical feature of these lakes is that fishes were diverse but small, in the 5–30 cm range.

3.8 Stereospondyls in refugia, lissamphibians on the rise

The Jurassic world. The main geographic change during the Jurassic was the breakup of Pangaea, which was accompanied by volcanism and the formation of intramontane basins. In North America, the fossiliferous basins of the Newark Supergroup accumulated a huge volume of sediments from Late Triassic to the Early Jurassic. During the Middle Jurassic (175–161 myr), new sea floor started to form between Europe, North Africa, and North America, leading to a partial separation of blocks within northern Pangaea. In eastern and southern Gondwana, rift valleys also formed between Africa, India, and Australia,

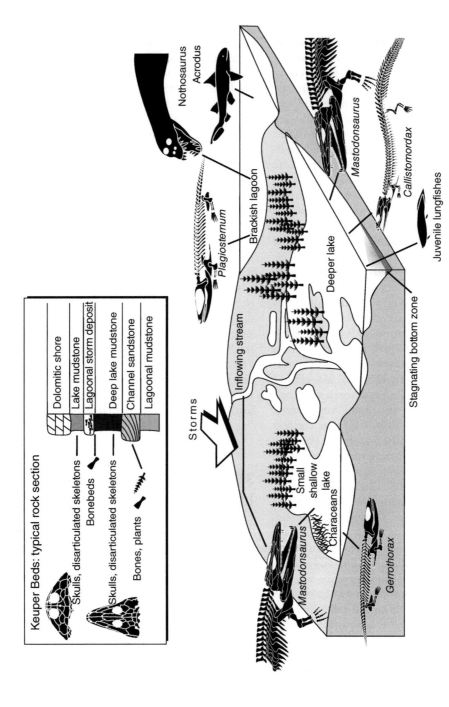

Figure 3.8 Environmental setting during the Middle Triassic in central Europe. The Keuper deposits formed in semi-arid lowland areas under subtropical conditions. The fish and aquatic temnospondyl faunas were rich, including the huge predator *Mastodonsaurus*.

initiating the birth of the Indian Ocean. Jurassic climates were warm up into the high latitudes, but the sea levels were rising steadily and many regions became more humid, especially along the northern Tethys margin as monsoonal influence intensified. Atmospheric oxygen levels were still low (15–18%), whereas carbon dioxide levels were still higher than today (Berner 2006). After two extinction events (Carnian–Norian, Triassic–Jurassic), the shallow marine and terrestrial faunas were substantially different by Early Jurassic time. In the sea, nothosaurs, pachypleurosaurs, placodonts, and thalattosaurs had disappeared, whereas ichthyosaurs and plesiosaurs diversified tremendously, reaching body sizes beyond 20 m. The small sphenodontids were also found in shallow marine habitats. On land, the dicynodonts and therocephalians had become extinct, as had the rauisuchians, phytosaurs, and aetosaurs. In turn, dinosaurs experienced a tremendous evolutionary radiation (Sues and Fraser 2010). Two cynodont groups became abundant: the tritylodontids (e.g., *Oligokyphus*) and mammals (e.g., *Morganucodon*). The Jurassic record of lissamphibians is much more substantial than the Triassic one, with gymnophionans and caudates making their first appearance, and salientians becoming more diverse. In addition, the enigmatic lissamphibian-like albanerpetontids make their first appearance. The large temnospondyls disappeared after the Rhaetian in most regions, with brachyopids persisting in refugia in Mongolia and chigutisaurids in Australia.

Kayenta Formation. The scenic landscapes in northern Arizona and Utah are composed of Triassic and Jurassic rocks that formed under terrestrial conditions. Among these, the Early Jurassic Kayenta Formation (196–183 myr) is famous for preserving small tetrapods in larger quantities, in addition to remains of dinosaurs. It includes mudstones and siltstones, formed in overbank deposits of larger rivers that alternate with aeolian rocks and smaller channel-fills (Sues *et al.* 1994). The siltstone-dominated facies of northern Arizona contains a rich fauna including two – in evolutionary terms – spectacular lissamphibian taxa: the earliest jumping salientian *Prosalirus* (Jenkins and Shubin 1998), and the limbed gymnophionan *Eocaecilia* (Jenkins and

Walsh 1993). Common amniotes include the cryptodiran turtle *Kayentachelys*, several early crocodile like reptiles, various dinosaur taxa, and the tritylodontid stem-mammals *Kayentatherium* and *Oligokyphus* (Sues *et al.* 1994).

British microvertebrate localities. In southern England, Middle Jurassic rocks preserve rich microvertebrate faunas. By the beginning of the Jurassic, large parts of Europe were flooded by the sea and thus formed a large shelf region. In southern Britain, three islands remained where terrestrial sediments were deposited. There, Middle Jurassic (Bathonian, 167–164 myr) rocks include marls and clays deposited in freshwater or under brackish conditions (Evans and Milner 1994). The fossiliferous soft marls formed in a coastal area with swampy environments including lagoons, creeks, and small freshwater lakes. A good modern analog appears to be the Florida Everglades. Apart from bony fishes (semionotids, pycnodontoids, and amiids), three clades of amphibians have been reported: (1) true frogs (*Eodiscoglossus*), (2) caudates (e.g., *Marmorerpeton*), and (3) the oldest record of the albanerpetontids. Small amniotes are also abundant, among them primitive lepidosauromorphs (*Marmoretta*), sphenodontians, and true lizards. Furthermore, the aquatic choristoderes are also present (*Cteniogenys*), along with stem-crocodiles.

Guimarota. At Leira (Portugal), the Guimarota coal mine has produced a wealth of vertebrate fossils in the last five decades (Krebs and Martin 2000). Chemical processing of the coal greatly facilitated the collection of small vertebrates. The Guimarota beds fall into the Late Jurassic (~152 myr) and consist of coal seams embedded in limestones. The sequence was deposited in a coastal swamp area, close to the North Atlantic rift valley, which already formed a seaway at the time. Abundant vegetation was deposited in peat lakes, indicating a tropical climate. Mollusk and ostracod faunas suggest the presence of both freshwater and coastal marine conditions in the area. Sharks and bony fishes are abundant, albanerpetontid amphibians (*Celtedens*) very common (Wiechmann 2000), and the reptile fauna includes squamates, turtles, stem crocodiles, pterosaurs, and dinosaurs. Most diverse are the mammals, with famous finds of now-extinct multituberculates, docodonts, and dryolestids (Krebs and Martin 2000).

Central Asia. Inner Asia was a safe haven for some tetrapods during the Mesozoic, among which were the temnospondyl amphibians. Altogether, diagnostic but tantalizingly fragmentary remains of brachyopids have been reported from Mongolia, Kyrgyzstan, Xinjiang, Sichuan, and Thailand (Shishkin 2000). Another deposit produced no temnospondyls, but caudate material occurs in the laminated freshwater limestones of Karatau (Kazakhstan), where the exquisite skeleton of *Karaurus* was found (Shishkin 2000).

3.9 Batrachians diversify, stereospondyls disappear

The Cretaceous world. Spanning some 80 myr, the Cretaceous was a long period full of substantial changes in global geography, faunas, and floras. The end of this period is often more highlighted than the duration itself, because of the extinction of iconic dinosaur taxa such as *Tyrannosaurus* and *Triceratops*, the last ammonites, and many species of marine invertebrates and plankton. It is also true, however, that the terrestrial ecosystems, and especially the flora, underwent major reorganization that paved the way for the modern world. During the Cretaceous, the Atlantic Ocean increased in size, first separating North Africa and North America, followed by rifting between Brazil and West Africa. The Indian Ocean expanded more slowly, initially flooding rift valleys between India and East Africa, then between India and Australia plus Antarctica. India started to drift northwards, still separated from Eurasia by the several thousand kilometer-wide Tethys Ocean. In the course of the Cretaceous, shelf areas expanded further and many shallow epicontinental seas formed: most of central and eastern Europe were marine from the Early Cretaceous on, and in North America a large seaway formed connecting the Arctic Sea with the Gulf of Mexico; vast areas of North Africa and Arabia were marine as well. The climate was warm-temperate up to high latitudes, the CO_2 levels high but steadily declining in the Late Cretaceous, while atmospheric oxygen reached a peak in the early Late Cretaceous (25%) (Berner 2006). The extinction event at the end of the Cretaceous (K–Pg boundary) has been studied extensively, with two "smoking guns" remaining after much discussion: (1) the evidence for climate change is substantial, suggesting a cooling of oceans that led to the extinction of marine plankton; (2) a crater over 100 km wide in Yucatán (Mexico) has been identified and dated, indicating the impact of a meteor shortly before the Cretaceous–Paleogene boundary (Schulte *et al.* 2010). Extinction was very pronounced in some groups, but also selective: for instance, most dinosaurs disappeared, but one group of small theropod dinosaurs, birds, survived. Likewise, some mammal clades (multituberculates) were affected, but not others. However, the amphibians that survived into the Late Cretaceous, namely lissamphibians and albanerpetontids, did not experience significant extinction (MacLeod *et al.* 1997). During the Cretaceous, salamanders and frogs diversified extensively, and the number of fossil deposits preserving these increased. The last known temnospondyl was the chigutisaurid *Koolasuchus* from the Aptian (112–125 myr) of Victoria, Australia (Warren *et al.* 1997).

The Jehol biota. Few fossil localities have been more surprising and enlightening than the Early Cretaceous limestones and tuffs from northeast China. These finely laminated rocks formed in lakes that covered a large area in the Chinese provinces of Liaoning, Inner Mongolia, and Hebei. Sedimentation in the lakes was under the influence of repeated volcanic eruptions in the vicinity, and limestone deposition was regularly interrupted by tuffs (ash layers). These fine-grained beds permitted even the most delicate structures to be preserved. Numerous quarries are worked by farmers who trade in fossils, but scientific excavations have also been undertaken in the last decade (Wang and Zhou 2003). In addition to the breathtaking fossils themselves, they provide a wealth of data on the formation of these exceptional conservation *Lagerstätten*. The Jehol fauna combines several exceptional features: a high concentration of vertebrate skeletons, the preservation of both lake and terrestrial faunas, and especially the exquisite quality of preservation. Fine structures, such as lizard scales, mammal fur, pterosaur "hair," and most notably dinosaurian proto-feathers and true feathers have made these deposits world-famous (Wang and Zhou 2003). The fauna includes water insects, fishes, caudates, anurans, turtles, mammals, choristoderes, squamates, pterosaurs, and a diverse assemblage of dinosaurs including early birds. The caudates are speciose, among them crown urodeles (*Chunerpeton*, *Beiyanerpeton*) and several stem taxa (*Sinerpeton*, *Laccotriton*, *Jeholotriton*, *Liaoxitriton*). The salientians encompass a stem taxon (*Mesophryne*) and the discoglossid-like *Callobatrachus* (Wang and Gao 2003). In sum, the lissamphibian finds add substantially to our knowledge of stem lineages and testify to the diversification of crown groups by the time of the Early Cretaceous.

Las Hoyas. Another conservation *Lagerstätte* is located near Cuenca, Spain. The Las Hoyas deposits formed under stagnating conditions in freshwater lakes, preserving elements of an inland flora and a rich vertebrate fauna with fishes, amphibians, stem crocodiles, and many dinosaurs. The albanerpetontid *Celtedens* and the crown salamander *Valdotriton* are known from complete skeletons (McGowan and Evans 1995; Evans and Milner 1996).

3.10 Lissamphibians expand into diverse habitats

The Cenozoic world. After the K–T extinction, the world climate was generally cooler, the temperature gradient from equator to poles higher, and the climate overall less stable. What was formerly known as Tertiary has recently been subdivided in two separate periods, the Paleogene (65–23 myr) and Neogene (23–1.8 myr). By the Eocene, the continents had attained their modern distribution, with Australia finally separated from Antarctica and India colliding with Eurasia. This collision formed part of a much larger process in which Africa and India crushed into Eurasia, closed the Tethys Ocean, and folded a 10 000 km long mountain range that today spans the Pyrenees, Alps, Balkans, Turkey, Iran, Afghanistan, Tibet, and southeast Asia. During that time, North and South America approached each another, and when the Central American land bridge was

established some 3 myr ago, a major faunal exchange started (Great American Interchange). This led to the migration of northern placental mammals into southern domains and the extinction of many marsupials in South America.

The evolution, distribution, and ecology of amphibians during the Cenozoic Era can be studied in much greater depth than for the Mesozoic because of the much greater number of fossil deposits. For instance, a substantial evolutionary radiation of anurans must have taken place (Báez 2000), whereas salamanders appear to have been already diverse by the Cretaceous (Milner 1983, 2000). Albanerpetontids became extinct only very recently, ~3 myr ago (Gardner and Böhme 2008).

The Messel crater lake. During the Eocene Epoch (55–33 myr), Europe was differentiated into an archipelago with numerous isolated regions and faunas. In the Franco-German Rhine Valley, then at the latitude of present-day Sicily, dense forests flourished in a subtropical climate. Various volcanoes shaped the landscape, and at Messel, near Frankfurt, a crater lake existed some 47 myr ago. A geological drilling project revealed the crater structure, indicating that the crater lake formed after a massive volcanic explosion. In this lake, dark bituminous mudstones (oil shales) formed in stagnating freshwater, preserving both the fishes and various terrestrial animals and plants that lived along its shore. The Messel oil shale was quarried in a pit some 900 m wide and 60 m deep, for almost a century (Franzen and Schaal 2000). Excavations over several decades have produced tremendous amounts of fossils, ranging from insects with original colors, exquisite fishes, amphibians, diverse lizards and snakes, crocodiles, and birds, as well as bats, primates, and primitive small horses. The amphibian fauna consists of crown-group lissamphibians: anurans (*Eopelobates*, palaeobatrachians) and the large salamandrid urodele *Chelotriton*. The Messel finds are exceptionally well preserved, which includes skin, but preparation is made difficult by the fast deterioration that starts shortly after the fossil-bearing rock is exposed and starts to dry out. To this end, a complicated preparation method had to be developed that removes the sediment completely and embeds the fossil in artificial resin.

References

Báez, A.M. (2000) Tertiary anurans from South America. In: H. Heatwole & R.L. Carroll (eds.), *Amphibian Biology. Volume 4. Palaeontology*. Chipping Norton, NSW: Surrey Beatty, pp. 1388–1401.

Baird, G.C., Sroka, S.D., Shabica, C.W., & Kuecher, G.J. (1986) Taphonomy of middle Late Carboniferous Mazon Creek area fossil localities, northeast Illinois: significance of exceptional fossil preservation in syngenetic concretions. *Palaios* **1**, 271–285.

Bakker RT. (1982) Juvenile–adult habitat shift in Permian fossil reptiles and amphibians. *Science* **217**, 53–55.

Benton, M.J. (2003) *When Life Nearly Died*. London: Thames & Hudson.

Berman, D.S., Reisz, R.R., Scott, D., Henrici, A., Sumida, S., & Martens, T. (2000) Early Permian bipedal reptile. *Science* **290**, 969–972.

Berner, R.A. (1990) Atmospheric carbon dioxide levels over Phanerozoic time. *Science* **249**, 1382–1386.

Berner, R.A. (2006) GEOCARBSULF: a combined model for Phanerozoic atmospheric O_2 and CO_2. *Geochimica et Cosmochemica Acta* **70**, 5653–5665.

Blom, H., Clack, J.A., Ahlberg, P.E., & Friedman, M. (2007) Devonian vertebrates from East Greenland: a review of faunal composition and distribution. *Geodiversitas* **29**, 119–141.

Bolt, J.R. (1969) Lissamphibian origins: possible protolissamphibian from the Lower Permian of Oklahoma. *Science* **166**, 888–891.

Borsuk-Białynicka, M. & Evans, S.E. (2009) Early Triassic vertebrate assemblage from karst deposits at Czatkowice, Poland. *Palaeontologia Polonica* **65**, 1–332.

Boy, J.A. (1977) Typen und Genese jungpaläozoischer Tetrapoden-Lagerstätten. *Palaeontographica A* **156**, 111–167.

Boy, J.A. (1990) Über einige Vertreter der Eryopoidea (Amphibia: Temnospondyli) aus dem europäischen Rotliegend (? höchstes Karbon – Perm). 3. *Onchiodon*. Paläontologische Zeitschrift **64**, 287–312.

Boy, J.A. & Schindler, T. (2000) Ökostratigraphische Bioevents im Grenzbereich Stefanium/Autunium (höchstes Karbon) des Saar-Nahe-Beckens (SW-Deutschland) und benachbarter Gebiete. *Neues Jahrbuch für Geologie und Palaontologie, Abhandlungen* **216**, 89–152.

Boy, J.A. & Sues, H.-D. (2000) Branchiosaurs: larvae, metamorphosis and heterochrony in temnospondyls and seymouriamorphs. In: H. Heatwole & R.L. Carroll (eds.), *Amphibian Biology. Volume 4. Palaeontology*. Chipping Norton, NSW: Surrey Beatty, pp. 973–1496.

Case, E.C. (1935) Description of a collection of associated skeletons of *Trimerorhachis*. *Contributions, Museum of Paleontology, University of Michigan* **4**, 227–274.

Clack, J.A. (2007) Devonian climate change, breathing, and the origin of the tetrapod stem group. *Integrative and Comparative Biology* **47**, 510–523.

Clack, J.A. (2012) *Gaining Ground: the Origin and Evolution of Tetrapods*, 2nd edition. Bloomington: Indiana University Press.

Eberth, D.A., Berman, D.S., Sumida, S.S., & Hopf, H. (2000) Lower Permian terrestrial paleoenvironments and vertebrate paleoecology of the Tambach Basin (Thuringia, Central Germany): the Upland Holy Grail. *Palaios* **15**, 293–313.

Erwin, D.H. (2001) Lessons from the past: biotic recoveries from mass extinctions. *Proceedings of the National Academy of Sciences* **98**, 5399–5403.

Evans, S.E. & Milner, A.R. (1994) Middle Jurassic microvertebrate assemblages from the British Isles. In: N.C. Fraser & H.-D. Sues (eds.), *In the Shadow of the Dinosaurs: Early Mesozoic Tetrapods*. New York: Cambridge University Press, pp. 303–321.

Evans, S.E. & Milner, A.R. (1996) A metamorphosed salamander from the Early Cretaceous of Las Hoyas, Spain. *Philosophical Transactions of the Royal Society B* **351**, 627–646.

Falcon-Lang, H.J. (1999) The Early Carboniferous (Asbian–Brigantian) seasonal tropical climate of Northern Britain. *Palaios* **14**, 116–126.

Falcon-Lang, H.J., Benton, M.J., Braddy, S.J., & Davies, S.J. (2006) The Late Carboniferous tropical biome reconstructed from the Joggins Formation of Nova Scotia, Canada. *Journal of the Geological Society of London* **163**, 561–576.

Fischer, A.G. (1986) Climatic rhythms recorded in strata. *Annual Review of Earth and Planetary Sciences* **14**, 351–376.

Fluteau, F., Besse, J., Broutinand, J., & Berthelin, M. (2001) Extension of Cathaysian flora during the Permian: Climatic and paleogeographic constraints. *Earth and Planetary Science Letters* **193**, 603–616.

Franzen, J.L. & Schaal, S. (2000) Der eozäne See von Messel. In: G. Pinna (ed.), *Europäische Fossillagerstätten*. Berlin: Springer, pp. 177–183.

Fröbisch, N.B. & Reisz, R.R. (2008) A new lower Permian amphibamid (Dissorophoidea, Temnospondyli) from the Fissure Fill Deposits near Richards Spur, Oklahoma. *Journal of Vertebrate Paleontology* **28**, 1015–1030.

Gardner, J.D. & Böhme, M. (2008) Review of the Albanerpetontidae (Lissamphibia), with comments on the paleoecological preferences of European Tertiary albanerpetontids. In: J.T. Sankey & S. Baszio (eds.), *Vertebrate Microfossil Assemblages: Their Role in Paleoecology and Paleobiogeography*. Bloomington: Indiana University Press, p. 178–218.

Grauvogel-Stamm, L. & Ash, S.R. (2005) Recovery of the Triassic land flora from the end-Permian life crisis. *Comptes Rendus Palevol* **4**, 593–608.

Harland, W.B. (1997) *The Geology of Svalbard*. London: Geological Society.

Hentz, T.F. (1988) Lithostratigraphy and paleoenvironments of Upper Paleozoic continental red beds, North-Central Texas: Bowie (new) and Wichita (revised) Groups. *University of Texas Bureau of Economic Geology* **170**, 1–55.

Jenkins, F. Jr. & Shubin, N. (1998) *Prosalirus bitis* and the anuran caudopelvic mechanism. *Journal of Vertebrate Paleontology* **18**, 495–510.

Jenkins, F.A., Jr. & Walsh, D.M. (1993) An Early Jurassic caecilian with limbs. *Nature* **365**, 246–250.

Klembara, J. (1995) The external gills and ornamentation of skull-roof bones of the Lower Permian Discosauriscus (Kuhn 1933) with remarks to its ontogeny. *Paläontologische Zeitschrift* **69**, 265–281.

Krebs, B. & Martin, T. (2000) *Guimarota: a Jurassic Ecosystem*. Munich: Pfeil.

Lindemann, F.J. (1991) Temnospondyls and the Lower Triassic paleogeography of Spitsbergen. In: Z. Kielan-Jaworowska, N. Heintz, & H.A. Nakrem (eds.), Fifth Symposium on Mesozoic Terrestrial Ecosystems and Biota. *Contributions from the Paleontological Museum Oslo* **364**, 39–40.

Link, P.K. (2009) "Icehouse" (cold) climates. In: V. Gornitz (ed.), *Encyclopedia of Paleoclimatology and Ancient Environments*. Heidelberg: Springer, pp. 463–471.

MacLeod, N., Rawson, P.F., Forey, P.L., *et al.* (1997) The Cretaceous–Tertiary biotic transition. *Journal of the Geological Society* **154**, 265–292.

Marshall, J.E.A. & Stephenson, B.J. (1997) Sedimentological responses to basin initiation in the Devonian of East Greenland. *Sedimentology* **44**, 407–419.

Martens, T. (1989) First evidence of terrestrial tetrapods with North American faunal elements in the red beds of Upper Rotliegendes (Lower Permian, Tambach beds) of the Thuringian Forest (G.D.R.) – first results. *Acta Musei Reginae Hradensis Scientiae Naturales* **22**, 99–104.

Martens, T. (2005) First burrow cast of tetrapod origin from the Lower Permian (Tambach Formation) in Germany. *New Mexico Museum of Natural History and Science Bulletin* **30**, 207.

McGowan, G. & Evans, S.E. (1995) Albanerpetontid amphibians from the Cretaceous of Spain. *Nature* **373**, 143–145.

Milner, A.R. (1980) The tetrapod assemblage from Nýřany, Czechoslovakia. In: A.L. Panchen (ed.), *The Terrestrial Environment and the Origin of Land Vertebrates*. London and New York: Academic Press, pp. 439–496.

Milner, A.R. (1982) Small temnospondyl amphibians from the Middle Late Carboniferous of Illinois. *Palaeontology* **25**, 635–664.

Milner, A.R. (1983) The biogeography of salamanders in the Mesozoic and early Cenozoic: a ladistic vicariance model. In: R.W. Sims, J.H. Price, & P.E.S. Whalley (eds.), *Evolution, Time and Space: the Emergence of the Biosphere*. London: Academic Press, pp. 431–468.

Milner, A.R. (1987) The Westphalian tetrapod fauna; some aspects of its geography and ecology. *Journal of the Geological Society of London* **144**, 495–506.

Milner, A.R. (2000) Mesozoic and Tertriary Caudata and Albanerpetontidae. In: H. Heatwole & R.L. Carroll (eds.), *Amphibian Biology. Volume 4. Palaeontology*. Chipping Norton, NSW: Surrey Beatty, pp. 1412–1444.

Milner, A.R. (2007) *Mordex laticeps* and the base of the Trematopidae. *Journal of Vertebrate Paleontology* **27**, 118A.

Olsen, H. (1993) Sedimentary basin analysis of the continental Devonian sediments on North-East Greenland. *Bulletin of the Grønlands Geologiske Undersøgelse* **165**, 1–108.

Olson, E.C. (1958) Fauna of the Vale and Choza: 14. Summary, review, and integration of the geology and the faunas. *Fieldiana: Geology* **10**, 397–448.

Olson, E.C. (1967) Early Permian vertebrates of Oklahoma. *Circular of the Oklahoma Geological Survey* **74**, 1–111.

Parrish, W.C. (1978) Paleoenvironmental analysis of a Lower Permian bonebed and adjacent sediments, Wichita County, Texas. *Palaeogeography, Palaeoclimatology, Palaeoecology* **24**, 209–237.

Paszkowski, M. (2009) The Early Triassic karst of Czatkowice 1, southern Poland. *Palaeontologia Polonica* **65**, 7–16.

Polley, B.P. & Reisz, R.R. (2011) A new Lower Permian trematopid (Temnospondyli: Dissorophoidea) from Richards Spur, Oklahoma. *Zoological Journal of the Linnean Society* **161**, 789–815.

Pruss, S.B. & Bottjer, D.J. (2005) The reorganization of reef communities following the end-Permian mass extinction. *Comptes Rendus Palevol* **4**, 485–500.

Pujol, F., Berner, Z., & Stüben, D. (2006) Palaeoenvironmental changes at the Frasnian/Famennian boundary in key European sections: chemostratigraphic constraints. *Palaeogeography, Palaeoclimatology, Palaeoecology* **240**, 120–145.

Rees, P.M., Ziegler, A.M., Gibbs, M.T., *et al.* (2002) Permian phytogeographic patterns and climate data/model comparisons. *Journal of Geology* **110**, 1–31.

Reisz, R.R. (2007) Terrestrial vertebrate fauna of the Lower Permian cave deposits near Richards Spur, Oklahoma with emphasis on dissorophoids. *Journal of Vertebrate Paleontology* **27**, 133A.

Reisz, R.R., Schoch, R.R., & Anderson, J.S. (2009) The armoured dissorophid *Cacops* from the Early Permian of Oklahoma and the exploitation of the terrestrial realm by amphibians. *Naturwissenschaften* **96**, 789–796.

Retallack, G.J. (1997) *A Color-Guide to Paleosols*. Chichester: Wiley.

Rolfe, W.D.I., Durant, G.P., Baird, W.J. (1994) The East Kirkton Limestone, Viséan, West Lothian, Scotland: an introduction and stratigraphy. *Transactions of the Royal Society of Edinburgh: Earth Sciences* **84**, 177–188.

Romer, A.S. (1928) Vertebrate faunal horizons in the Texas Permo-Carboniferous red beds. *University of Texas Bulletin* **2801**, 67–108.

Romer, A.S. (1935) Early history of Texas redbeds vertebrates. *Bulletin of the Geological Society of America* **46**, 1597–1658.

Rubidge, B.S. (1995) *Biostratigraphy of the Beaufort Group (Karoo Supergroup)*. South African Committee for Stratigraphy, Biostratigraphic Series 1. Pretoria: Council of Geoscience.

Sander, P.M. (1989) Early Permian depositional environments and pond bonebeds in Central Archer County, Texas. *Palaeogeography, Palaeoclimatology, Palaeoecology* **69**, 1–21.

Saunders, A.D. & Reichow, M.K. (2009) The Siberian Traps and the End-Permian mass extinction: a critical review. *Chinese Science Bulletin* **54**, 20–37.

Schneider, J. (1993) Environment, biotas and taphonomy of the lacustrine Niederhäslich Limestone, Döhlen Basin, Germany. *Transactions of the Royal Society of Edinburgh: Earth Sciences* **84**, 453–464.

Schulte, P., Alegret, L., Arenillas, I., *et al.* (2010) The Chicxulub asteroid impact and mass extinction at the Cretaceous–Paleogene boundary. *Science* **327**, 1214–1218.

Schultze, H.-P. (2009) Interpretation of marine and freshwater paleoenvironments in Permo-Carboniferous deposits. *Palaeogeography, Palaeoclimatology, Palaeoecology* **281**, 126–136.

Schultze, H.-P. & Soler-Gijon, R. (2004) A xenacanth clasper from the ?uppermost Carboniferous-Lower Permian of Buxières-les-Mines (Massif Central, France) and the palaeoecology of the European Permo-Carboniferous basins. *Neues Jahrbuch für Geologie und Paläontologie Abhandlungen* **232**, 325–363.

Shishkin, M.A. (2000) Mesozoic amphibians from Mongolia and the Central Asian republics. In: M.J. Benton, M.A. Shishkin, D.M. Unwin, & E.N. Kurochkin (eds.), *The Age of Dinosaurs in Russia and Mongolia*. Cambridge: Cambridge University Press, pp. 297–308.

Smith, R. & Botha, J. (2005) The recovery of terrestrial vertebrate diversity in the South African Karoo Basin after the end-Permian extinction. *Comptes Rendus Palevol* **4**, 555–568.

Smith, R.M.H. (1995) Changing fluvial environments across the Permian-Triassic boundary in the Karoo Basin, South Africa and possible causes of tetrapod extinctions. *Palaeogeography, Palaeoclimatology, Palaeoecology* **117**, 81–104.

Smithson, T. R. (1994) *Eldeceeon rolfei*, a new reptiliomorph from the Viséan of East Kirkton, West Lothian, Scotland. *Transactions of the Royal Society of Edinburgh: Earth Sciences* **84**, 377–82.

Streel, M., Caputo, M.V., Loboziak, S., & Melo, J.H.G. (2000) Late Frasnian–Famennian climates based on palynomorph analyses and the question of the Late Devonian glaciations. *Earth Science Reviews* **52**, 121–173.

Sues, H.-D. & Fraser, N.C. (2010) *Triassic Life on Land: the Great Transition*. New York: Columbia University Press.

Sues, H.-D., Clark, J.M., & Jenkins, F.A., Jr. (1994) A review of Early Jurassic tetrapods from the American Southwest. In: N.C. Fraser & H.-D. Sues (eds.), *In the Shadow of the Dinosaurs: Early Mesozoic Tetrapods*. New York: Cambridge University Press, pp. 284–294.

Taylor, E.L., Taylor, T.N., & Cuneo, N.R. (1992) The present is not the key to the past: a Permian petrified forest in Antarctica. *Science* **257**, 1675–1677.

Wang, X.L. & Zhou, Z.H. (2003) Mesozoic Pompeji. In: M.M. Chang (ed.), *The Jehol Biota*. Shanghai: Scientific and Technical, pp. 19–35.

Wang, Y. & Gao, K.Q. (2003) Amphibians. In: M.M. Chang (ed.), *The Jehol Biota*. Shanghai: Scientific and Technical, pp. 76–85.

Ward, P.D., Montgomery, D.R. and Smith, R.M.H., (2000) Altered river morphology in South Africa related to the Permian–Triassic extinction. *Science* **289**, 1740–1743.

Warren, A.A., Rich, T.H., & Vickers-Rich, P.V. (1997) The last last labyrinthodonts? *Palaeontographica A* **247**, 1–24.

Werneburg, R. (2001) Die Amphibien- und Reptilfaunen im Permokarbon des Thüringer Waldes. *Beiträge zur Geologie von Thüringen, Neue Folge* **8**, 125–152.

Wiechmann, M.P. (2000) The albanerpetontids from the Guimarota mine. In: T. Martin & B. Krebs (eds.), *Guimarota: a Jurassic Ecosystem*. Munich: Pfeil, pp. 51–54.

Wiman, C. (1914) Über die Stegocephalen aus der Trias Spitzbergens. *Bulletin of the Geological Institute Upsala* **13**, 1–34.

Woods, A.D. (2005) Paleoceanographic and paleoclimatic context of Early Triassic time. *Comptes Rendus Palevol* **4**, 463–472.

Ziegler, P. (1989) *The Evolution of Laurussia*. Dordrecht: Kluwer.

4 The Amphibian Soft Body

The staple diet of paleontologists is the study of skeletons, the only body parts to be preserved in the great majority of fossils. Yet there are exceptions, in which soft tissues have been partially preserved. These are often unexpected in occurrence and depend on the type of fossil deposit. In the case of extinct amphibians, a rich body of evidence has accumulated from numerous geological formations preserving traces of gills, skin, eye pigments, gut contents, and even the outline of intestines. These data, although highly fragmentary and selective, shed light on otherwise unknown aspects of amphibian paleobiology. They play an important role in providing anatomical information for evolutionary hypotheses.

Soft-body preservation forms only one line of evidence. In order to understand the evolution of a group, primary fossil evidence must be supplemented by anatomical data from extant taxa. The present chapter outlines how this is done in the least hypothetical way. To this end, several body regions are discussed which permit the reconstruction of organ systems that played a significant role in the evolution of land vertebrates. For the early tetrapods, the existence of three surviving amphibian clades – caecilians, salamanders, and frogs – can be regarded as fortunate. They provide much anatomical information inaccessible to paleontology.

Amphibian Evolution: The Life of Early Land Vertebrates, First Edition. Rainer R. Schoch.
© 2014 Rainer R. Schoch. Published 2014 by John Wiley & Sons, Ltd.

4.1 How to infer soft tissues in extinct taxa

The crucial aspect of the reconstruction procedure is that it relies on two different lines of evidence which are independent from each other: (1) direct fossil evidence (skeleton, soft-tissue preservation) and (2) phylogenetic reasoning. On closer inspection, fossil evidence is not always as "hard" as one might like. Bones are often not adequately preserved, and hardly any two deposits preserve fossils in the same way. The most common situation is the presence of disarticulated material, which needs to be identified bit by bit, referred to the same taxon (which requires knowledge of more complete finds from other deposits), and

finally reconstructed in three dimensions. Most early tetrapod taxa are known from incomplete finds, but many are represented by more than one specimen – again a fortunate case for paleontology. This permits incomplete finds to complement one another, but also grants insight into individual variation, development, and geographical variation or evolutionary patterns on a small scale. Provided that all these criteria are fulfilled, a fossil taxon may provide rich data on the anatomy of an extinct species.

The second line of evidence (phylogenetic reasoning) is derived from phylogenetic systematics and employs anatomical (or other) data gathered from extant taxa (Figure 4.1). It forms an indirect means of assessing the soft-anatomical features that a given fossil taxon *may* have had – but it will

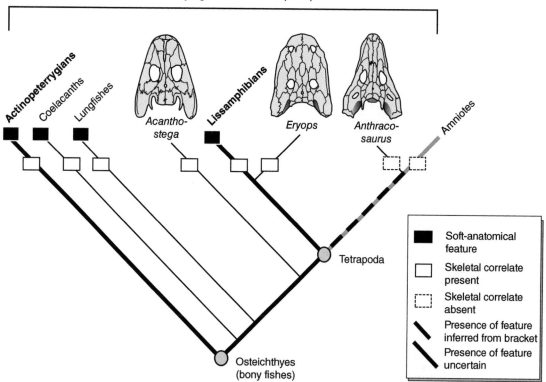

Figure 4.1 The extant phylogenetic bracket (EPB) exemplified by early tetrapods. To trace the presence of soft-anatomical features, their distribution among extant bracket groups is first assessed (bracket in bold). In a second step, skeletal correlates of these features are examined and their distribution mapped on the cladogram. Together, these lines of evidence indicate the plausibility that a given structure was present in a given taxon.

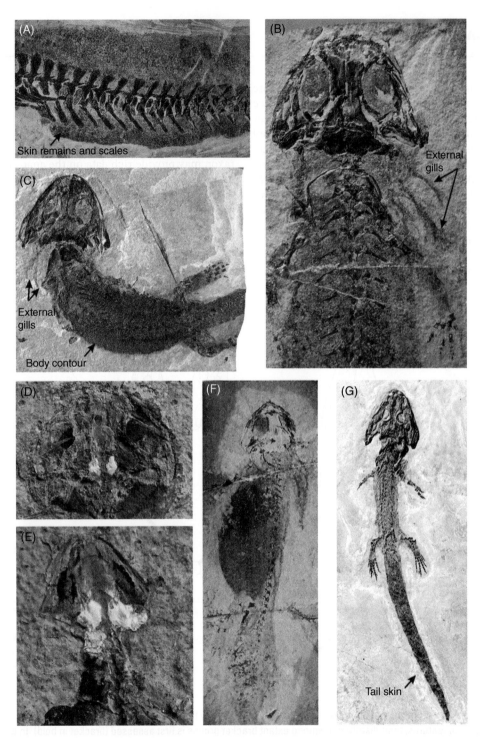

Figure 4.2 Features of the soft body are sometimes preserved in early tetrapods. This ranges from skin and tail fins (A, G, temnospondyl *Sclerocephalus*) over whole-body contours (B, C, branchiosaurid *Apateon*) and external larval gills (B, C) to fillings of ear capsules (D, E, larval newt *Chelotriton*), intestines (F, branchiosaurid *Apateon*), and eye pigments (D).

never reach the degree of certainty that a well-preserved fossil skeleton does. That said, phylogenetic reasoning can be a powerful tool that *supplements* direct fossil evidence (Bryant and Russell 1990; Bryant and Seymour 1992). The approach has been coined "extant phylogenetic bracketing" or EPB (Witmer 1995). It makes use of two particular aspects of cladistics: outgroup comparison and parsimony.

The EPB requires substantial knowledge of relationships in a studied group. That is, a sound phylogenetic hypothesis must exist before the approach can be attempted at all. Then, two extant clades that *contain* the fossil taxon are to be found. What does "contain" mean here? Suppose we want to reconstruct an unfossilized structure in the stem-tetrapod *Acanthostega* (it is irrelevant here whether we call it a tetrapod or a tetrapodomorph). The most important question is whether we can find two living taxa that form a phylogenetic frame into which *Acanthostega* can be placed – this frame will be the extant phylogenetic bracket. In the present case, lungfishes and crown tetrapods are considered the two closest extant relatives. In other words, *Acanthostega* nests higher than dipnoans but below tetrapods, thus forming the sister group of Tetrapoda in the present three-taxon statement.

Although the dipnoan–tetrapod monophyly is only one among several hypotheses, we may choose this as our preferred framework (Mickoleit 2004). The impact of an alternative, e.g., the actinistian–tetrapod hypothesis of Schultze (1991), can always be compared at a later stage. Unlike extant dipnoans (three genera), tetrapods are highly speciose and diverse. Therefore it is advisable to select a representative tetrapod clade that is likely to retain the plesiomorphic tetrapod condition in the studied aspect. In most of the present cases, salamanders are almost always the best choice.

The EPB procedure takes three steps: (1) one needs to find a skeletal (osteological) correlate for a soft-anatomical feature. For instance, this may be the attachment site for a muscle or cartilage on a bone surface that is preserved in fossils; (2) the similarities of the soft-anatomical feature between the extant taxa are hypothesized to be inherited from a common ancestor; (3) this hypothesis is

tested by searching for a skeletal correlate of the soft-anatomical character in the fossil taxon (Figure 4.1). When the hypothesis passes the test, the soft-anatomical feature can be inferred in the extinct taxon with a relatively high degree of confidence.

In many cases, the EPB has to be expanded because the available outgroup taxa lack the studied feature altogether. Unfortunately, this is often the case in early tetrapod anatomy. In fact, both lungfishes and the living coelacanth (*Latimeria*) are too modified to serve as a reasonable guide for soft-anatomical inference in the head and pectoral girdle. In *Latimeria*, the skull and cranial musculature have been greatly modified, and many skull elements shared by crown tetrapods and fossil tetrapod taxa are absent. Lungfishes are even less suited, because the living forms have highly reduced skeletons, and the Devonian ancestors had various bones that cannot be homologized with other groups. In both cases, the long separate evolution has remodeled the head extensively. Thus, the EPB would be incomplete on the fish side, unless another, more distant outgroup taxon is chosen. Fortunately, this is possible. Basal ray-finned fishes (*Polypterus*, *Amia*, *Lepisosteus*) preserve the closed skull of early bony fishes that characterized most Paleozoic tetrapods (Allis 1897, 1922; Lauder 1980). They retain the full set of bones primitive for stem-tetrapods. The cranial and visceral muscles associated with basal actinopterygian skulls are therefore likely to represent the primitive condition.

4.2 Fossil evidence: soft tissue preservation

Skin. The most common soft-part preservation in early tetrapods consists of faint traces of the skin (Figure 4.2A). In most cases, they form dark shadows with no clearly defined shape. In *Rotliegend* deposits of Europe, black silhouettes contouring the body outline are common in some fine-grained mudstones (Boy 1972). Similar finds are known from much younger deposits in the Paleogene, such as from Messel (Franzen and Schaal 2000). The bony scales, which were very thin in larval temnospondyls and microsaurs, are often embedded in such dark matrices. Willems and

Wuttke (1987) have shown that at least some of this skin preservation in the *Rotliegend* does not preserve the skin itself but rather a sheet of bacteria that were feeding on the skin and themselves became petrified. In the branchiosaurid *Apateon* from Odernheim, such fossilized microbes have produced very well-preserved dark halos.

Gills. In some rare cases of skin preservation, remains of gills are also present (Bulman and Whittard 1926; Boy 1974; Milner 1982). These are usually only contours that do not preserve fine structures (Figure 4.2B,C). Imprints of internal gills of *Eusthenopteron* were described by Jarvik (1980), but otherwise only external larval gills are known from direct preservation (Witzmann 2004). They cover a wide range of taxa, known from several seymouriamorphs (Klembara 1995) and temnospondyls (Boy 1974; Werneburg 1991; Witzmann 2006b).

Braincase. Calcareous fillings of the ear capsules are common in branchiosaurids and known in modern anuran analogs (Boy 1972). They indicate the position and size of the otic capsules, which were cartilaginous and not themselves preserved (Figure 4.2D,E). In exceptional cases, even other parts of the braincase were filled in and thus preserved, such as the endolymphatic sacs inside the ear capsules (Boy 1974).

Pigments. Werneburg (2007) described coloration patterns in the skin of branchiosaurids from the Permian of Thuringia, Germany. These are likely to contain regionally variable patterns of pigments. Round black patches, resembling skin preservation, are reported from inside the orbit in small temnospondyl larvae in *Rotliegend* sediments, and these have been interpreted as eye pigments (Boy 1974). They are clearly distinct from scleral rings, and both co-occur occasionally (Schoch 1992).

Intestine fillings. In carbonate nodules from Mazon Creek, Milner (1982) reported the exceptional case of larval amphibamid temnospondyls with intestine outlines. These are caused by the filling of these organs by a matrix different from that of the surrounding sediment.

Cartilage. Usually, only skeletal elements that contained some bony tissue or enamel are preserved in fossils. However, histology has revealed that cartilage may be enclosed in bone and preserved

with it (de Ricqlès 1975; Sanchez *et al.* 2010). Imprints of cartilaginous ceratobranchials occur in *Archegosaurus* (Witzmann 2006a) and *Glanochthon latirostre* (Schoch and Witzmann 2009)

Early bone formation. In various Carboniferous–Permian *Lagerstätten*, small larvae are preserved in which the skeleton was only partially ossified. Preservation of tiny bone primordia, early stages of bone formation, permitted the study of ossification sequences and direction of bone growth (Boy 1974; Schoch 1992, 2002, 2004).

4.3 Head and visceral skeleton

In contrast to the lightly built lissamphibians, the closed and heavy skulls of Paleozoic tetrapods are a substantial constraint on the reconstruction of muscle arrangements in the head. Such skulls are also typical of extant basal actinopterygians (*Polypterus*, *Lepisosteus*, *Amia*) and all known fossil tetrapodomorphs, as exemplified by *Eusthenopteron* and *Acanthostega*. The primitive condition for crown tetrapods is therefore a closed skull with head musculature attaching along the internal side of the cheek, the skull table, and the lateral wall of the braincase (Figure 4.3). Indeed, many Paleozoic tetrapods preserve muscle attachment sites that pass the EPB test when compared with such skeletal correlates in extant tetrapods and bony fishes.

Epaxial musculature. Elevation of the head is usually the first movement in the feeding process, and throughout bony fishes and tetrapods it is mediated by epaxial muscles (EA).

Adductor mandibulae. The jaw-closing muscles of extant bony fishes and tetrapods are relatively consistent in number and arrangement, although the latter have modified skulls with large openings. These muscles always insert inside the fossa and along the medial side of the mandible. Many taxa share three branches of the adductor mandibulae (AM), but in *Polypterus*, *Amia*, and teleosts, different terminologies are in use from those of tetrapods (Allis 1897, 1922; Luther 1914; Jarvik 1980; Diogo *et al.* 2008). In bony fishes, the AM attaches to the braincase, parasphenoid, cheek (quadrate and preoperculum), and hyomandibula; in tetrapods, it originates from the squamosal,

Figure 4.3 Reconstruction of musculature relies on the EPB, here shown by (A–D) the actinopterygian fish *Polypterus* (adapted from Allis 1922) and (E–H) the salamander *Dicamptodon* (Schoch, unpublished data). Abbreviations are explained in text.

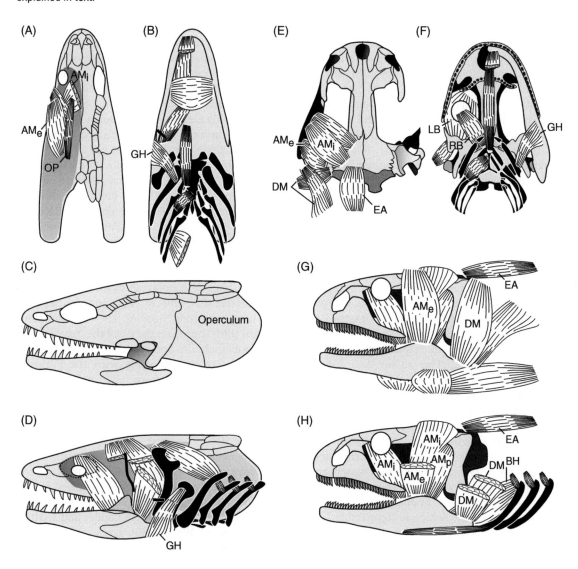

otic capsule, parietal, and frontal (Luther 1914; Carroll and Holmes 1980; Iordansky 1990). In tetrapods, the adductor mandibulae has three main portions: external (AMe), internal (AMi), and posterior (AMp).

Following the EPB approach, a configuration similar to *Polypterus* is likely for stem-tetrapods and Paleozoic crown tetrapods (temnospondyls,

embolomeres, seymouriamorphs, many lepospondyls). A notable difference between actinopterygians and sarcopterygians is the possession of additional skull elements (jugal, squamosal), which make the cheek substantially longer in tetrapodomorphs (Janvier 1996). This probably corresponds to the different proportions of the adductor muscles.

Levator palatoquadrati. Where present, this muscle raises the cheek and upper jaw relative to the braincase. It is present in its plesiomorphic form in all bony fishes, best represented by *Polypterus* and *Amia*, which retain the moveable palatoquadrate. Although there are no direct skeletal correlates of this muscle preserved, the levator palatoquadrati (LPQ) was probably lost or had changed its role in taxa with basicranial articulation, such as in early stem-amphibians (temnospondyls). In anthracosaurs and chroniosuchians, the muscle may still have been present in its plesiomorphic state.

Arcus palatini musculature. Like the LPQ, the levator arcus palatini (LAP) raises the palatoquadrate against the braincase, with its antagonist being the adductor arcus palatini (AAP). The LAP has four portions in actinopterygians, of which only one is retained in lungfishes and tetrapods (Lubosch 1938). In tetrapods, the LAP still attaches along the pterygoid and braincase, but as these skulls are largely consolidated between braincase and palatoquadrate, they have adopted different functions. In adult salamanders and frogs, the LAP is large and raises the eye (Iordansky 1990), forming a further example of an exaptation (see below).

Subcranial muscle. The subdivided, kinetic braincase is an autapomorphy of sarcopterygians, and known also from numerous well-preserved tetrapodomorphs (Jarvik 1980; Janvier 1996). However, among extant taxa only *Latimeria* retains such a joint. In this taxon, the subcranial muscle mediates movement of the anterior braincase block (Thomson 1967). The intracranial joint disappeared in tetrapodomorphs, with *Acanthostega* already having a solid single-unit braincase (Clack 1998). The subcranial muscle, which is comparably large in *Latimeria*, has been homologized with the retractor bulbi (RB) of tetrapods (Janvier 1996).

Opercular muscles. In bony fishes, the operculum articulates with the underlying hyomandibula, and the opercular and branchiostegal elements are interconnected by a series of muscles on the medial side, the hyohyoideus superioris (HHS) (Allis 1897). Opening or closing of the opercular series – which permits water to flow out of the gill chamber – is mediated by rotation of the hyomandibula. This is made possible by muscles attaching along different sides of the hammer-shaped hyomandibula, the dilatator operculi (DOP). Another muscle, the adductor operculi (AOP), attaches directly along the operculum. The opercular muscles (OP) are not present in their original form in any tetrapod, and the opercular elements are completely absent. In stem-tetrapods, the opercular series was already absent in *Tiktaalik*, with *Panderichthys* being the last tetrapodomorph in which the gill cover worked in the plesiomorphic way. The muscles attaching to the hyomandibula underwent modification along with this element (see section 4.5, below).

Visceral muscles. In bony fishes, numerous muscles connect the hyoid and gill arches with the mandible and pectoral girdle. Some of these muscles are retained in larval lissamphibians (Lauder and Shaffer 1985), but substantially modified in their metamorphosing adults (Drüner 1901; Wake and Deban 2000). An important role in feeding is played by the sternohyoideus (SH), which connects the hypohyals with the pectoral girdle (in salamanders it is often called the rectus cervicis). In both osteichthyans and larval salamanders, this muscle ranks among the primary mouth-opening muscles. The branchiomandibularis (BM) runs from the tip of the mandible to the hypobranchials, and the coracomandibularis (CM) connects the mandible with the pectoral girdle. The geniohyoideus (GH) connects the mandible with the branchial arches. Finally, the branchiohyoideus (BH) unfolds the branchial basket in order to enlarge the buccal cavity (Deban and Wake 2000). Together with a range of others, these muscles form a complex network with interconnected skeletal elements, in concert mediating the depression of the lower jaw, hyoid arch, and branchial arches in bony fishes and larval salamanders (Lauder 1980; Lauder and Shaffer 1985; Deban and Wake 2000).

Depressor mandibulae. The jaw-opening depressor mandibulae (DM) is confined to dipnoans and tetrapods. Embryology reveals that it derives from a hyoid muscle (constrictor hyoideus, CH) in both groups, but only one of the two portions present in tetrapods is actually homologous in the two groups (Diogo *et al.* 2008). This is the anterior depressor mandibulae (DMa), which attaches to the squamosal and braincase in tetrapods and inserts on the mandible behind the jaw articulation. It is not difficult to imagine a slight shift from the hyoid arch to the mandible. The posterior portion (DMp), attaching along the epaxial musculature in salamanders, is not homologous to the dipnoan depressor (Diogo *et al.* 2008).

In Paleozoic tetrapods, the presence of a depressor mandibulae is indicated by a retroarticular process, a bony projection behind the jaw articulation. In temnospondyls, such a process is generally present, albeit of a different length. It often preserves muscle scars pointing dorsally and posteriorly, which is consistent with the alignment of the DM in salamanders and frogs (Lubosch 1938).

Eye musculature. The eye-raising muscle of tetrapods, the levator bulbi (LB), is the homolog of the palatoquadrate muscle (LAP) of bony fishes. Its antagonist is the retractor bulbi (RB), which is a tetrapod character judged by its function and attachment, but derived from the subcranial muscle (SM) of sarcopterygians. In batrachians, this muscle is large and originates along the margin of the parasphenoid where the anterior process merges into the quadrangular plate. Similar muscle attachments are found in temnospondyls. The slit-like palatal windows of stem-tetrapods and stem-amniotes permitted such a muscle to attach in a similar way to the parasphenoid. This is consistent with the presence of the RB throughout tetrapods (Mickoleit 2004).

The retention of several visceral muscles in larval salamanders that are otherwise unknown from tetrapods highlights the importance of studying all phases of development. Here, salamanders can indeed be viewed as a fortunate case in which crucial functional components of bony fishes have been retained in tetrapods (see Chapter 5).

4.4 Respiratory organs

When tetrapods left the water they had to tackle numerous problems, but the physical properties of air also provided some huge advantages: it is much easier to take up oxygen from air than from water (Schmidt-Nielsen 1997). There are three main reasons for this: (1) one liter of air contains 209 milliliters of O_2, whereas the same amount of water has only 0.7 milliliters of dissolved oxygen;

(2) pumping air through a respiratory organ requires much less energy in air than in the more viscous water; and (3) the diffusion rate in air is 10 000 times higher than in water. This suggests that once the appropriate organs were available, respiration on land could be made an effective process – and indeed early tetrapods made use of two different organs.

Small animals rely entirely on diffusion of oxygen and carbon dioxide, but in vertebrates specialized organs and a blood circulatory system evolved to transport respiratory gases through the voluminous body. There are three different respiratory organs: gills, lungs, and epithelial surfaces. Gills and lungs are structurally opposite solutions to the problem of surface increase: gills are inversions, lungs protuberances. It is true that gills evolved under water and are not used in air in modern vertebrates, but there is no reason in principle why they could not work on land. Lungs, in turn, evolved under water as well, and were ready to work on land. However, in contrast to the water-processing lungs of some invertebrates, vertebrate lungs were air-breathing from the start. The best-suited tissues for respiration purpose are epithelia, such as the outer layer of the skin (epidermis) or the internal layer of the mouth, pharyngeal, and intestinal cavities. Respiratory organs have consequently evolved in both body regions, and they did so repeatedly. The plesiomorphic condition of bony fishes is respiration with gills, which form in pouches between the head and pectoral girdle. They require a water current running from the mouth cavity over the gills to the gill slits, the openings of the gill pouches within the body wall. Any respiratory epithelia inside the pouches are called *internal gills*, while those outside the wall are *external gills*. At this stage, these terms are only descriptive, without reference to homology.

Both types of gills are associated with the gill arch skeleton, which is homologous throughout gnathostomes (Janvier 1996). These arches are composed of curved bows, primitively five arranged in a series, each consisting of several rod-like elements (ceratobranchials, epibranchials, pharyngobranchials) (Figure 4.4). They articulate with unpaired elements in the midline of the pharyngeal floor (basibranchials). Internally,

Figure 4.4 The extant fish *Polypterus* (A–C, adapted from Allis 1922) serves as a guide in reconstructing the musculature of the extinct tetrapodomorph fish *Eusthenopteron* (D–F, adapted from Jarvik 1980 and Schoch unpublished data).

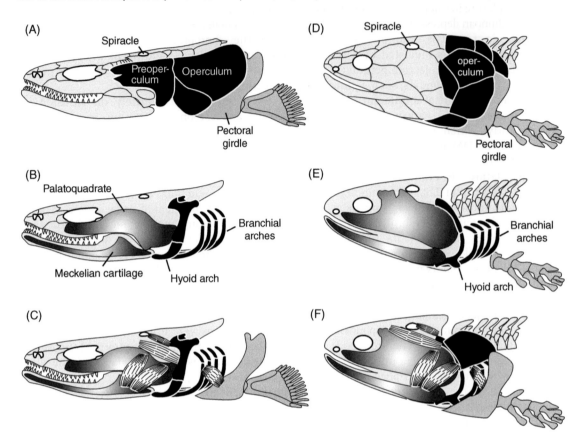

facing into the pharynx, two sets of cartilaginous thorns are attached to the arch (gill rakers). On the posterior face of each gill arch, the epithelium covering the skeletal elements forms a large sheet that functions as respiratory surface – this is the gill proper. These surfaces are not identical in the two gill types: internal gills form paired lamellae that are divided by a septum. In external gills, the septum itself forms the respiratory sheet, and there are no lamellae. However, the morphological outcome is very similar: in external gills, the end of the septum is partitioned into numerous lamella-like lobes, which are arranged in pairs like the lamellae of internal gills (Schoch and Witzmann 2011). Thus, when comparing internal and external gills, the septum is probably homologous, but the lamellae are not.

Internal gills. At first sight, the phylogenetic distribution of gill types appears to be clear-cut: internal gills (Figure 4.5) are present in all "fishes" (= fish-like gnathostomes) and absent in all crown tetrapods. Internal gills are present in both *Latimeria* and dipnoans, and thus form the primitive condition of stem-tetrapods such as *Acanthostega*. But where and when were the internal gills lost? In a simple functional scenario, the loss should have occurred in the first terrestrial tetrapods. However, I have already shown how difficult it is to infer lifestyle in many Paleozoic taxa.

In fact, the story turned out to be more complicated – and resulted in an unexpected picture. Skeletal correlates of internal gills were first mentioned by Coates and Clack (1991), who

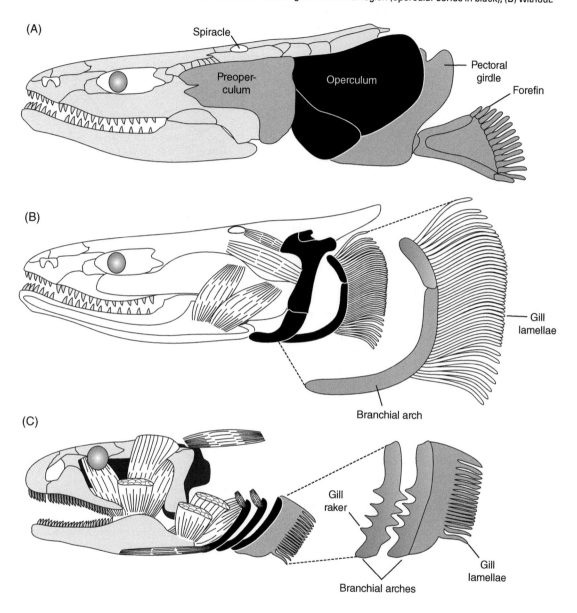

Figure 4.5 Anatomy of the gill region in (A, B) a bony fish (*Polypterus*, adapted from Allis 1922) and (C) a salamander (*Dicamptodon*, unpublished data). (A) With dermal bones covering the branchial region (opercular series in black); (B) without.

(A)

Spiracle

Preoper-
culum

Operculum

Pectoral
girdle

Forefin

(B)

Gill
lamellae

Branchial arch

(C)

Gill
raker

Gill
lamellae

Branchial arches

discovered grooves on the posterior side of the gill arch elements in *Acanthostega*. Such grooves, they argued, are only found in bony fishes with internal gills, but not in salamander larvae, which have external gills. Schoch and Witzmann (2011) found the reason for this: the gill arteries lie close to the gill arch in all internal gills, running in grooves along the skeleton (Figure 4.6A,C). In salamanders, the septum bifurcates (into septal "lamellae") at a considerable distance from the skeletal element, and there lie the arteries (Figure 4.6B,D). They are far away from the gill arch and consequently do not leave traces on the bone like the grooves in bony fishes. The

Figure 4.6 The two different types of gills in bony fishes. (A, C) The internal adult gills, common to all bony fishes, are formed by two sheets of lamellae separated by a septum. (B, D) The external larval gills of lissamphibians are instead formed by the septum, and there are no homologs of fish lamellae. (A, B) Lateral view; (C, D) cross-section. Adapted from Schoch and Witzmann (2011).

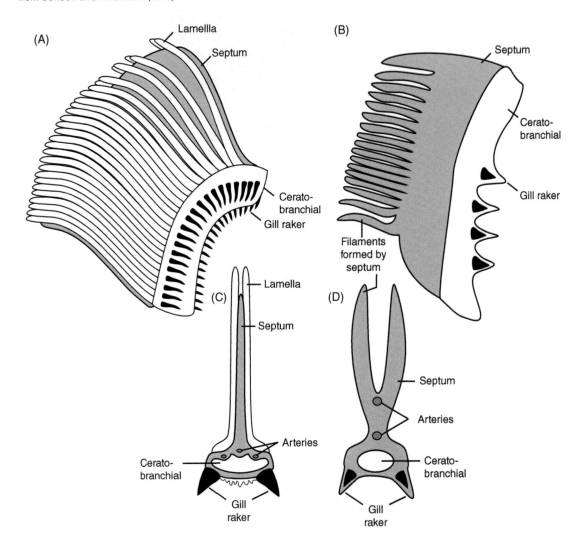

discovery of such grooves, along with other correlates of internal gills, may not be surprising in an aquatic stem-tetrapod like *Acanthostega*. In the meantime, they were also found in *Ichthyostega*, which has also increasingly been viewed as water-dwelling (Clack 2012). However, evidence of internal gills also comes from a very different group: Schoch and Witzmann (2011) recently highlighted that such grooves exist in temnospon-

dyls. These were recognized by Bystrow (1938), but at the time were interpreted as support for external gills.

External gills. External gills are only present in larvae, and indeed the "larval stage" is often defined by the presence of external gills in lissamphibians. External gills in larvae of bony fishes are exceptional, and are certainly not homologous to those of lissamphibians (Figure 4.7) (Witzmann

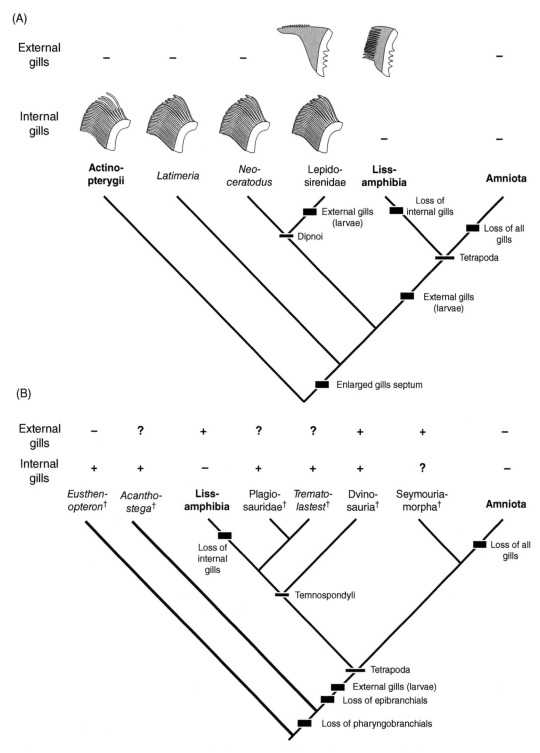

Figure 4.7 EPB and skeletal correlates of gill vessels provide insight into the evolution of gills. (A) Bracket showing distribution of internal and external gills. (B) Cladogram with major events in gill evolution mapped. Adapted from Schoch and Witzmann (2011).

2004). As mentioned above, the external gills are formed by the septum, which together with the skeletal element is the only homologous part between external and internal gills. In caecilians and salamanders, the larval gills develop on branchial arches II, IV, and V; in anurans the posterior one is usually absent (Duellman and Trueb 1994). Caecilians have three external larval gills, forming expanded sheets in typhlonectid embryos (Wake 1977) plus fimbriate ones in embryos of *Ichthyophis* (Dünker *et al.* 2000). Salamanders have three external gills of various shapes and sizes, correlating with properties of the water body (stream type, pond type). In plethodontid salamanders, the encapsulated larvae undergo direct development but still retain larval gills, and in some species they are leaf-like; in viviparous *Salamandra*, gill fimbriae are elongated, presumably to take up oxygen within the oviducts (Duellman and Trueb 1994). Finally, in anuran tadpoles external larval gills are overgrown by a flap of skin. Hence, in this clade, external larval gills become secondarily internal (Schmalhausen 1968).

4.5 Lateral lines, electroreception, and ears

The capacity to hear is an ancient trait of vertebrates, and the hearing organs are diverse. Both hearing and balancing senses rely on receptor cells that develop locally from the ectoderm. Based on their possession of hair-like structures, they are called hair cells. They are arranged in clusters and the hairs are sensitive to deflection, generating an electrical response in the cell. Depending on the organ, these receptors are called neuromasts (lateral sense), maculae and cristae (vestibular or balancing sense), or papillae (auditory sense). Strictly speaking, only the auditory sense is referred to as hearing, but functionally the lateral-line system of fishes is a hearing organ as well.

Lateral line. The lateralis organs (lateral-line system) form an ancient trait of vertebrates (Mickoleit 2004). They consist of numerous separate mechanoreceptors located in the skin. Each sensory organ (neuromast) consists of a group of receptor cells bearing sensitive hairs (cilia) that are enclosed in a gelatinous capsule (cupula). Neuromasts may be located as single units or arranged in lateral lines. In bony fishes, they are located within the dermal bones and connected to the outer surface by means of pores; in tetrapods they lie in open grooves or simply within the dermis. The lateral-line neuromasts are sensitive to changes in velocity and permit orientation under water independent of sight. Based on their anatomical and functional consistency, the homology between lateral-line organs of bony fishes and lissamphibians is generally accepted (Mickoleit 2004). Occurring throughout ontogeny in fishes, they are confined to larval stages in amphibians, with the exception of neotenic species, where they persist in aquatic adults (Figure 4.8), and a few aquatic anurans retaining them in the adult stage (*Pipa, Xenopus*). Lateral-line organs only function in organisms that return to the water regularly. They were evidently present in stem-tetrapods, where they were located in bony canals (Clack 2012). That is, anatomical correlates in dermal skull bones indicate the presence of the lateral-line system, confirming the presence of lateral lines in the bracket taxa (bony fishes and lissamphibians). In Paleozoic tetrapods, both stem-amphibians and stem-amniotes, lateral lines were located in grooves aligned in exactly the same pattern as the closed canals of bony fishes. This indicates that lateral lines were not re-invented in lissamphibians, and that they were finally lost in the stem-group of amniotes, where they persisted in seymouriamorphs and lepospondyls. Clack (2012) pointed out that the open lateral-line sulci in tetrapods are a pedomorphic trait with respect to the enclosed canals of their fish-like ancestors. In bony fishes, the canal neuromasts form superficially in the epidermis, and sink into a furrow formed by dermis and epidermis.

Electroreception. A second group of sensory organs of use under water are the electroreceptors of sharks and bony fishes, which are similar in receptor anatomy to the lateralis organs. Electrosensory organs help in the detection and identification of conspecifics and prey items. In addition to orientation, electrosensory organs may also be used to generate electric fields, a feat accomplished by specialized electroreceptors. Certain rays, eels, and catfishes have independently evolved this capacity in order to threaten

Figure 4.8 Many Paleozoic tetrapods were more or less aquatic. Lateral lines, homologous to those of fishes, are found as closed canals or open grooves in many stem-tetrapods, anthracosaurs, and temnospondyls. (A) Skull roof of neotenic temnospondyl *Micromelerpeton*. (B) Hyobranchial skeleton (black, ossified; white, unossified; inferred from relatives in which these structures are preserved), branchial dentition, and external gills. Adapted from Schoch (2009a).

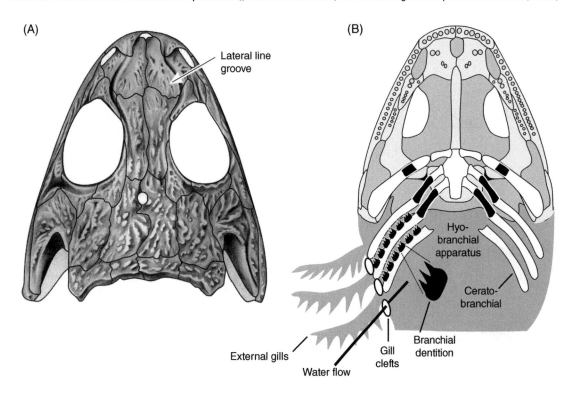

(A)

Lateral line groove

(B)

Hyo-branchial apparatus

Cerato-branchial

External gills

Gill clefts

Branchial dentition

Water flow

enemies or paralyze or even kill prey. Electrosensory organs are present in larval salamanders and caecilians (Fritzsch and Wahnschaffe 1983), but absent in other tetrapods. Klembara (1994) suggested that depressed, densely pitted regions in the skull roof of Permian seymouriamorphs (*Discosauriscus*) may have housed electroreceptors.

Balance and sound organs. The organs for sound perception and balance are both located in the inner ear. Together, they are referred to as the stato-acoustic sense. The static or vestibular organ is an autapomorphy of vertebrates, which use it for maintaining balance in the water. Whereas hagfishes and lampreys have only two semicircular canals, gnathostomes have three, corresponding with the three dimensions of space. In contrast to all other sense organs, the vestibular apparatus does not provide information on the environment, but on the orientation and movement of the body itself. The vestibular system has not essentially changed with the fish–tetrapod transition. The receptors for the vestibular sense are called maculae and cristae, and they are sensitive to displacement occurring when the body changes its orientation.

The second, acoustic, system involves receptors (papillae) sensitive to pressure changes. As in the lateral-line organs, papillae are capable of detecting vibrations in the water. In terrestrial tetrapods, airborne vibrations are perceived, but because of their much smaller amplitude an impedance-matching system evolved: the middle ear. In tetrapods, the acoustic organ system thus falls in two separate components: (1) the sensory receptor (papilla) in the water-filled inner ear cavity and (2) the middle ear, an air-filled canal housing the ear ossicle, which acts as sound transmitter (Figure 4.9).

Figure 4.9 Anatomy of the amphibian ear. (A) Stapes and batrachian operculum in the extant goliath frog (*Conraua*, unpublished data). (B) Operculum and opercularis muscle in a salamander (adapted from Duellman and Trueb 1994). (C, D) Inner and middle ear of a frog in cross-section (adapted from Wever 1985). Pap am, papilla amphibiorum; pap bas, papilla basilaris.

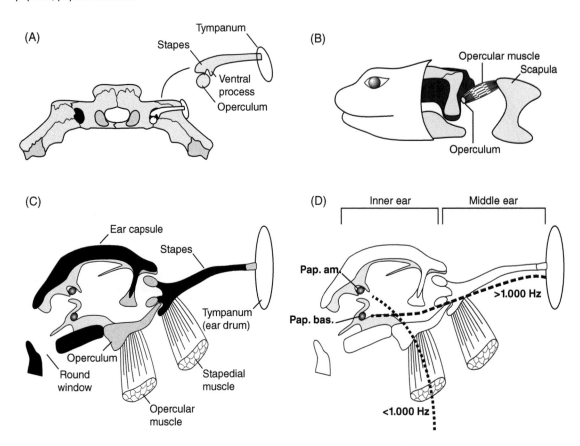

In bony fishes, hearing in the inner ear is performed by the maculae, which are covered by a gelatinous cupula that contains mineralized bodies. In actinopterygians, *Latimeria*, and dipnoans, the mineralized parts, called otoliths (ear stones), are large and formed of aragonite (Nolf 1985). In tetrapods, the same organs contain small calcite crystals. In addition to these receptors, tetrapods also have an acoustic sense, formed by the already mentioned papillae. Common to most tetrapods is the papilla basilaris. As this is absent in dipnoans and other bony fishes, it had long been considered a tetrapod autapomorphy. However, Fritzsch (1987) reported a papilla basilaris in *Latimeria*, and it is therefore likely that this papilla was lost in dipnoans (Mickoleit 2004).

This reasoning indicates that the papilla basilaris was the first receptor of the auditory sense and can be inferred to have existed in tetrapodomorphs. A second receptor (papilla amphibiorum) is present only in lissamphibians (Parsons and Williams 1963; Duellman and Trueb 1994). Amniotes are thus considered to retain the plesiomorphic condition, with a single papilla covering the entire range of frequencies. Only in modern amphibians has auditory processing been divided into low- and high-frequency streams, with the ear drum of frogs associated with the papilla basilaris, mediating the high-frequency end of the spectrum. Hearing mediated by the papilla basilaris thus evolved under water, first confined to low-frequency sound, transmitted by vibrations of the whole skull (Christensen-Dalsgaard and Carr 2008).

Spiracle and middle ear. The spiracle is a canal connecting the pharyngeal cavity with the dorsal side of the skull in jawed vertebrates. It is associated with the hyoid arch, specifically with one component (hyomandibula) that forms part of its lateral wall. Like the gill pouches, the spiracle is primitively water-filled, and has often been considered a modified gill slit. However, its orientation is different from the branchial arches and water flows through it in reversed fashion, from dorsal to ventral. A spiracle is present in sharks and rays, in *Polypterus* and other basal actinopterygians, as well as in *Latimeria*, but it is only vestigial, without openings, in lungfishes (Rauther 1930; Bartsch 1994). In many sharks and all rays, the spiracle controls the influx of respiratory water (von Wahlert 1966). In bony fishes, its role is less clear. Budgett (1903) and Magid (1966) observed the intake of air through the spiracle in *Polypterus*, when the fish is at the water surface. This happened on many occasions, but especially during phases of excitement or raised activity, or in water that was short of oxygen. This confirms that the spiracle is used as a respiratory canal for the lung in some bony fishes, in contrast to its use in sharks. A bony canal consistent with the features of the spiracle has been identified in many tetrapodomorph fishes, where it is largely similar to that of *Polypterus* (Jarvik 1980). It is therefore generally accepted that the spiracular canal was present in stem-tetrapods, and a similar anatomy is known from temnospondyls and anthracosaurs (Clack 1993).

The spiracle is considered a homolog of the middle ear cavity in tetrapods (Clack 1993). In tetrapods, the hyomandibula (stapes) is not attached to the spiracular wall, but enclosed in the spiracular canal, which is always air-filled. Like the spiracle, this middle ear cavity opens ventrally into the pharynx, by means of a narrow channel known as the eustachian tube. Dorsally, the middle ear is closed by a membrane, referred to as the ear drum (tympanum). The ear drum holds the same position as the dorsal spiracular opening in bony fishes, a region known as the temporal notch (squamosal embayment). A middle ear cavity of

this type is present in frogs and most amniotes and may be considered a synapomorphy of tetrapods, although other evidence contradicts this (see below).

The amphibian ear. The evolutionary transformation of the fish hyomandibula into the tetrapod stapes ranks among the most interesting topics in vertebrate evolution. The hyomandibula is a massive bone that tightly integrates numerous anatomical structures (muscles, ligaments, the gill-covering operculum). Movement of the hyomandibula is mediated by several muscles, contributing to the opening of the operculum, changing the shape of the spiracle, and constraining movements of the mandible, palatoquadrate, and braincase. In tetrapods, however, the stapes is not involved in any such role – cranial mobility has been largely reduced, the spiracle has become the middle ear cavity that *contains* the stapes, and the opercular bones are lost. The massive hyomandibula is thus a feature found in groups that primarily feed and breathe under water: actinopterygians, *Latimeria*, and stem-tetrapods (Jarvik 1980; Janvier 1996). By contrast, extant lungfishes have a small rudimentary hyomandibula and the opercular region is largely soft with a reduced operculum (Bartsch 1994), but Devonian stem taxa are more consistent with other bony fishes in this set of characters.

In tetrapods, the stapes is shorter than the hyomandibula and largely freed from connections to other skeletal elements, except for its articulation with the otic capsule. The reduced importance and connectivity of the hyomandibula/stapes in lungfishes and tetrapods is considered a convergence: fossils show that both stem-dipnoans and stem-tetrapods retained the primitive condition of bony fishes, encompassing a complete hyoid arch. The hyomandibula of *Eusthenopteron* was still a large and solid element with numerous muscle attachments (Jarvik 1980; Brazeau and Ahlberg 2006).

The tetrapod stapes attaches to the margin of an opening in the ear capsule, the oval window. This round opening evolved from a slit-like fontanelle in bony fishes, but the morphology of the oval window and the mode of attachment are exclusive to and found throughout tetrapods. Distally, the stapes is thin and lightly built, attaching to the tympanum in frogs and amniotes. In salamanders and caecilians, the stapes is more robust and rudimentary, attaching either to the quadrate or the squamosal. In both groups the tympanum, middle ear cavity, and eustachian tube are consistently absent.

Hence, there are two divergent types of middle ears in lissamphibians – the anuran and salamander–caecilian types. The similarities between the anuran and amniote ears are usually interpreted as convergences (Lombard and Bolt 1979). This is concluded from inconsistent anatomical structures in anurans and amniotes, especially the course and position of nerves and blood vessels relative to the middle ear and tympanum, which indicate a convergent origin of tympanum and middle ear cavity. However, the absence of these structures in salamanders and caecilians is probably a derived state rather than inherited from stem-tetrapods. This conclusion is based on an entirely phylogenetic argument: the most likely stem-group of all three lissamphibian clades are the dissorophoid temnospondyls, which all possessed a large tympanum, a delicate anuran-like stapes, and a middle ear cavity similar to that of extant frogs (Bolt and Lombard 1985; Maddin et al. 2012). If the temnospondyl origin of lissamphibians is accepted, the primitive condition of the amphibian ear should therefore be the possession of a tympanum, middle ear cavity, and eustachian tube, with the stapes completely enclosed within the air-filled middle ear cavity. Here, this set of structures is referred to as the tympanic ear. As stem-amniotes lack evidence of a tympanum and middle ear cavity, they are generally not considered to have possessed a tympanic ear – this indicates the independent evolution of such ears in lissamphibians and amniotes. This hypothesis is supported by the presence of massive stapes in stem-amphibians and stem-amniotes, which often articulated with the quadrate or squamosal.

Among lissamphibians, anurans and salamanders have a second ear ossicle that is formed by an isolated piece of the ear capsule (Figure 4.9). It is often bony, but may also be cartilaginous. Unfortunately, this element is referred to as the operculum, although it is neither homologous nor functionally comparable to the gill-covering elements of bony fishes. It is an endoskeletal element, in contrast with the dermal origin of the fish operculum. To avoid confusion, I refer to this element as the batrachian operculum. This element is located posterior to the oval window and forms the origin of a muscle that attaches to the scapula. Thus, the batrachian operculum and the so-called opercularis muscle connect the inner ear with the shoulder girdle and forelimb, forming an independent hearing apparatus from that of the stapes. This apparatus transmits low-frequency vibrations from the ground to the inner ear, which are perceived by the papilla amphibiorum (Wever 1985). The fact that the papilla amphibiorum and the opercular apparatus are functionally coupled suggests that the ancestors of caecilians probably possessed an operculum, although the extant taxa lack it; the massive footplate of the caecilian stapes might well include an operculum.

References

Allis, P. (1897) The cranial muscles and cranial and first spinal nerves in *Amia calva*. *Journal of Morphology* **12**, 487–762.

Allis, P. (1922) The cranial anatomy of *Polypterus*, with special reference to *Polypterus bichir*. *Journal of Anatomy* **56**, 189–291.

Bartsch, P. (1994) Development of the cranium of *Neoceratodus forsteri*, with a discussion of the suspensorium and the opercular apparatus in Dipnoi. *Zoomorphology* **114**, 1–31.

Bolt, J.R. & Lombard, R.E. (1985) Evolution of the amphibian tympanic ear and the origin of frogs. *Biological Journal of the Linnean Society* **24**, 83–99.

Boy, J.A. (1972) Die Branchiosaurier (Amphibia) des saarpfälzischen Rotliegenden (Perm, SW-Deutschland). *Abhandlungen des Hessischen Landesamts für Bodenforschung* **65**, 1–137.

Boy, J.A. (1974) Die Larven der rhachitomen Amphibien (Amphibia: Temnospondyli, Karbon-Trias). *Paläontologische Zeitschrift* **48**, 236–268.

Brazeau, M. D. & Ahlberg, P. E. (2006) Tetrapod-like middle ear architecture in a Devonian fish. *Nature* **439**, 318–321.

Bryant, H.N. & Russell, A.P. (1992) The role of phylogenetic analysis in the inference of unpreserved attributes of extinct taxa. *Philosophical Transactions of the Royal Society of London B* **337**, 405–418.

Bryant, H.N. & Seymour, K.L. (1990) Observations and comments on the reliability of muscle reconstruction in fossil vertebrates. *Journal of Morphology* **206**, 109–117.

Budgett, J.S. (1903) Notes on the spiracles of *Polypterus*. *Proceedings of the Zoological Society of London* **1903**, 10.

Bulman, O.M.B. & Whittard, W.F. (1926) On Branchiosaurus and allied genera (Amphibia). *Proceedings of the Zoological Society of London* **1926**, 533–579.

Bystrow, A.P. (1938) *Dvinosaurus* als neotenische Form der Stegocephalen. *Acta Zoologica* **19**, 209–295.

Carroll, R.L. & Holmes, R. (1980) The skull and jaw musculature as guides to the ancestry of salamanders. *Zoological Journal of the Linnean Society* **68**, 1–40.

Christensen-Dalsgaard, J. & Carr, C.E. (2008) Evolution of a sensory novelty: tympanic ears and the associated neural processing. *Brain Research Bulletin* **75**, 365–370.

Clack, J.A. (1993) Homologies in the fossil record: the middle ear as a test case. *Acta Biotheoretica* **41**, 391–410.

Clack, J.A. (1998) The neurocranium of *Acanthostega gunnari* Jarvik and the evolution of the otic region in tetrapods. *Zoological Journal of the Linnean Society* **122**, 61–97.

Clack, J.A. (2012) *Gaining Ground: the Origin and Evolution of Tetrapods*, 2nd edition. Bloomington: Indiana University Press.

Coates, M.I. & Clack, J.A. (1991) Fish-like gills and breathing in the earliest known tetrapod. *Nature* **352**, 234–236.

Deban, S.M. & Wake, D.B. (2000) Aquatic feeding in salamanders. In: K. Schwenk (ed.), *Feeding: Form, Function, and Evolution in Tetrapod Vertebrates*. Boston: Academic Press, pp. 65–94.

de Ricqlès, A. (1975) Quelques remarques paléo-histologiques sur le problème de la néotenie chez les stégocéphales. *CNRS Colloquium International* **218**, 351–363.

Diogo, R., Hinits, Y., & Hughes, S. (2008) Development of mandibular, hyoid and hypobranchial muscles in the zebrafish: homologies and evolution of these muscles in bony fishes and tetrapods. *BMC Evolutionary Biology* **8**, 24–46.

Drüner, L. (1901) Studien zur Anatomie der Zungenbein-, Kiemenbogen- und Kehlkopfmuskulatur der Urodelen. I. Theil. *Zoologisches Jahrbuch für Anatomie und Ontogenie* **15**, 435–622.

Duellman, W.E. & Trueb, L. (1994) *Biology of Amphibians*. Baltimore: Johns Hopkins University Press.

Dünker, N., Wake, M.H., & Olson, W.M. (2000) Embryonic and larval development in the caecilian Ichthyophis kohtaoensis (Amphibia, Gymnophiona): a staging table. *Journal of Morphology* **243**, 3–34.

Franzen, J.L. & Schaal, S. (2000) Der eozäne See von Messel. In: G. Pinna (ed.), *Europäische Fossillagerstätten*. Berlin: Springer, pp. 177–183.

Fritzsch, B. (1987) Inner ear of the coelacanth fish *Latimeria* has tetrapod affinities. *Nature* **327**, 153–154.

Fritzsch, B. & Wahnschaffe, U. (1983) The electroreceptive ampullary organs of urodeles. *Cell and Tissue Research* **229**, 483–503.

Iordansky, N.N. (1990) *Evolution of Complex Adaptations. The Jaw Apparatus of Amphibians and Reptiles*. Nauka, Moscow. (In Russian.)

Janvier, P. (1996) *Early Vertebrates*. Oxford Monographs on Geology and Geophysics 33. Oxford: Oxford Univesity Press.

Jarvik, E. (1980) *Basic Structure and Evolution of Vertebrates*. Vols. **1–2**. London and New York: Academic Press.

Klembara J. (1994) Electroreceptors in the Lower Permian Discosauriscus austriacus. *Palaeontology* **37**, 609–626.

Klembara, J. (1995) The external gills and ornamentation of skull-roof bones of the Lower Permian Discosauriscus (Kuhn 1933) with remarks to its ontogeny. *Paläontologische Zeitschrift* **69**, 265–281.

Lauder, G.V. (1980) Evolution of the feeding mechanism in primitive actinopterygian fishes: a functional anatomical analysis of *Polypterus*, *Lepisosteus*, and *Amia*. *Journal of Morphology* **163**, 283–317.

Lauder, G.V. & Shaffer, H.B. (1985) Functional morphology of the feeding mechanism in aquatic ambystomatid salamanders. *Journal of Morphology* **185**, 297–326.

Lombard, R.E. & Bolt, J.R. (1979) Evolution of the tetrapod ear: an analysis and reinterpretation. *Biological Journal of the Linnean Society* **11**, 19–76.

Lubosch, W. (1938) Muskeln des Kopfes. Viscerale Muskulatur. In: L. Bolk, E. Göppert, E. Kallius, & W. Lubosch (eds.), *Handbuch der vergleichenden Anatomie der Wirbeltiere*. Berlin: Urban & Schwarzenberg, pp. 1011–1106.

Luther, A. (1914) Über die vom N. trigeminus versorgte Muskulatur der Amphibien. *Acta Societatis Scientarum Fennicae* **44**, 1–151.

Maddin, H., Jenkins, F.A., & Anderson, J.S. (2012) The braincase of *Eocaecilia micropodia* (Lissamphibia, Gymnophiona) and the origin of caecilians. *PloS ONE* **7**, e50743.

Magid, A.M.A. (1966) Breathing and function of the spiracles in *Polypterus senegalus*. *Animal Behavior* **14**, 530–533.

Mickoleit, G. (2004) *Phylogenetische Systematik der Wirbeltiere*. Munich: Pfeil.

Milner, A.C., Milner, A.R., & Walsh, S.A. (2009) A new specimen of *Baphetes* from Nýřany, Czech Republic and the intrinsic relationships of the Baphetidae. *Acta Zoologica* **90**, 318–334.

Milner, A.R. (1982) Small temnospondyl amphibians from the Middle Late Carboniferous of Illinois. *Palaeontology* **25**, 635–664.

Milner, A.R. (2007) *Mordex laticeps* and the base of the Trematopidae. *Journal of Vertebrate Paleontology* **27**, 118A.

Milner, A.R. & Sequeira, S.E.K. (1994) The temnospondyl amphibians from the Viséan of East Kirkton, West Lothian, Scotland. *Transactions of the Royal Society of Edinburgh: Earth Sciences* **84**, 331–361.

Nolf, D. (1985) Otolithi piscum. In: H.-P. Schultze (ed.), *Handbook of Paleoichthyology*. Stuttgart: Gustav Fischer, Vol. 10, pp. 1–145.

Parsons, T.S. & Williams, E.E. (1963) The relationships of the modern Amphibia: a re-examination. *Quarterly Review of Biology* **38**, 26–53.

Rauther, W. (1930) Kiemen der Anamnier: Kiemenderivate der Cyclostomen und Fische. In: L. Bolk, E. Göppert, E. Kallius, & W. Lubosch (eds.), *Handbuch der vergleichenden Anatomie der Wirbeltiere*, vol. 3, pp. 211–276.

Sanchez, S., de Ricqlès, A., Schoch, R.R., & Steyer, J.S. (2010) Developmental plasticity of limb bone microstructural organization in *Apateon*: histological evidence of paedomorphic conditions in branchiosaurs. *Evolution and Development* **12**, 315–328.

Schmalhausen, I.I. (1968) *The Origin of Terrestrial Vertebrates*. London and New York: Academic Press.

Schmidt-Nielsen, K. (1997) *Animal Physiology: Adaptation and Environment*. Cambridge: Cambridge University Press.

Schoch, R.R. (1992) Comparative ontogeny of Early Permian branchiosaurid amphibians from southwestern Germany. Developmental stages. *Palaeontographica A* **222**, 43–83.

Schoch, R.R. (2002) The early formation of the skull in extant and Paleozoic amphibians. *Paleobiology* **28**, 378–396.

Schoch, R.R. (2004) Skeleton formation in the Branchiosauridae as a case study in comparing ontogenetic trajectories. *Journal of Vertebrate Paleontology* **24**, 309–319.

Schoch, R.R. (2009) The evolution of life cycles in early amphibians. *Annual Review of Earth and Planetary Sciences* **37**, 135–162.

Schoch, R.R. & Witzmann, F. (2009) Osteology and relationships of the temnospondyl *Sclerocephalus*. *Zoological Journal of the Linnean Society London* **157**, 135–168.

Schoch, R.R. & Witzmann, F. (2011) Bystrow's paradox: gills, forssils, and the fish-to-tetrapod transition. *Acta Zoologica* **92**, 251–265.

Schultze, H.-P. (1991) A comparison of controversial hypotheses on the origin of tetrapods. In: H.-P. Schultze & L. Trueb (eds.) *Origins of the Higher Groups of Tetrapods: Controversy and Consensus*. Ithaca: Cornell University Press, pp. 29–67.

Thomson, K.S. (1967). Mechanisms of intracranial kinesis in fossil rhipidistian fishes (Crossopterygii) and their relatives. *Zoological Journal of the Linnean Society* **46**, 223–253.

von Wahlert, G. (1966) Atemwege und Schädelbau der Fische. *Stuttgarter Beiträge zur Naturkunde A* **159**, 1–10.

Wake, D.B. & Deban, S.M. (2000) Terrestrial feeding in salamanders. In: K. Schwenk (ed.), *Feeding: Form, Function, and Evolution in Tetrapod Vertebrates*. Boston: Academic Press, pp. 95–116.

Wake, M.H. (1977) The reproductive biology of caecilians: an evolutionary perspective. In: D.H. Taylor & S.I. Guttman (eds.), *The Reproductive Biology of Amphibians*. New York: Plenum, pp. 73–101.

Werneburg, R. (1991) Die Branchiosaurier aus dem Unterrotliegend des Döhlener Beckens bei Dresden. *Veröffentlichungen des Naturhistorischen Museums Schleusingen* **6**, 75–99.

Werneburg, R. (2007) Timeless design: colored pattern of skin in Early Permian branchiosaurid (Temnospondyli: Dissorophoidea). *Journal of Vertebrate Paleontology* **27**, 1047–1050.

Wever, E.G. (1985) *The Amphibian Ear*. Princeton: Princeton University Press.

Willems, H. & Wuttke, M. (1987) Lithogenese lakustriner Dolomite und mikrobiell induzierte "Weichteilerhaltung" bei Tetrapoden des Unter-Rotliegenden (Perm, Saar-Nahe-Becken, SW-Deutschland). *Neues Jahrbuch für Geologie und Paläontologie Abhandlungen* **174**, 213–238.

Witmer, L. M. (1995). The extant phylogenetic bracket and the importance of reconstructing soft tissues in fossils. In: J. J. Thomason (ed.), *Functional Morphology in Vertebrate Paleontology*. Cambridge: Cambridge University Press, pp. 19–33.

Witzmann F. (2004) The external gills of Palaeozoic amphibians. *Neues Jahrbuch für Geologie und Paläontologie Abhandlungen* **232**, 375–401.

Witzmann F. (2006a). Morphology and palaeobiology of the Permo-Carboniferous temnospondyl amphibian *Archegosaurus decheni* Goldfuss, 1847 from the Saar-Nahe Basin, Germany. *Transactions of the Royal Society of Edinburgh: Earth Sciences* **96**, 131–162.

Witzmann F. (2006b). Developmental patterns and ossification sequence in the Permo-Carboniferous temnospondyl *Archegosaurus decheni* (Saar-Nahe Basin, Germany). *Journal of Vertebrate Paleontology* **26**, 7–17.

5 Evolution of Functional Systems

A major goal of paleobiology is to understand not only the basic functions of extinct organisms, but also the evolutionary changes that organs have undergone. The study of early tetrapod anatomy has reached a phase in which morphology, phylogeny, and functional data derived from extant groups can be integrated to trace major evolutionary transformations. The fish–tetrapod transition had a profound impact on almost all organs, but only some can be studied in the fossil record. Feeding, breathing, and hearing exemplify cases in which many new data have become available recently. Mechanical properties of skeletons, evidence of muscles, and phylogenetically bracketed traits of function and behavior come together in this area. Although early tetrapods retained many structural features of their fish ancestors, they also remodeled essential parts of the skeleton. When bony fishes are compared with salamanders, surprisingly few differences are found in the distribution of jaw and branchial muscles and the way they operate during feeding and breathing. How did the tetrapodomorph fishes feed and breathe, and which successive modifications occurred to their skeletons? How did the limbed stem-tetrapods differ in these body regions? What impact did the loss of the opercular bones, the disintegration of the hyoid arch, and the separation of skull and shoulder girdle have on feeding and breathing? How did the middle ear emerge from these complex changes? What impact did the origin of amphibian metamorphosis have on these organ systems?

Amphibian Evolution: The Life of Early Land Vertebrates, First Edition. Rainer R. Schoch.
© 2014 Rainer R. Schoch. Published 2014 by John Wiley & Sons, Ltd.

5.1 How paradigms and brackets give a functional scenario

When discussing amphibian soft-tissue structures in Chapter 4, extant phylogenetic brackets (EPBs) were used on various occasions. These discussions were all centered on static morphological traits, such as musculature and its skeletal correlates. In a further step, functional morphology is now considered. This procedure follows the same reasoning as in other brackets, here dealing with patterns of spatiotemporal muscle activity and biomechanical properties of body parts (Figure 5.1). These brackets use data derived from experimental approaches such as high-speed cinematography and electromyography of feeding strikes in living animals (Lauder 1980a, 1980b).

A second approach that has delivered new insights does not strictly follow the EPB protocol: experimental data on skull sutures (Markey *et al.* 2006). As bones and their sutures are universal properties of vertebrates, analysis in extant taxa permits inference in extinct taxa. Rather than phylogenetic

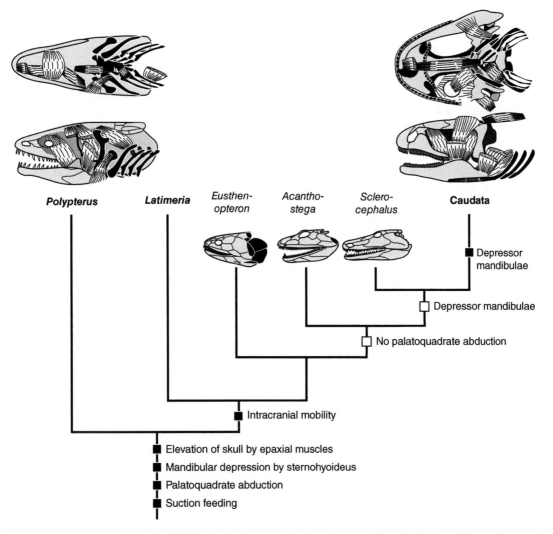

Figure 5.1 Extant phylogenetic bracket (EPB) inferring important functions in the feeding strike in bony fishes and tetrapods.

reasoning, this is an example of actualism, which focuses on material properties that have not changed with time or by evolution. Experiments by Markey *et al.* (2006) showed that, on a gross scale, interdigitating sutures are subject to tension, whereas abutting bones respond to strain. With these insights at hand, suture morphologies of fossil taxa can be analyzed to find out the major forces that acted on skulls of long-extinct fishes and tetrapods. Following this line of evidence, Markey and Marshall (2007) examined sutures of *Eusthenopteron*, *Acanthostega*, and the dissorophoid *Phonerpeton*. They found that between the *Eusthenopteron* and *Acanthostega* nodes a shift from suction feeding to a jaw-prehensive mode of prey ingestion must have occurred.

The two lines of evidence outlined here differ from what paleontologists often refer to as "functional morphology": rather than providing a theoretical paradigm in order to explain the functional role of a feature in an extinct taxon (Rudwick 1964), a phylogenetic bracket of functions deals with experimental data and seeks to detect the universal aspects of biomechanical and behavioral properties found in the extant bracket taxa. However, both the paradigm and EPB approaches share the premise that the study of extant exemplars – either functional analogs or biomechanical homologs – can be guides to understanding evolutionary history. The fact that both approaches have been successful in various cases highlights that paleontology and evolutionary biology are not historical sciences *per se*, but also employ aspects of experimental and theoretical sciences, which fall in the domain of ahistoric disciplines.

Lauder (1990) has outlined the integration of functional and morphological data within a phylogenetic frame. Shared patterns among functional traits form helpful guides for an evolutionary scenario that focuses on some key features of early tetrapod life. The starting point will be the mode of underwater feeding employed by modern bracket taxa, in order to form a frame for reconstructing evolutionary transformations in the skull and pectoral girdle. The major problem here is that the bones and muscles of the skull, hyoid arch, branchial arches, and pectoral girdle are so tightly interconnected that a separate discussion of feeding, breathing, and hearing is impossible. Instead, I shall discuss these traits as aspects of a single scenario in successive sections. It is also crucial to understand that bones are important but their roles in biomechanics can only be understood when their often complex relations to muscles and ligaments are known – to this end, the EPB is the only source, and is indispensable.

Prey capture in *Polypterus* and related fishes. Lauder (1980a) showed that *Polypterus*, *Lepisosteus*, and *Amia* share numerous motor patterns of muscles involved in feeding. These muscles include the ones discussed in Chapter 4, most of which are conserved in at least some extant tetrapods that feed in the water (salamanders).

In *Polypterus*, the feeding strike involves the following actions in succession: (1) elevation of the head (EA), (2) depression of the hyoid arch (SH, HY), (3) adduction of the operculum (AOP), (4) elevation of the palatoquadrate (LAP), (5) abduction (opening) of the operculum (DOP), and finally (6) closure of the mouth (AM). These patterns are shared with other actinopterygians, and Lauder (1982) concluded that they represent the primitive condition of all bony fishes. Characteristically, mouth opening is achieved by raising the neurocranium and depressing the hyoid arch. Because there is no depressor mandibulae (DM), the mandible is lowered by means of its connection to the hyoid arch. Thus, the hyomandibula and ceratohyal are essential components not only for moving the operculum or manipulating the spiracle, but also for opening the mouth, accomplished by a ligament connecting the two units. The operculum is held closed during the gape, but opened when the jaw-closing phase has been initiated by the mandibular adductors (AM). The elevation of the palatoquadrate plays an important role in the later part of mouth opening, maximizing mouth width during the expansive phase. Most of the skeletal components and muscular correlates are present in finned stem-tetrapods, as preserved in *Eusthenopteron*, *Osteolepis*, and *Panderichthys*.

Prey capture in *Latimeria*. Although extant lungfishes and *Latimeria* are more closely related to tetrapodomorphs than *Polypterus* is, their skulls are generally considered too modified to be guides to the primitive condition of tetrapods.

A notable exception is the intracranial joint, a consistent feature of tetrapodomorph fishes that is preserved in only one extant taxon, *Latimeria chalumnae* (Thomson 1967). The division of the braincase into two components characterizes sarcopterygians and is believed to be derived from a partial fissure of early osteichthyans, which is conserved in actinopterygians (Janvier 1996). This joint and the associated large muscle (subcranial or subcephalic muscle, SM) had long been regarded as questionably homologous and not necessarily a reliable guide to the primitive sarcopterygian condition, but Lauder (1980b) showed that *Latimeria* shares the essential features of the osteichthyan feeding apparatus, and that the intracranial joint fits rather easily into this frame. In *Latimeria*, the intracranial joint forms part of a four-bar linkage mechanism (jaw articulation–intracranial joint–hyomandibula-braincase joint–symplectic-mandible joint). In this system, mandibular depression is initiated in the typical osteichthyan fashion by the sternohyoideus (SH) muscle, which in *Latimeria* elevates the hyoid arch (Lauder 1980b). When the mandible is depressed, the two braincase blocks are elevated by contraction of the epaxial muscles. This is enabled by the stabilization of the pectoral girdle, accomplished by the hypaxial muscles, which in turn constrains the effect of sternohyoideus contraction to the hyoid arch. During the compressive phase, the adductors raise the mandible and the subcranial muscle lowers the anterior braincase, thus closing the mouth. The subcranial muscle is therefore the antagonist of the sternohyoideus in *Latimeria* (Lauder 1980b). It is noteworthy that the above-cited biomechanical properties of feeding in *Latimeria* were largely derived from mechanical models rather than cinematography, and thus are based on a paradigm. The implications for tetrapodomorphs are that the intracranial joint was at least involved in mouth closure – considering the fate of the subcranial muscle in lissamphibians, this has interesting implications for evolutionary changes in the skull, as discussed below.

Aquatic feeding in salamanders. Larval and neotenic salamanders are the only tetrapods to retain a large complement of structures and muscles that perform an aquatic feeding strike similar to bony fishes. Lauder and Shaffer (1985)

and Reilly and Lauder (1990) accumulated many data on shared patterns of muscle acticity, their timing, and the anatomical framework. Interestingly, even in cases where morphology has been substantially altered, muscle activities and their roles in the feeding strike have been much more precisely conserved than the morphology (Lauder and Shaffer 1985). Deban and Wake (2000) more recently summarized the facts and opinions about aquatic feeding in salamanders. As studied in *Ambystoma mexicanum* (Lauder and Shaffer 1985), the feeding strike includes the following steps, with involved muscles given in brackets: (1) elevation of neurocranium (EP), (2) depression of the mandible (DM), (3) retraction of the ceratohyal (SH), (4) stabilization of the pectoral girdle by means of the hyomandibularis muscle (HM), and (5) closure of the mouth (AM).

In comparison to bony fishes, the palatoquadrate is not substantially moved against the braincase and the musculature that moves the two units in fishes (LAP) has been recruited by the eye in tetrapods. There is also no equivalent of the intracranial joint, with the braincase forming a single unit. Furthermore, only the ventral portion of the hyoid arch is involved in feeding in salamanders: the ceratohyal (the dorsal portion, of course, is a sound-transmitter and called stapes). Interestingly, the so-called hyomandibular ligament connects the mandible with the ceratohyal, mediating jaw depression when the sternohyoideus muscles fires (Lauder and Shaffer 1985). The close muscular connection between the surviving ventral portions of hyoid and branchial arches is referred to as the hyobranchial apparatus. The hyomandibular ligament – and by that the mechanical coupling of lower jaw and hyobranchium – is shared with osteichthyans, and thus likely to have been present in tetrapodomorphs. Although the phases of muscular activity are similar to those of *Polypterus*, the total number of muscles and biomechanical units involved is smaller. Aquatic feeding in salamanders also relies on an enhanced kind of suction, driven by the explosive expansion of the buccal cavity. The mechanical apparatus behind this powerful suction is a simple four-bar system: the parallel ceratohyal and first ceratobranchial articulate with the basibranchial. Before feeding commences, this apparatus is folded together. During suction feeding, the sternohyoideus and branchiohyoideus unfold it by pulling the hyoid and branchial bars into an upright position, which pulls the mandible back and greatly enlarges the buccal cavity. The geniohyoideus finally pulls the whole apparatus back into its resting position, closing the mouth and folding the hyobranchium (Lauder and Shaffer 1985).

Why experimental data are indispensable. Functional considerations based on skeletal features alone miss an important aspect: they will often not be sufficient to grasp the complete set of components of an extinct mechanical apparatus. Only by inference of data on muscles and ligaments does a complete picture emerge. For instance, in extant taxa the mandibulohyoid ligament leaves practically no trace (skeletal correlate) on the hard parts that it connects. Therefore, it is unlikely to be detected in fossil taxa, especially because potential correlates may also be interpreted in an alternative way. Therefore, the EPB is the only guide at hand, adding significant information to the reconstruction of a long-extinct feeding apparatus.

Multiplicity of components and functions. The study of feeding strikes in bony fishes and tetrapods reveals another point that is worth a moment of thought. In some cases, the same set of muscles is used for different purposes – this is *functional multiplicity*. Depending on the particular situation, a muscle may stabilize a body region at one time – for instance, to form an anchor for other muscles – and move body parts at another time. That the same muscle may perform rather different, sometimes even opposite functions parallels the role that genes play in the current understanding of developmental genetics. Rather than "coding for" particular traits, the same gene may be active in numerous entirely different situations, delivering products (proteins) required under diverse conditions. The parallel shows that biological functions are usually much more complex and multifaceted than they first appear, and it should remind us that there is no one-to-one relation between a structure and a function. Returning to the muscle example, the adductor mandibulae has been found to be active

not only during mouth closure, but also at the beginning of the feeding strike, and the hyobranchial muscles perform very complex actions during the feeding cycle (Lauder 1980a, 1980b; Lauder and Shaffer 1985).

A second lesson to be learned from the study of aquatic feeding is the advantage gained by a *multiplicity of components*. In bony fishes, there is always more than one muscle performing a particular function in the strike (e.g., mouth opening, expansion of buccal cavity, opening of the gill chamber, and mouth closure). For instance, the mouth may be opened by action of the sternohyoideus (which pulls back the hyoid and with it the mandible) *and* by raising the braincase through the epaxial muscles – but the mouth is *also* opened by the depressor mandibulae in lungfishes and tetrapods. Likewise, mouth closing is achieved not only by the adductor mandibulae, but also by the subcranial muscle in taxa having an intracranial joint. Thus, the possession of several separate components performing the same function (in different ways) not only forms an insurance against default but also, more importantly, allows functional fine-tuning of these multiple components. This is not restricted to muscles, but also concerns skeletal elements. The result is exemplified by *Polypterus*, which shows a complex succession of muscle activities and movements of jaw and branchial elements.

Not surprisingly, the example of salamanders shows that an evolutionary reduction of some osteichthyan muscles and bones did not affect the functionality of the apparatus – even the spatiotemporal patterns of muscular activity were conserved. The loss of dermal bones in the mandible and gill cover (the gular plates and opercular bones) has opened an avenue for expanding the hyobranchium far beyond the narrow limits of the rigid bony fish skeleton. Structural multiplicity contrasts with mechanical freedom in this case, probably forming trade-offs that are "re-negotiated" anew in each new species. Paralleling aquatic salamanders, extreme suction feeding has also evolved in teleosts, but not by hyobranchial expansion; instead, modification of the mouth margin has been the key innovation (Lauder and Liem 1989).

5.2 Feeding and breathing under water

This section discusses a scenario of feeding and respiration in tetrapodomorphs, based on the above-described bracket taxa, supplemented by direct osteological information from the fossil taxa and the phylogenetic succession of taxa. In the Devonian lobe-finned fishes, exemplified by *Eusthenopteron*, feeding and breathing were tightly coupled. The two-unit braincase, palatoquadrate, mandible, hyoid arch, branchial arches, and opercular series were all interconnected by joints (Jarvik 1954, 1980; Thomson 1967). These data are confirmed and supplemented by a bracket including *Polypterus*, *Amia*, and *Latimeria*, and dipnoans on the one side of the bracket and aquatic salamanders on the other. For instance, there is little ground to doubt that a mandibulohyoid ligament connected the mandible and hyoid arch. In *Eusthenopteron*, the ventral part of the hyomandibula probably directly attached to the palatoquadrate.

The importance of the hyoid arch. As in the extant bracket taxa, the hyoid arch played a pivotal role in the integration of the skull, gill cover, and cranial musculature in *Eusthenopteron* (Figure 5.2). This is reflected by the numerous muscles attaching to the hyomandibula, as exemplified by *Polypterus* and *Amia*. These are: the posterior portion of the adductor mandibulae, the spiracular muscles (which manipulate the shape of the spiracle), and two specific hyomandibula muscles, the adductor and retractor hyomandibularis (AHM, RHM). The hyomandibula operates the movements of the opercular bone, which mediates water breathing: rotating the hyomandibula opens the operculum. Furthermore, it also mechanically couples the palatoquadrate with the operculum in *Polypterus*, coordinating movements between the cheek and operculum: elevating the cheek ultimately affects opening of the gill chamber. This is consistent with the observation that shortly after the LAP has started to be active, the DOP joins it (Lauder 1980a). This is also apparent from the anatomy of *Polypterus*, where the LAP, DOP, and AM are all connected with each other (Allis 1897). Hence, movement of one muscle has an impact on the action of others.

The palatoquadrate problem. At this stage it is necessary to comment on a debate about cranial kinesis in *Eusthenopteron*. Jarvik (1954) reported a series of joints between the palatoquadrate and braincase, which would have prevented the two units from moving against each other (Figure 5.3). Thomson (1967) found no such tight connection in other lobe-finned fishes, even close relatives of *Eusthenopteron*. Later, Jarvik (1980) reiterated his point without referring to Thomson or other papers. Whereas the number of joints between the palatoquadrate and braincase are debated in that taxon, close relatives of *Eusthenopteron* had only two points of attachment between the anterior braincase (ethmoid) and palatoquadrate: one behind the nasal capsule and one shortly anterior to the intracranial joint (basipterygoid facet). As in *Latimeria*, there was no direct contact between the posterior braincase and palatoquadrate (Thomson 1967). Even if the skull of *Eusthenopteron* was akinetic, other lobe-finned fishes evidently retained the intracranial and palatoquadrate joints. This forms an important cornerstone for the following scenario.

The feeding strike in osteolepiform fishes. Bracketed by *Polypterus* and *Latimeria* on the fish side, tetrapodomorph fishes are likely to have retained the mobile cheek, which is basically confirmed by the fossil anatomical data. As laid out by Thomson (1967), movement of the anterior braincase was linked to mobility of the cheek. Indeed, mechanical models (paradigms) show that in a skull like that of *Osteolepis* or *Gogonasus*, lifting the ethmoid portion automatically raises the cheek and vice versa. In turn, lifting the palatoquadrate in such a system pushes the hyomandibula back, which contributes to a compression of the hyoid and branchial arches. This movement would have forced water and prey further posterior, and subsequent opening of the operculum would let the water flow out.

In sum, the fossilized parts of the stem-tetrapod jaw and hyoid apparatus indicate that the units were linked in a similar way as in other bony fishes. The intracranial joint and palatoquadrate were probably moved only after peak gape had been achieved, which would have required the subcranial muscle to keep the endocranial components together before that point was reached.

Figure 5.2 Transformation of important organs during the fish–tetrapod transition. (A) *Sclerocephalus*; (B) *Acanthostega*; (C) *Eusthenopteron*. Based on anatomical data adapted from Jarvik (1980) and Clack (2002a). A central role played the fragmentation of the hyopid arch, by which the hyomandibular was freed from numerous connections ("roles") and ready to serve as ear ossicle. The spiracle, probably water-filled and *adjacent to* the hyomandibular, transformed into an air-filled cavity that *contains* the stapes.

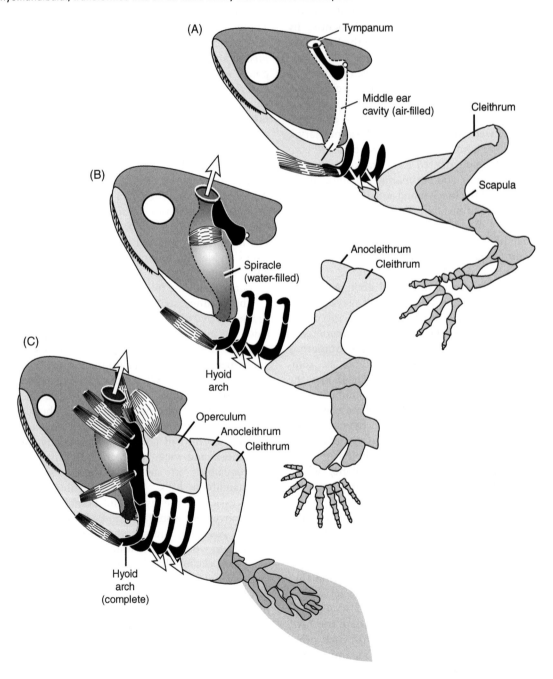

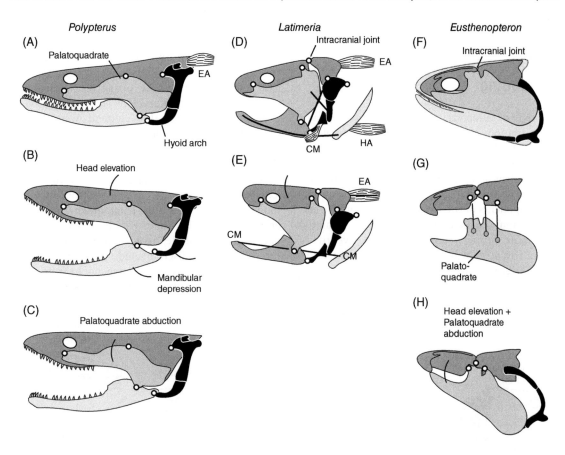

Figure 5.3 Skull mechanics, bracketed by fishes and lissamphibians and inferred for *Eusthenopteron*. (A–C) *Polypterus* (adapted from Allis 1922 and Lauder 1980a); (D, E) *Latimeria* (adapted from Lauder 1980b); (F–H) *Eusthenopteron* (adapted from Thomson 1967 and Jarvik 1980). *Latimeria* and *Eusthenopteron* share the intracranial joint, which was lost in tetrapods.

Therefore, both EPB and mechanical models indicate that the feeding strike in osteolepiforms was as follows: (1) the head was lifted by action of epaxial muscles, (2) depression of the sternohyoideus rotated the hyoid arch back, and (3) by transmission through the mandibulohyoid ligament and in concert with the depressor mandibulae the mandible was lowered; (4) when peak gape had been reached, the levator arcus palatini muscles raised the palatoquadrate, which by linkage with the ethmoid also lifted the snout; (5) this affected a rotation of the hyoid arch and an opening of the gill-covering opercular bones, permitting water to flow out, accompanied and enforced by (6) closure of the mouth by action of the subcranial muscle (pulling the snout back in line with the posterior braincase) and the jaw adductors. This scenario can only form the core of a much more complicated story, because our knowledge of ligaments is limited and at best indirectly assessed by EPB.

Feeding and breathing. Primitively, feeding and breathing employed the same mechanical actions in the skull and gill region. Breathing was operated, as it still is in *Polypterus*, *Amia*, and *Latimeria*, by the opercular suction pump: mouth opening and hyoid retraction sucked in water, which was finally pumped through the gills. From this perspective, the feeding strike is an extended version of the breathing cycle, with the action of the ventral hyoid and branchial musculature wedged in between closing and opening of the gill chamber. Thus, in bony fishes the muscular activities and mechanical processes involved in feeding and breathing are tightly coupled, particularly by the hyoid arch and its dermal bones, the opercular and gular series.

5.3 Decoupling breathing and feeding

The origin of tetrapods involved at least three major transformations that can be traced in skeletal features: (1) feeding and breathing, (2) hearing, and (3) locomotion. Here, I focus on the linkage between the modification of the feeding apparatus and middle ear. The middle ear of tetrapods evolved from two components of the hyoid arch, the spiracle and hyomandibula, which were successively separated from their former connections and eventually coupled in a novel way, performing a novel function and playing a new biological role. This was permitted by the breakup of the hyoid arch and the reduction

of the opercular pump. So, without this change in the breathing mechanism, there would have been no platform for the evolution of the middle ear.

The breakup of the hyoid arch. The hyoid arch played a crucial role in upholding connections and controlling movements of the opercular water pump in osteolepiforms. The hyomandibula was the crucial element, which is apparent by its size and complexity not only in osteolepiforms but also in extant actinopterygians and coelacanths.

In *Panderichthys*, this element was shorter than in *Eusthenopteron*, having lost the ventral part (Brazeau and Ahlberg 2006). This indicates that the hyoid arch was already partitioned into a dorsal portion ("proto-stapes") and a ventral one (ceratohyal). Likewise, the hyomandibula no longer articulated with the palatoquadrate (Downs *et al.* 2008). The tight coupling between the operculum, mandible, hyoid arch, and palatoquadrate was thus disconnected, and these units became successively more independent from each other (Figure 5.4). Judging from the structure of

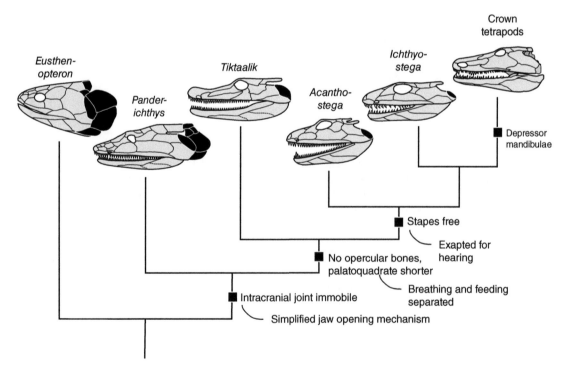

Figure 5.4 Major events in the evolution of feeding mechanics and skull mobility mapped onto a cladogram of the fish–tetrapod transition.

cranial joints, feeding still involved palatoquadrate abduction, a feature retained in some stem-tetrapods (whatcheeriids, *Crassigyrinus*, baphetids–but not in colosteids and temnospondyls) and in the anthracosaur stem-amniotes. Modification of the hyomandibula must therefore be seen in a different functional context, of which two components are apparent: (1) the skull became markedly flattened between the *Eusthenopteron* and *Panderichthys* nodes, changing the anatomical frame considerably (Downs *et al.* 2008), and (2) the braincase was increasingly consolidated, with the neurocranium of *Acanthostega* forming a single unit (Clack 1998).

The crucial transformation of the hyomandibula started with the disconnection of the hyoid arch and palatoquadrate. This is consistent with the continued use of palatoquadrate abduction in feeding but the loss of kinesis within the braincase. The fusion of the two braincase units therefore opened the door for a stepwise disconnection of the hyomandibula from feeding and aquatic breathing.

A second step was the loss of the bony opercular and gular bones, which is first seen at the *Tiktaalik* node (Figure 5.4). By analogy with lungfishes, an opercular fold was probably still present, but composed of soft tissue instead of bony elements. The loss of the bony gill cover is unlikely to have been caused by changes in breathing – internal gills and branchial arches were not substantially changed between the *Panderichthys* and *Acanthostega* nodes. However, the opercular pump was evidently weakened. Although it is not directly apparent from the fossil taxa which structure replaced the opercular pump, salamanders provide a hint: in these, the ceratobranchials bear large spike-like projections (gill rakers) that act as a zipper to close the gill slit. Coordinated opening and closing of the gills slits of course forms a pumping apparatus ("ceratobranchial pump") not unlike that of the operculum. The difference is that the pump is here composed of endoskeletal components, and that each slit can be controlled independently. The muscles driving this pump are purely visceral muscles connecting the branchial arches.

A third step was the re-orientation of the hyomandibula, first apparent in *Acanthostega*, and this point marks the transformation into the tetrapod stapes: rather than the ventral alignment of the hyomandibula, the stapes faces laterally in most early tetrapods and is only connected to the braincase and cheek, ready to form a brace between the two but also to transmit vibrations from the outside of the skull to the inner ear. This does not require a tympanum or middle ear cavity, as salamanders and caecilians exemplify.

The decoupling of water breathing and feeding thus paved the way for hearing. That said, it is important to stress that we need an evolutionary explanation not only for the origin of the middle ear, but also for the decoupling of the hyomandibula and ceratohyal in the first place. This problem, however, is a puzzle with some pieces remaining unknown.

Modularity. The breakup of the hyoid arch and the recruitment of its constituents for entirely different functions sheds some light on a new hot topic in evolutionary biology: modularity. Recent years have brought an increased interest in the phenotype also from disciplines that were traditionally uninterested in morphology – especially genetics. Modularity, along with a suite of other concepts, represents the new research fields dealing with how phenotypes develop and evolve. The basic idea is simple: organisms are integrated wholes, but they can only grow and develop because they fall into well-defined components, each of which can evolve with the required degree of autonomy. Modularity is thus an essential property for both development and evolution. As a concept, it guides the search for such units, and the present case exemplifies the idea neatly. Wagner and Schwenk (2000) have called this "evolutionarily stable configurations" or ESCs. Once detected, it is hoped that such ESCs will not only reveal the *building blocks* of development and evolution, but also shed new light on *phylogenetic characters*.

The hyoid arch was a tightly integrated component in the mechanical system of muscles, ligaments, and bones. When it broke up, somewhere between the *Panderichthys* and *Acanthostega* nodes, it not only decoupled feeding from breathing, but also opened a new avenue for hearing. The recruitment of hyomandibula and spiracle for hearing means the creation of a new evolutionary module, defined by the novel arrangement in which the air-filled spiracle contains the stapes. Module formation thus required two steps: (1) decoupling of pre-existing connections and (2) coupling and novel integration of two former sub-components. This reveals that modularity forms one aspect of a more inclusive theme, organismal integration.

5.4 Hearing: exapting the spiracle and hyomandibula

Hearing is an old heritage of vertebrates, and tetrapods only modified the existing receptors and sound-transmitting devices they inherited from bony fishes. The generally accepted scenario is that lateral lines were retained in early tetrapods but that the tympanic ear evolved convergently in lissamphibians and amniotes (Lombard and Bolt 1979; Clack 1992). This not only implies that the stapes is homologous throughout tetrapods but that the middle ear and tympanum evolved convergently several times (Lombard and Bolt 1979; Mickoleit 2004). A repeated evolution of the middle ear cavity from the spiracular canal is not difficult to imagine, whereas the enclosure of the stapes within that cavity requires a set of parallel events in lissamphibians and amniotes. That this occurred convergently is indicated by differences in anatomical details (Lombard and Bolt 1979). Further, whereas there is no doubt concerning the general homology of the hyomandibula with the tetrapod stapes, uncertainties remain about the identity and conservation of its various processes and muscle attachment sites across the fish–tetrapod transition. Here, I focus on the exaptation of the hyomandibula–stapes for hearing and a scenario for the origin of the middle ear cavity and tympanum. Any such hypothesis has to explain how a water-filled spiracle, supported by a tightly interconnected hyomandibula, evolved into an air-filled tympanic ear that contains a free-moving stapes. It should also explain why salamanders and caecilians lack such an ear, and specifically what makes their stapes appear so similar to that of early tetrapods. To meet these demands, the scenario requires the integration of developmental, paleontological, and functional data in a novel way (Figure 5.5).

Clack (1992) developed a scenario in which the tetrapod stapes evolved in two major steps: (1) it was freed from the duties of controlling opercular movement, and (2) it was freed from connecting the mandible with the ceratohyal and braincase, which enabled it to be included within a middle ear cavity. In the course of these changes, the spiracle transformed into the middle ear cavity. This implies that the spiracle persisted for much longer than was traditionally thought, and was retained in various stem-tetrapods (colosteids, baphetids) and stem-amniotes (anthracosaurs). This may have sounded heretical when it was first proposed, but now that the persistence of internal gills has been demonstrated in early tetrapods and even some Triassic temnospondyls (Schoch and Witzmann 2011), it adds to a more consistent picture of the primarily aquatic habits of early tetrapods.

The changing role of the spiracle. Clack (1992) suggested that the spiracle formed part of a specific air-breathing mechanism, by which air was taken from the dorsal surface of the skull via the spiracle and pharynx into the lungs. This is based on the observations of Budgett (1903) that extant *Polypterus* inhales air through the spiracle. Although it is likely that such a mechanism was also present in stem-tetrapods, there is some seemingly contradicting evidence: in *Eusthenopteron*, the spiracular canal contained numerous denticulate ossicles (Jarvik 1980). By analogy with the pharyngeal dentition of gill slits, this indicates that the spiracle was at least sometimes water-filled in *Eusthenopteron*, as these denticles serve as a filter preventing larger particles intruding into the gill pouches. Von Wahlert (1966) observed that the spiracle is "cleaned" in bony fishes by flooding it with water, but this does not mean that it has anything to do with a water-breathing mechanism. In combination, these observations are not necessarily in conflict: as in *Polypterus*, *Eusthenopteron* and other stem-tetrapods may well have used the spiracle for inhaling fresh air to supply the lungs with extra oxygen whenever required. The consistent presence of the spiracle in all stem-tetrapods highlights its importance for these still-aquatic animals.

The stem-tetrapod stapes. The osteolepiform hyomandibula articulated with the posterior braincase by a hinge joint with two vertically

Figure 5.5 The tetrapod stapes has a complicated evolutionary history. Starting with a small and massive bone in stem-tetrapods (D, *Acanthostega*, adapted from Clack 1998), it persisted as such in the amniote stem lineage (E, anthracosaur *Proterogyrinus*, adapted from Clack 2012; F, microsaur *Asaphestera*, adapted from Carroll and Gaskill 1978). In temnospondyls (C, *Sclerocephalus*, adapted from Schoch and Witzmann 2009), the stapes was much longer and more lightly built, with an additional joint to the floor of the braincase (ventral process). In batrachian lissamphibians, an additional element (operculum) was added: (A) caudate *Ranodon sibiricus*; (B) anuran *Conraua goliath*.

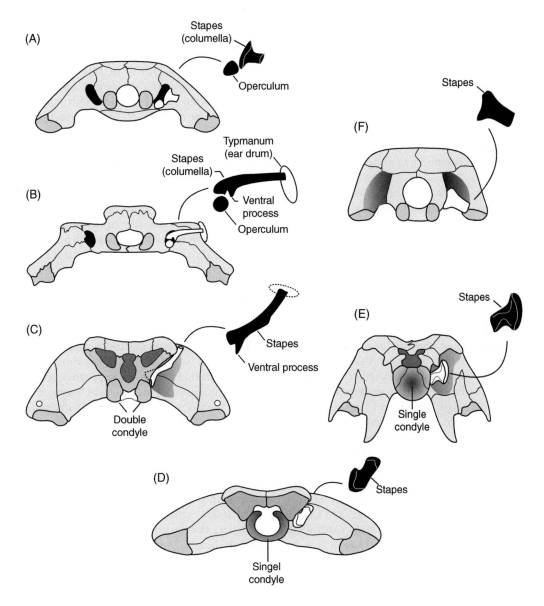

aligned facets (Jarvik 1980; Carroll 1980). Ventral to these, a slit-like opening (vestibular fontanelle) exposed the inner ear. In *Acanthostega*, the stapes had attained the characteristic shape of the ear ossicle in stem-tetrapods and anthracosaurs: a single articulation with the braincase by a large footplate, a short stylus, and a fan-shaped distal end, obviously forming attachment for ligaments and/or musculature (Figure 5.5D). Rather than hinging at the lateral wall of the braincase, the footplate of this stapes was connected to the margins of the vestibular fontanelle. This opening was enlarged in *Acanthostega* and had assumed a round outline. Two formerly separate components – hyomandibula and vestibular fontanelle – were thus linked in tetrapods, where they are known as the stapes and oval window. Only by this coupling could the stapes become a transmitter of vibrations to the inner ear.

This stapes – here referred to as the *cheek-anchored* type – resembles the ear ossicle of many extant salamanders by sharing the following features: (1) footplate without a second process, (2) absence of dorsal rod-like process correlating with the lack of a tympanum, and (3) broadened distal end facing the squamosal or quadrate. In salamanders, this cheek-anchored stapes transmits low-frequency vibrations to the inner ear (Wever 1985). The presence of a similar type of ear ossicle, with marked attachment sites along the broadened distal end, suggests that the early tetrapod stapes was also connected to the squamosal and served as a low-frequency transmitter. Stapes of this type are found in colosteids (*Greererpeton*), baphetids (*Kyrinion*), anthracosaurs (*Palaeoherpeton*, *Pholiderpeton*), and the whatcheeriids (*Pederpes*) (Clack 2003).

This scenario adds the low-frequency hearing function to other roles already suggested for the early tetrapod stapes. Carroll (1980) held that this type of stapes supported the braincase, and Clack (1992) added that it might also have controlled movements of spiracular air breathing. There is no reason why the cheek-anchored stapes might not have performed all these functions together. In salamanders at least, it acts both as a brace between the cheek and braincase and as an ear ossicle (Iordansky 1990 reported slight mobility of cheek and braincase in salamanders). This line of thought requires further considerations of the salamander stapes: its similarity to that of stem-tetrapods is not readily apparent from phylogeny. Below I will outline a hypothesis explaining the "re-appearance" of the cheek-anchored stapes in salamanders and caecilians, which appears to be in strong contradiction to the hypothesized dissorophoid ancestry of lissamphibians.

Clack's (1992) interpretation of the stapes and spiracle in *Acanthostega* is supported by recent findings on the middle ear region of *Ichthyostega* reported by Clack (2012). In this slightly more tetrapod-like taxon, the stapes was unique, with a huge blade-like distal portion that apparently attached to the medial wall of the spiracle. Such a construction resembles the Weberian ear ossicles of ostariophysean teleost fishes, which use their air-filled swim bladder as a hearing organ. If this analogy holds, then *Ichthyostega* had an air-filled spiracle that was already employed as a hearing organ. In contrast to anurans and amniotes, the stapes was not inside the spiracle (which was therefore not yet a middle ear cavity) but attached to its medial margin. If this functional interpretation is correct, then *Ichthyostega* would testify to the air-filled status of the spiracle in limbed tetrapodomorphs. However, the functional solution in *Ichthyostega* is best considered a unique condition, because *Acanthostega* and more crownward stem-tetrapods had a range of different stapes which were all cheek-anchored. Such a stapes is unlikely to have supported spiracular hearing, because the ossicle was more massive than in *Acanthostega* and probably only suited to the transmission of low-frequency sounds.

Lepospondyl stapes. Following Carroll and Gaskill (1978), the stapes of lepospondyls was largely comparable with the cheek-anchored stapes of stem-tetrapods. It was always short and stout, with a single head articulating with the oval window, and a single distal end usually contacting the quadrate (Figure 5.5 F). As lepospondyls lack a temporal notch, this condition recalls the situation in salamanders and caecilians. Thus, lepospondyls appear to have abandoned the spiracle and had no middle ear either. By analogy with salamanders and caecilians, the short stapes acted as a brace between cheek and braincase and transmitted low-frequency vibrations.

Temnospondyl ears. In temnospondyls, the stapes attained a different morphology from that of all other early tetrapods and must have performed a different function. Three features (Figure 5.5C) characterize the great majority temnospondyl ears: (1) there is a ventral process, clearly offset from the footplate, which articulates with the braincase or its floor; (2) the main body of the ossicle is elongate, delicate, and anteroposteriorly compressed; and (3) the distal end is rounded and ends in or near the temporal notch (Bolt and Lombard 1985). A survey of the well-known temnospondyl stapes confirms that most taxa share all three features, albeit showing a wide range of structural diversity. These three features are also found in anuran stapes (Bolt and Lombard 1985), and it is hard to envision how such a degree of anatomical consistency could have evolved by convergence, as suggested by some authors (Laurin 1998; Laurin and Soler-Gijón 2006). In temnospondyls, the cited features are present in temnospondyls of very different size, ranging from 1.4 cm long skulls of *Doleserpeton* to 1.4 m long skulls of *Mastodonsaurus*. In the large stereospondyls, the stylus is very long and oriented dorsally, pointing into the large circular temporal notch.

Unlike the situation in stem-tetrapods, the stapes was different in taxa that had lost the temporal notch: the Permian genus *Dvinosaurus* had a short, blade-like stapes with a single head and a cheek-anchored distal end. The (probably ligamentous) connection to the squamosal was maintained by the quadrate process, which was aligned laterally. The dorsal process, or stylus proper, was short and did not reach the temporal region. Lacking a temporal notch, *Dvinosaurus* recalls the situation in salamanders, which also lack a temporal notch and have a cheek-anchored stapes. Similar stapes are present in the stereospondyls *Batrachosuchus* and *Gerrothorax*, both also lacking a temporal notch. These data suggest that there was a link between the morphology of the stapes and the presence of a temporal notch. If the notch was present, the stapes was rod-like and pointed into the notch (the basal temnospondyl *Edops* forming an exception).

Origin of the tympanic ear: dissorophoids and frogs. In amphibamid dissorophoids, the stapes is especially similar to that of anurans. This may reflect the tiny size of both taxa but also documents shared derived characters: rather than dorsally, the stapes is directed more laterally. The relatively short stylus and its attachment to the proportionally very large tympanum are further shared features. In amphibamids, it is plausible to consider the temporal notch as having housed a tympanum. It is greatly enlarged, very similar to that of frogs, and preserves traces of soft tissue that attached to it. In addition, the quadrate forms a dorsal extension that appears to have supported a roundish structure that encircled the temporal region. In anurans, such a structure is present and referred to as the tympanic annulus. Interestingly, this cartilaginous ring develops from the quadrate, recalling the dorsal extension of dissorophoids (Bolt and Lombard 1985). Together with the pedicellate dentition, this set of features provides the most convincing evidence for lissamphibian relationships.

Poor preservation in the braincase region has so far precluded the study of the ear capsule. In salamanders and frogs, it houses a large opening in which a second ear ossicle is located, the *batrachian operculum*. As mentioned earlier, this cartilaginous element ossifies during metamorphosis and is connected by a muscle to the scapula. This second and independent hearing apparatus has not been found in any Paleozoic tetrapod. In most temnospondyls, the ear capsule was concealed from the occiput by the exoccipital, and there was no room for the attachment of an opercularis muscle. In amphibamids the condition may have been different (Sigurdsen and Bolt 2010), but this region is usually heavily crushed in the delicate fossils. The present state of knowledge indicates that a batrachian operculum was not present in any amphibamid. In the putative stem-batrachian *Gerobatrachus*, the braincase is mostly absent (J.S. Anderson, personal communication 2012).

At any rate, the dissorophoid stapes is likely to have formed part of a tympanic ear. The remaining problem is whether the various other temnospondyls (1) were tympanate as in dissorophoids, (2) were atympanate with a spiracular breathing apparatus as in stem-tetrapods, or (3) possessed some other kind of spiracular system. The problem can be constrained by the observation that in many temnospondyls the stapes is associated with the temporal notch – a feature distinguishing them from other early tetrapods. Milner and Sequeira (1994) and Robinson *et al.* (2005) have shown that the stapes of the early temnospondyls *Balanerpeton* and *Dendrerpeton* were already delicate and consistent with that of dissorophoids, suggesting they served as sound transmitters in a manner similar to the dissorophoid stapes.

This may be only half the story, however. Temnospondyls provide much insight into development, which had an important impact on the morphology of the stapes. In *Sclerocephalus*, larvae had a short and undifferentiated stapes, resembling that of adult stem-tetrapods (Boy 1988). The larval stapes had a prominent quadrate process and a single-headed proximal end, and it probably was cheek-anchored. In contrast, the adult stapes was elongate and delicate, with a ventral process articulating with the parasphenoid, and the distal end pointing into the temporal notch (Schoch and Witzmann 2009). Although restricted to a single taxon, this evidence indicates that temnospondyl stapes underwent ontogenetic modification, and this will be of importance for the interpretation of salamander stapes.

Loss of the tympanic ear: salamanders and caecilians. The absence of the middle ear cavity and tympanum poses a substantial problem for any evolutionary scenario of tetrapod hearing. One reason is that it involves the loss of an apparently hard-won set of characters that are otherwise "good" or "convincing" tetrapod autapomorphies. Another reason is usually not highlighted in studies confined to extant tetrapods: many salamander stapes resemble the primitive tetrapod condition in the morphology of the ear ossicle and its connection to the cheek.

Given that salamander ears evolved from dissorophoid ears, there are two possible scenarios in which the salamander condition might have evolved: either (1) by a complete loss of the middle ear for functional reasons (because it disturbed other important functions or became obsolete through some unknown behavior) or (2) by a slow-down of its development (reduction by pedomorphosis), facilitated by the presence of the opercular apparatus which took over the functional properties of the tympanic ear. In other words, either the tympanic ear *had to* be reduced or its loss

was a by-product of some other change that was readily compensated by an alternative mechanism.

The pedomorphosis scenario has two advantages: it provides an evolutionary mechanism by which the reduction might have proceeded, and it takes account of the resemblance between salamander and stem-tetrapod stapes. The loss hypothesis is supported only by the fact that all salamanders lack the middle ear cavity, eustachian tube, and tympanum. There is no intermediate condition between the salamander and anuran/amniote conditions. This suggests that salamanders passed through an evolutionary stage in which the middle ear had to be abandoned – perhaps as in the burrowing amphisbaenians. Thus, it remains unknown how the reduction occurred, and which steps it involved. This weakens the second hypothesis, which is otherwise more elegant than the loss scenario. Pedomorphosis could explain why salamanders have a stapes but no middle ear or tympanum, because in frogs and amniotes the stapes starts to form relatively early in development, whereas the middle ear develops only during metamorphosis in anurans. Some anurans have also lost the tympanum (Smirnov and Vorobyeva 1988). Furthermore, the rudimentary appearance of the stapes in salamanders is consistent with early ontogenetic stages of stapes in anurans and other groups. This could imply that the middle ear developed at a slower rate than the rest of the body in salamanders.

Whereas pedomorphosis would give a neat picture of developmental evolution, the adaptive reason behind such a heterochronic shift remains completely unclear. The pedomorphosis scenario is supported by the general pedomorphic appearance of many salamanders when compared to Paleozoic tetrapods or amniotes, and by the frequent occurrence of neoteny, an adaptive strategy involving pedomorphosis. If salamanders originated by neoteny, the "incomplete" status of their skeleton would be easier to understand: the absence of skull and girdle bones is consistent with the absence of middle ear structures.

Caecilians are a second group that retain the stapes but lack all other middle ear components (Maddin *et al.* 2012). However, considering their burrowing mode of life, the reduction of the middle ear is easier to understand than in salamanders. By analogy with amphisbaenians, the burrowing lifestyle required a massive skull and the reduction of sound perception to low-frequency vibrations in the ground. The vocalization that characterizes frogs plays no role in caecilians or salamanders.

5.5 Respiration in early tetrapods

It is no coincidence that the loss of the opercular and gular elements marks the climax of the fish–tetrapod transition: it signals the decoupling of feeding and breathing mechanisms. At about the same time, other organs of breathing appeared or were modified from those that already existed: (1) external gills for aquatic breathing in early tetrapod larvae, (2) cutaneous respiration employed preferably in small tetrapods, (3) the evolution of more efficient lungs, and (4) the establishment of two divergent air-pumping mechanisms for lung ventilation: costal inhalation in stem-amniotes and buccal pumping in stem-amphibians.

The buccal pump, in its most primitive version, was the original mode by which the first tetrapods breathed air, whereas the aspiration pump of amniotes is the derived mechanism (Brainerd 1994). Extant amphibians assume an intermediate position in using the buccal pump for inhalation (where the mouth cavity is compressed) and the trunk musculature for exhalation (Brainerd 1999). *Polypterus*, *Amia*, and *Lepisosteus* also use buccal pumping (Brainerd 1994), and because extant lungfishes do the same the mechanism is regarded as an osteichthyan autapomorphy. A major innovation between the dipnoan and tetrapod nodes was the use of nares for inhalation that can be closed when the air is pumped into the lung (Gans 1970). Tetrapods were the first to use the hypaxial musculature to force air out of the lungs; while lissamphibians retained this in combination with the buccal pump, amniotes largely replaced the buccal inhalation by movement of the ribs. However, buccal pumping was not entirely given up, because at least lepidosaurs still practice it in addition to rib movements (Brainerd 1999).

By employing external gills, vascularized epidermis and skin folds, enlarged lungs, and modified pumping mechanisms, the first tetrapods did not just *transform* the existing breathing mechanisms, but also *diversified* the options for air breathing. The huge benefit of this diversification was evolutionary flexibility, required to cope with the complicated habitats at the water–land interface. The cost of this flexibility was the loss of the very successful opercular pump, which in turn led to the morphological changes documented in the fish–tetrapod transition.

Stem-tetrapods. Although evidence is still scarce, most stem-tetrapods appear to have retained internal gills that were attached to the branchial arches. The dorsal part of these arches (pharyngobranchials, epibranchials) eventually disappeared in crown tetrapods (Schmalhausen 1968; Clack 2012), but atavistic re-appearances have been reported from salamanders (Reilly and Lauder 1988). Ossified branchial arches are well preserved in *Acanthostega* and *Ichthyostega*, where they share the skeletal correlates with the internal gills of bony fishes (Coates and Clack 1991; Schoch and Witzmann 2011). Colosteids had numerous dentigerous plates in the region where the gill slits were located, indicating at least a water-filled pharynx. Nothing is known about lungs – whose existence is inferred by the EPB – or the first origin of larval gills. Their presence in both stem-amniotes (seymouriamorphs) and stem-amphibians (temnospondyls) indicates their status as a tetrapod synapomorphy. However, such gills were not present in larvae of *Eusthenopteron* (Schultze 1984), which developed the opercular bones early in ontogeny. Dipnoans thus must have evolved their external larval gills independently (Witzmann 2004). This is also indicated by the absence of larval gills in *Neoceratodus*, the basalmost of the modern lungfishes (Schoch and Witzmann 2011). Probably stem-tetrapods developed their internal gills early in ontogeny and kept them throughout their aquatic lives.

Respiration in temnospondyls. Temnospondyls evolved terrestrial forms early in their phylogeny, probably starting with small taxa such as *Dendrerpeton*. Lung breathing must have played an important role in these early amphibious forms. Indeed, a major temnospondyl character – the large palatal windows – indicates an enhanced form of buccal pumping: as in modern amphibians

(Brainerd 1999), air was sucked in by the nares and swallowed by lowering the skin in the palatal windows (Clack 1992). The existence of such a *palatal buccal pump* is thus indicated by anatomical correlates, especially the extensive palatal openings and insertion sites for a large retractor bulbi muscle that retracts the eyeballs. Air breathing was thus a driving factor of temnospondyl morphology – but by no means the only method of respiration evolved by the group. The abundance of hyobranchial skeletons with skeletal correlates of internal gills in adults and the preservation of external gill filaments in many clades indicate that gills played an important role in temnospondyls (Witzmann 2004; Schoch and Witzmann 2011). Based on these data, various aquatic taxa retained internal gills (dvinosaurs, stereospondyls). Likewise, many temnospondyl larvae breathed with external gills (dvinosaurs, dissorophoids, eryopids, stereospondylomorphs), irrespective of the adult breathing mechanism. Finally, cutaneous respiration might have been practiced at least by the miniaturized amphibamids.

Apparently, each temnospondyl clade had its own mix of respiratory mechanisms: buccal respiration with lungs, skin breathing, external gills in larvae, and internal gills in adults of some groups. This situation may explain the taxonomic diversity (in terms of species numbers), in contrast to the much smaller diversity of stem-amniotes. If lissamphibians are indeed temnospondyls, the diversity of respiratory mechanisms in lissamphibians would simply conserve the condition of their temnospondyl stem-group. In the alternative case, temnospondyls and lissamphibians would have evolved respiratory diversity in parallel. The main restriction in lissamphibians is the loss of internal gills. Again, this evolutionary flexibility came at a price: the conservation of buccal pumping as an inhalation mechanism appears to have been a limiting factor for the evolution of higher and constant metabolic rates (Perry and Sander 2004) – probably one of the reasons why lissamphibians never evolved endothermy.

Skin breathing and the loss of lungs. Lissamphibians have repeatedly lost lungs in situations where they were in conflict with other organs or a particular mode of life. For instance, the speciose plethodontid salamanders all lack lungs and also abandoned the larval gill-breathing phase. Instead, they have put all their efforts into skin breathing. It is hypothesized that plethodontid ancestors invaded fast-flowing streams, in which the possession of lungs would have been a threat to swimming (Wake 2009). Consequently, lungs were completely reduced, which imposed a strong size constraint on these caudates. The frequent evolution of lunglessness in other salamanders, anurans, and caecilians highlights that skin respiration forms a strong attractor for selection, albeit at the cost of losing respiratory flexibility.

Stem-amniotes. Like the temnospondyls, stem-amniotes are defined (in part) by a skeletal correlate of breathing: the elongated ribs and the rib basket they span (Janis and Keller 2001). Inhalation of air by expansion of the rib cage is likely to have been practiced by anthracosaurs, chroniosuchids, seymouriamorphs, and lepospondyls. The postcrania of these forms were substantially more robust and completely ossified, even in the tiniest taxa, suggesting either longer land excursions (anthracosaurs) or a terrestrial mode of life (seymouriamorphs, microsaurs). At the same time, hyobranchial skeletons are absent in most stem-amniotes and external gills are known only in larval seymouriamorphs. This indicates that internal gills had been lost, external gills were confined to larvae, and most taxa had no clear-cut larval phase. Stem-amniotes were clearly less flexible regarding their "toolkit" of breathing mechanisms.

What was the driving factor of costal aspiration? Perhaps it was the loss of the internal gill option and the necessity for an effective air-breathing mechanism in more terrestrial taxa. But why did buccal pumping not suffice? A common correlation in stem-amniotes is that the bodies were much longer than those of most temnospondyls. This has significant functional implications for buccal pumping: as evidenced by caecilians, elongate bodies require more buccal pump cycles (Brainerd 1999). In small animals or taxa with a low metabolism, such as lissamphibians, this does not pose a problem. However, buccal pumping may not be sufficient in larger and/or more active animals, and costal aspiration becomes an attractive alternative for them.

5.6 The evolution of terrestrial feeding

Stem-tetrapods and anthracosaurs. In their study of cranial sutures, Markey and Marshall (2007) concluded that *Acanthostega* fed in a different way than *Eusthenopteron* and *Polypterus*. They did not question the aquatic mode of life in stem-tetrapods, but their suture data show that the skull roof of *Acanthostega* experienced similar forces to that of *Phonerpeton*, an undoubted terrestrial feeder among the dissorophoid temnospondyls. The terrestrial bite thus first evolved under water. This is not surprising, as the opercular pump was already absent in *Acanthostega*, whereas enhanced suction feeding by four-bar hyobranchial depression evolved only much later in salamanders. Thus, neither the hyobranchium nor the former opercular region were able to create sufficient suction for a suck-and-gape feeding strike. In contrast, jaw prehension appears to have been the mode by which early tetrapods fed, irrespective of their mode of life and preferred habitat. The feeding apparatus of anthracosaurs appears primitive in the retention of mobility between palatoquadrate and braincase and the closed, fully dentigerous bony palate. The skull– with the exception of the crocodile-like *Anthracosaurus* – was deep-flanked with substantial attachments for jaw adductors along the pterygoid and an exceptionally deep mandible. At the same time, no ossified hyobranchial elements have been found associated with anthracosaur skulls, suggesting that they had abandoned any significant involvement of the hyoid and branchial arches in feeding. This probably means that suction was even less important than in *Acanthostega*, but it is not clear what to conclude from that. Lateral lines and bodily proportions suggest that at least some anthracosaurs were aquatic, but did they all feed in the water? There is no reason why the moveable palatoquadrate could not have worked outside the water, although it is not clear what effect it might have had. Panchen (1970) suggested that they were feeding in the water.

Temnospondyls. The dentition was very conservative in this vast clade, consisting of large tusks and numerous smaller teeth. Unique for temnospondyls and lissamphibians is the open palate, in which the pterygoid, palatine, and ectopterygoid form thin strips of bone bordering large palatal openings. Extant frogs and salamanders withdraw the eyes when swallowing large prey items, which is made possible by these openings. In temnospondyls, small bony plates paved the palatal openings. These bear small teeth that are directed posteriorly, apparently assisting swallowing larger prey items by pushing them further into the pharynx. The palatal openings, the tooth-bearing plates, and a broad attachment site for the eye-retracting muscle (shared by batrachians and temnospondyls) constitute a form–functional complex. Originally, the combination of large eyes and a flat skull required the enlargement of the palatal opening in temnspondyls, but the driving factor was probably air breathing (buccal pump).

Dissorophoids and zatracheids, the most terrestrial temnospondyls as inferred from their postcranial morphology and associated tracks, also reveal interesting patterns in dentition and the morphology of hyobranchia. Dissorophids and trematopids had enlarged, markedly curved fangs with which they probably grasped larger prey items. Some amphibamids evolved pedicellate teeth, which in lissamphibians are used in feeding on small terrestrial invertebrates (Duellman and Trueb 1994). Dissorophoids had a fontanelle in the anterior palate similar to that of salamanders, which houses the intermaxillary gland that produces sticky secretions aiding in attaching prey to the tongue. Zatracheids had numerous tiny teeth and a huge fontanelle, indicating extensive use of such secretions. They also had an elaborate hyobranchium, which consisted of numerous thin rods. This skeletal structure was present only in metamorphosed specimens, and it forms the first evidence that the hyobranchial skeleton was involved in supporting the tongue – a common condition in salamanders, where a projectile tongue evolved in plethodontids.

Many temnospondyls preferred more or less aquatic modes of life, although it is difficult to specify what their habitats were like. A consistent feature of these taxa is the possession of partially ossified hyobranchial skeletons, notably a large and robust basibranchial. Phylogenetic bracketing indicates that this element is embedded in a sheet of muscles connecting the mandible with the pectoral girdle and hypaxial muscles (sternohyoideus, branchiohyoideus, and geniohyoideus). The sternohyoideus muscle acts as a mandibular depressor in both bony fishes and aquatic salamanders, and it attaches along the basibranchial (Allis 1897; Lauder and Shaffer 1985). The presence of this element does therefore not have implications for the existence of branchial arches and gills, as suggested earlier (Boy 1974), but it shows that the mouth-opening mechanism employed the visceral musculature. As this mechanism works primarily under water, it indicates aquatic feeding. Consistent with the occurrence of a robust basibranchial is the evidence of water-filled pharynx and gill slits, indicated by denticular plates in the gill region. This evidence is found in *Sclerocephalus*, *Glanochthon*, *Archegosaurus*, rhinesuchids, and various stereospondyls.

5.7 Transforming fins into limbs

Hall (1999) formulated the four main steps in the evolutionary transformation from fin to limb: (1) the dermal fin rays were lost, (2) the endoskeleton was modified, (3) the distal endoskeleton differentiated to form proper joints for wrist and ankle, and (4) new endoskeletal elements, the digits, appeared (Figure 5.6). The fourth step has been considered the most important by all authors, giving rise to the hand and foot of tetrapods, which are referred to as *autopodia* (Greek, meaning aptly "the foot itself"). Autopodia consist of five or more digits, and they differ from the distal portion of fish fins in a topological rather than a functional feature: their position relative to the main limb axis. This axis is called the "primary axis" (Shubin and Alberch 1986; Wagner and Larsson 2007). The distal elements that branch off the primary axis in fishes are called radials. Radials are located on the anterior margin of the primary axis (*pre-axial*), digits along the posterior margin (*post-axial*). By simple morphological standards, radials and digits cannot be homologized, but rather they form heterotopic structures. On closer inspection, however, the distinction becomes less clear, and the currently available evidence remains somewhat ambiguous.

The primitive condition. All gnathostomes share limbs, which first evolved as paired appendages containing a cartilaginous or bony endoskeleton and a bony, enamel-bearing exoskeleton. The question of homology between fins and limbs puzzled Geoffroy St.-Hilaire (1807), who studied *Polypterus*. Today, it is beyond any doubt that paired fins are the homologs of tetrapod limbs, but the homology between the skeletal elements of actinopterygians and sarcopterygians remains controversial (Janvier 1996). The first, proximal endoskeletal element is probably homologous throughout sarcopterygians (humerus/femur in tetrapods). The other parts of the endoskeleton are more controversial.

Digits or not? *Eusthenopteron, Panderichthys, and Tiktaalik*. It is still controversial at which phylogenetic node the digits originated, how digits can be identified at all, and whether individual digits can be homologized between tetrapodomorphs (Johanson *et al.* 2007; Wagner and Larsson 2007; Clack 2009; Swartz 2012). *Eusthenopteron* and other tristichopterids appear not to have had digits; the limb contains only pre-axial radials. Recently, Boisvert *et al.* (2008) reported evidence for rudimentary digits from CT scans of the forelimb in *Panderichthys*, where they identified four small ossicles in the distal part of the limb as digit-like structures. In contrast to more basal taxa, *Tiktaalik* appears to have had both pre- and post-axial radials branching off the primary limb axis (Wagner and Larsson 2007). The post-axial radials might be homologous with digits, but it remains unclear whether they can be homologized with those of more crownward stem-tetrapods.

Digits first evolved before the full complement of wrist and ankle bones was reached (Johanson *et al.* 2007). This forms a discontinuity between the proximal and distalmost elements of the limb, and such a gap is also known to exist in salamander limb development (Fröbisch 2008). The radials in the forelimb of *Tiktaalik* were studied in great detail by Shubin *et al.* (2006), who concluded that individual elements permitted flexion and rotation similar to the wrist joint in tetrapods. This suggests that the wrist might have preceded the elbow in acquiring the required flexibility for terrestrial locomotion.

Figure 5.6 Transition from fins to limbs in tetrapodomorphs: (A) hindlimb; (B) forelimb. Adapted from Clack (2012) and Boisvert *et al.* (2008). The autopod (hand and foot skeleton) is probably a new structure, without homologs in most bony fishes.

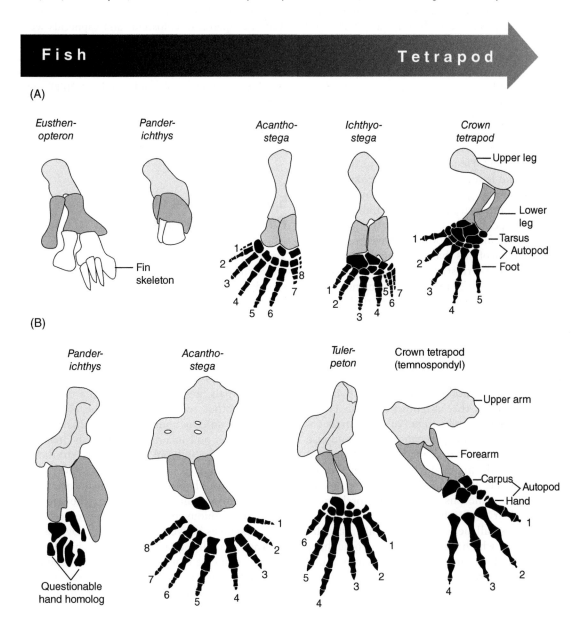

Polydactyl stem-tetrapods. In contrast to the apparently aquatic *Acanthostega*, *Ichthyostega* has always been viewed as an animal capable of crossing land bridges (Gregory and Raven 1941; Jarvik 1996). It had fully ossified, robust limbs and girdles, heavy ribs, and regionally differentiated neural arches. However, it now appears to have been an aquatic animal, and a recent analysis of three-dimensional limb joint mobility sheds light on its walking capabilities. Pierce *et al.* (2012)

found that *Ichthyostega* did not employ typical tetrapod locomotory behaviors such as lateral side walking. It lacked the necessary rotary motions in its limbs to push the body off the ground and move the limbs in an alternating sequence. The hindlimbs were not suited for locomotion on land, whereas the forelimb movements at best permitted mudskipper-like "crutching" motions. The forelimb thus appears to have taken the lead in the evolution of land locomotion, with the hindlimb first forming a propulsive adjunct of the tail for swimming before it was recruited to enable walking on land. Hence, the ability to rotate the humerus and femur along their long axis and to use symmetrical gaits appears to have evolved above the *Ichthyostega* node. This indicates that the limbs, specifically the polydactyl autopodia, were not used for locomotion on land, but formed a novel version of fins that were only later exapted to permit longer terrestrial excursions.

However, as is often the case in science, there is an alternative perspective on the problem, this time arising from the study of locomotory behavior in lungfishes (King *et al.* 2011). These authors found that the African lungfish *Protopterus* uses pelvic fin-driven, tetrapod-like gaits, including walking and bounding, under water. This finding agrees with the locomotory model of *Ichthyostega* in that it reveals the assignment of the Polish Middle Devonian tracks to polydactyl tetrapodomorphs as unwarranted, because they could also stem from more basal tetrapodomorphs. However, it disagrees with this model in the role played by the pelvic girdle and hindlimb in stem-tetrapods. Either tetrapodomorphs had the behavioral capacity to use their hindlimbs in walking under water, despite the small pelvis and its weak muscular support – in which case, behavior would have taken the lead and anatomy followed only much later (above *Ichthyostega*) – or, alternatively, lungfishes evolved this capacity independently from tetrapods, and stem-tetrapods were indeed not capable of using their hindlimbs for walking.

Pentadactyl stem-tetrapods. In contrast to temnospondyls and lissamphibians, the five-digited (pentadactyl) stem-tetrapods had relatively well-ossified limbs and girdles, providing attachments for muscles involved in terrestrial locomotion. Nevertheless, they had elongate and high swimming tails and lateral-line systems, indicating a primarily aquatic existence. Although very diverse in body shape, limbed stem-tetrapods from *Tulerpeton* to colosteids and baphetids were probably much better swimmers than crawlers. This is consistent with the occurrence of mass accumulations of *Whatcheeria* in sediments of a small water body (Bolt and Lombard 2000), and similar but less numerous finds were reported for *Acanthostega* (Clack 2012). This indicates that the animals preferred to remain in the water even when the habitat was shrinking.

5.8 Locomotion of Paleozoic tetrapods

Temnospondyls. Various types of aquatic and terrestrial locomotion can be inferred for temnospondyls, which managed to explore numerous habitats in different pulses of evolutionary radiation. *Dendrerpeton* and *Balanerpeton* probably best exemplify the primitive condition, in which small, salamander-like taxa dwelled in Carboniferous forests and wetlands. The fully ossified girdles and limbs include the coracoids, pubes, carpals, and tarsals – elements unossified in many other temnospondyls. The absence of lateral-line sulci is consistent with this inferred mode of life, and locomotion probably involved moderate body and tail undulations similar to those of caudates. *Cochleosaurus* and *Edops* exemplify an early offshoot, with the former being an able tail-undulating swimmer, the latter a heavy amphibious predator capable of crossing short distances between water bodies. Dvinosaurs returned to a fully aquatic mode of life, highlighted by their elongate trunks (> 28 vertebrae) and tails and feeble limbs. Marked lateral undulations are the most likely mode of locomotion. The Permian genera *Archegosaurus*, *Sclerocephalus*, *Onchiodon*, and *Eryops* form a wide range of skeletal types, with *Archegosaurus* having the most incompletely ossified and gracile skeleton, *Eryops* the most heavily ossified one. Their considerable size of 1–2 m suggests that they were crocodile-like predators with variable capacity to leave the water. Although developing slowly from gill-bearing larvae in lakes, *Onchiodon* and *Eryops* had fully ossified limbs and girdles and a complicated humerus structure, indicating powerful forelimb muscles (Miner 1925). The terrestrial dissorophoids were not just miniature versions of eryopids, but really form a novel and unique locomotory morphotype. The dissorophids are perhaps the most interesting in terms of locomotion, as their trunks were very foreshortened, the skull disproportionally large, and the axial skeleton covered by a carapace of double osteoderms. Their limbs were more slender and substantially longer than in *Eryops*. Dilkes and Brown (2007) analyzed this remarkable construction in the dissorophid *Cacops*, suggesting that vertical flexion was more likely than lateral undulation, supported by the mobility of the double osteoderm series. They proposed that the 50–80 cm long animals ran in short spurts, forming an analog to the extant Natterjack toad *Bufo calamita*. Other taxa with heavier carapaces (*Dissorophus, Broiliellus*) might have used different modes of walking. The tiny amphibamids also had short trunks, but more slender limbs and no osteoderms, and probably practiced a symmetrical walk like salamanders and crocodiles, where the body is supported by diagonally opposite movements of the fore- and hindlimbs (Figure 5.7). The Mesozoic stereospondyls returned to a permanently aquatic existence, evolving (1) large and elongate crocodile-like predators (rhinesuchids, capitosaurs, metoposaurids), (2) deep-bodied newt- to eel-like forms (trematosaurids), and (3) flat- and short-bodied forms similar to extant giant salamanders and flatfishes (brachyopoids, plagiosaurids). All these taxa had poorly ossified limb and girdle bones, lacked coracoids and pubes, often lacked carpals and tarsals, and the humerus had poorly differentiated muscle attachment sites. Groups 1 and 3 had extraordinarily heavy skeletons, by both morphological and histological measures. This suggests that they were aquatic bottom-dwellers, and indeed they appear to have been poor swimmers. Instead, group 2 was lightly built and probably swam by lateral undulation of the elongate tail. Voigt (2012) has summarized the existing knowledge of putative temnospondyl tracks, the larger of which are readily identified by their four-digited hand impressions. These match

the anatomy of eryopids (e.g., *Onchiodon*), while the smaller ones are consistent with the anatomy of amphibamids and metamorphosed branchiosaurids. Large Triassic tracks, known as *Capitosauroides*, are rare and suggest very slow and sluggish motion of stereospondyls.

Lepospondyls. This assemblage is remarkable, because four out of six lepospondyl clades evolved eel-like bodies with limbs either greatly reduced (lysorophians) or entirely lost (aïstopods, adelospondyls, acherontiscids). Their divergent skull morphologies indicate that these groups are probably not intimately related within Lepospondyli, and the two large-scale phylogenetic analyses of Anderson (2001) and Vallin and Laurin (2004) have found radically different hypotheses of relationships between them. Anderson (2002) suggested that limblessness probably evolved convergently in these groups, highlighting general problems with phylogenetic analysis of such taxa. Assuming that lepospondyls really were a natural group, such parallel evolution may have been triggered by a common developmental–evolutionary framework: perhaps the general tendency of microsaurs and nectrideans to have disproportionately small limbs and elongate trunks with a high vertebral number was an easy starting point for the addition of vertebrae and further reduction of limbs by decreasing their growth rate (negative allometry). At any rate, the number of trunk vertebrae varies substantially more than in temnospondyls or other early tetrapod clades, suggesting that there was no obvious constraint on the number of vertebral segments in lepospondyls.

However, there were also quite different locomotory patterns in lepospondyls, represented by two important Permian groups: microsaurs and nectrideans. Microsaurs repeatedly evolved taxa with body shapes like those of modern land salamanders. Others resemble lizards, such as the heavy *Pantylus*, which had massive limbs and girdles. Consistent features of all these forms were their elongate trunks and tails, suggesting that they all practiced some form of lateral undulation. Nectrideans were short-trunked but had tall neural spines in their tails, which served as main propulsors in swimming. Diplocaulids form an exception, having dorsoventrally flattened bodies with a boomerang-like skull that might have acted as a hydrofoil.

Figure 5.7 Comparison of skeletons and tracks. (A) Late Devonian *Acanthostega* and eight-digited tracks (adapted from Clack 1997a, reproduced with permission of Elsevier). (B–D) Late Permian tetrapods: (B) seymouriamorph tracks of *Amphisauropus* in comparison to *Seymouria* (adapted from Voigt 2012 and Berman *et al.* 2000); (C) large temnospondyl tracks of *Limnopus* in comparison to *Sclerocephalus* (adapted from Voigt 2012 and Schoch and Witzmann 2009); (D) small temnospondyl tracks of *Batrachichnus* in comparison to an amphibamid (adapted from Voigt 2012 and Schoch and Rubidge 2005).

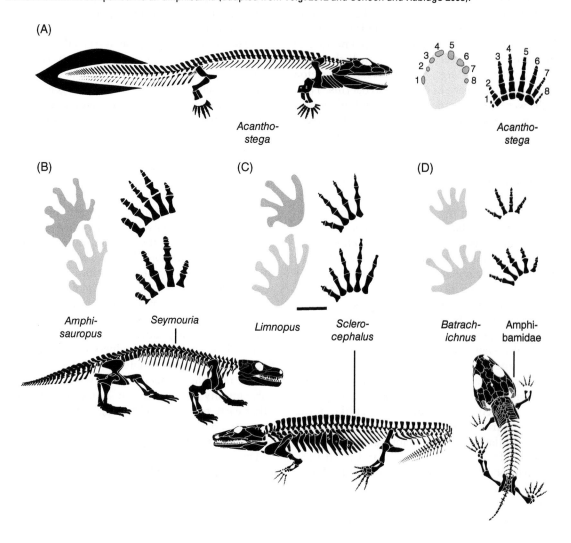

References

Allis, P. (1897) The cranial muscles and cranial and first spinal nerves in *Amia calva*. Journal of Morphology **12**, 487–762.

Allis, P. (1922) The cranial anatomy of *Polypterus*, with special reference to *Polypterus bichir*. *Journal of Anatomy* **56**, 189–291.

Anderson, J.S. (2001) The phylogenetic trunk: maximal inclusion of taxa with missing data in an analysis of the Lepospondyli (Vertebrata, Tetrapoda). *Systematic Biology* **50**, 170–193.

Anderson, J.S. (2002) Revision of the aïstopod genus *Phlegethontia* (Tetrapoda: Lepospondyli). *Journal of Paleontology* **76**, 1029–1046.

Berman, D.S., Reisz, R.R., Scott, D., Henrici, A., Sumida, S., & Martens, T. (2000) Early Permian bipedal reptile. *Science* **290**, 969–972.

Boisvert, C., Mark-Kurik, E., & Ahlberg, P. (2008) The pectoral fin of *Panderichthys* and the origin of digits. *Nature* **456**, 636–638.

Bolt, J.R. & Lombard, R.E. (1985) Evolution of the amphibian tympanic ear and the origin of frogs. *Biological Journal of the Linnean Society* **24**, 83–99.

Bolt, J.R. & Lombard, R.E. (2000) Palaeobiology of *Whatcheeria deltae*, a primitive Mississippian tetrapod. In: H. Heatwole & R.L. Carroll (eds.), *Amphibian Biology. Volume 4. Palaeontology*. Chipping Norton, NSW: Surrey Beatty, pp. 1044–1052.

Boy, J.A. (1974) Die Larven der rhachitomen Amphibien (Amphibia: Temnospondyli, Karbon-Trias). *Paläontologische Zeitschrift* **48**, 236–268.

Boy, J.A. (1988) Über einige Vertreter der Eryopoidea (Amphibia: Temnospondyli) aus dem europäischen Rotliegend (? höchstes Karbon – Perm). 1. *Sclerocephalus. Paläontologische Zeitschrift* **62**, 107–132.

Brainerd, E.L. (1994) The evolution of lung-gill bimodal breathing and the homology of vertebrate respiratory pumps. *American Zoologist* **34**, 289–299.

Brainerd, E.L. (1999) New perspectives on the evolution of lung ventilation mechanisms in vertebrates. *Experimental Biology Online* **4**, 11–28.

Brazeau, M. D. & Ahlberg, P. E. (2006) Tetrapod-like middle ear architecture in a Devonian fish. *Nature* **439**, 318–321.

Budgett, J.S. (1903) Notes on the spiracles of *Polypterus. Proceedings of the Zoological Society of London* **1903**, 10.

Carroll, R.L. (1980) The hyomandibular as a supporting element in the skull of primitive tetrapods. In: A.L. Panchen (ed.), *The Terrestrial Environment and the Origin of Land Vertebrates*. London and New York: Academic Press, pp. 293–317.

Carroll, R.L. & Gaskill, P. (1978) The order Microsauria. *Memoirs of the American Philospohical Society* **126**, 1–211.

Clack, J.A. (1992) The stapes of *Acanthostega gunnari* and the role of the stapes in early tetrapods.

In: D.B. Webster, R.R. Fay, & A.N. Popper (eds.), *The Evolutionary Biology of Hearing*. New York and Berlin: Springer, pp. 405–420.

Clack, J.A. (1997) Devonian tetrapod trackways and trackmakers; a review of the fossils and footprints. *Palaeogeography, Palaeoclimatology, Palaeoecology* **130**, 227–250.

Clack, J.A. (1998) The neurocranium of *Acanthostega gunnari* Jarvik and the evolution of the otic region in tetrapods. *Zoological Journal of the Linnean Society* **122**, 61–97.

Clack, J.A. (2003) A new baphetid (stem tetrapod) from the Upper Carboniferous of Tyne and Wear, U.K., and the evolution of the tetrapod occiput. *Canadian Journal of Earth Sciences* **40**, 483–498.

Clack, J. A. (2009) The fin to limb transition: new data, intepretations, and hypotheses from paleontology and developmental biology. *Annual Reviews of Earth and Planetary Sciences* **37**, 163–179.

Clack, J.A. (2012) *Gaining Ground: the Origin and Evolution of Tetrapods*, 2nd edition. Bloomington: Indiana University Press.

Coates, M.I. & Clack, J.A. (1991) Fish-like gills and breathing in the earliest known tetrapod. *Nature* **352**, 234–236.

Deban, S.M. & Wake, D.B. (2000) Aquatic feeding in salamanders. In: K. Schwenk (ed.), *Feeding: Form, Function, and Evolution in Tetrapod Vertebrates*. Boston: Academic Press, pp. 65–94.

Dilkes D.W. & Brown, L.E. (2007) Biomechanics of the vertebrae and associated osteoderms of the Early Permian amphibian *Cacops aspidephorus. Journal of Zoology* **271**, 396–407.

Downs, J.P., Daeschler, E.B., Jenkins, F.A., & Shubin, N.H. (2008) The cranial endoskeleton of *Tiktaalik roseae. Nature* **455**, 925–929.

Duellman, W.E. & Trueb, L. (1994) *Biology of Amphibians*. Baltimore: Johns Hopkins University Press.

Fröbisch, N.B. (2008) Ossification patterns in the tetrapod limb: conservation and divergence from morphogenetic events. *Biological Reviews* **83**, 571–600.

Gans, C. (1970) Respiration in early tetrapods: the frog is a red herring. *Evolution* **24**, 723–734.

Geoffroy St.-Hilaire, E. (1807) Premier mémoire sur les poissons, òu l'on compare les piéces osseuses de leurs nagoires pectorales avec les os de l'extrémité antérieure des autres animaux à vertèbres. *Annales du Muséum* **9**, 357–370.

Gregory, W.K. & Raven, H.C. (1941) Origin of paired fins and limbs. *Annals of the New York Academy of Sciences* **42**, 273–360.

Hall, B.K. (1999) *Evolutionary Developmental Biology*, 2nd edition. Dordrecht: Kluwer.

Iordansky, N.N. (1990) *Evolution of Complex Adaptations. The Jaw Apparatus of Amphibians and Reptiles.* Nauka, Moscow. (In Russian.)

Janis, C.M. & Keller, J.C. (2001). Modes of ventilation in early tetrapods: costal aspiration as a key feature of amniotes. *Acta Palaeontologica Polonica* **46**, 137–170.

Janvier, P. (1996) *Early Vertebrates.* Oxford Monographs on Geology and Geophysics 33. Oxford: Oxford Univesity Press.

Jarvik, E. (1954) On the visceral skeleton of *Eusthenopteron* with a discussion of the parasphenoid and palatoquadrate in fishes. *Kunglik Svenska Vetenskapsakademien Handlingar* **5**, 1–104.

Jarvik, E. (1980) *Basic Structure and Evolution of Vertebrates.* Vols. **1–2**. London and New York: Academic Press.

Jarvik, E. (1996) *The Devonian Tetrapod Ichthyostega.* Fossils and Strata 40. Oslo: Scandinavian University Press.

Johanson, Z., Joss, J., Boisvert, C.A., Ericsson, R., Sutija, M., & Ahlberg, P.E. (2007) Fish fingers: digit homologues in sarcopterygian fish fins. *Journal of Experimental Zoology* **308B**, 757–768.

King, H.M., Shubin, N.H., Coates, M.I., & Hale, M.E. (2011) Behavioral evidence for the evolution of walking and bounding before terrestriality in sarcopterygian fishes. *Proceedings of the National Academy of Sciences* **108**, 21146–21151.

Lauder, G.V. (1980a) Evolution of the feeding mechanism in primitive actinopterygian fishes: a functional anatomical analysis of *Polypterus, Lepisosteus,* and *Amia. Journal of Morphology* **163**, 283–317.

Lauder, G.V. (1980b) The role of the hyoid apparatus in the feeding mechanism of the coelacanth *Latimeria chalumnae. Copeia* **1980** (1), 1–9.

Lauder, G.V. (1982) Patterns of evolution in the feeding mechanism of actinopterygian fishes. *American Zoologist* **22**, 275–285.

Lauder, G.V. (1990) Functional morphology and systematics: studying functional patterns in an historical context. *Annual Review of Ecology and Systematics* **21**, 317–340.

Lauder, G.V. (1981) Form and function: structural analysis in evolutionary morphology. *Paleobiology* **7**, 430–442.

Lauder, G.V. & Shaffer, H.B. (1985) Functional morphology of the feeding mechanism in aquatic ambystomatid salamanders. *Journal of Morphology* **185**, 297–326.

Laurin, M. (1998) The importance of global parsimony and historical bias in understanding tetrapod evolution. Part I. Systematics, middle ear evolution, and jaw suspension. *Annales des Sciences naturelles* **19**, 1–42.

Laurin, M. & Soler-Gijón, R. (2006) The oldest known stegocephalian (Sarcopterygii: Temnospondyli) from Spain. *Journal of Vertebrate Paleontology* **26**, 284–299.

Lombard, R.E. & Bolt, J.R. (1979) Evolution of the tetrapod ear: an analysis and reinterpretation. *Biological Journal of the Linnean Society* **11**, 19–76.

Maddin, H., Jenkins, F.A., & Anderson, J.S. (2012) The braincase of *Eocaecilia micropodia* (Lissamphibia, Gymnophiona) and the origin of caecilians. *PloS ONE* **7**, e50743.

Markey, M.J. & Marshall, C.R. (2007) Terrestrial-style feeding in a very early aquatic tetrapod is supported by evidence from experimental analysis of suture morphology. *Proceedings of the National Academy of Sciences* **104**, 7134–7138.

Markey, M.J., Main, R.P., & Marshall, C.R. (2006) *In vivo* cranial function and suture morphology in the extant fish *Polypterus*: implications for inferring skull function in living and fossil fish. *Journal of Experimental Biology* **209**, 2085–2102.

Mickoleit, G. (2004) *Phylogenetische Systematik der Wirbeltiere.* Munich: Pfeil.

Miner, R.W. (1925) The pectoral limb of *Eryops* and other primitive tetrapods. *Bulletin of the American Museum of Natural History* **51**, 145–312.

Panchen, A.L. (1970) Anthracosauria. In: O. Kuhn (ed.), *Encyclopedia of Paleoherpetology*, Vol. 5A. Stuttgart: Fischer.

Perry, S.F. & Sander, M. (2004) Reconstructing the evolution of the respiratory apparatus in tetrapods. *Respiratory Physiology and Neurobiology* **144**, 125–39.

Pierce, S.E., Clack, J.A., & Hutchinson, J.R. (2012) Three-dimensional limb joint mobility in the early tetrapod *Ichthyostega*. *Nature* **486**, 523–526.

Reilly, S.M. & Lauder, G.V. (1988) Atavisms and the homology of hyobranchial elements in lower vertebrates. *Journal of Morphology* **195**, 237–245.

Reilly, S.M. & Lauder, G.V. (1990) The evolution of tetrapod feeding behavior: kinematic homologies in prey transport. *Evolution* **44**, 1542–1557.

Robinson, J., Ahlberg, P.E. & Koentges, G. (2005) The braincase and middle ear region of Dendrerpeton acadianum (Tetrapoda: Temnospondyli). *Zoological Journal of the Linnean Society* **143**, 577–597.

Rudwick, M.J.S. (1964) The inference of function from structure in fossils. *British Journal for the Philosophy of Science* **15**, 27–40.

Schmalhausen, I.I. (1968) *The Origin of Terrestrial Vertebrates*. London and New York: Academic Press.

Schoch, R.R. & Rubidge, B.S. (2005) The amphibamid *Micropholis* from the Lystrosaurus Assemblage Zone of South Africa. *Journal of Vertebrate Paleontology* **25**, 502–522.

Schoch, R.R. & Witzmann, F. (2009) Osteology and relationships of the temnospondyl *Sclerocephalus*. *Zoological Journal of the Linnean Society London* **157**, 135–168.

Schoch, R.R. & Witzmann, F. (2011) Bystrow's paradox: gills, forssils, and the fish-to-tetrapod transition. *Acta Zoologica* **92**, 251–265.

Schultze, H.-P. (1984) Juvenile specimens of *Eusthenopteron foordi* Whiteaves, 1881 (osteolepiform rhipidistian, Pisces) from the Upper Devonian of Miguasha, Quebec, Canada. *Journal of Vertebrate Paleontology* **4**, 1–16.

Shubin, N.H. & Alberch, P. (1986) A morphogenetic approach to the origin and basic organisation of the tetrapod limb. *Evolutionary Biology* **20**, 319–387.

Shubin, N.H., Daeschler, E.B., & Jenkins, F.A., Jr. (2006) The pectoral fin of *Tiktaalik roseae* and the origin of the tetrapod limb. *Nature* **440**, 764–771.

Sigurdsen, T. & Bolt, J.R. (2010) The Lower Permian amphibamid *Doleserpeton* (Temnospondyli: Dissorophoidea), the interrelationships of amphibamids, and the origin of modern amphibians. *Journal of Vertebrate Paleontology* **30**, 1360–1377.

Smirnov, S.V. & Vorobyeva, E.I. (1988) Morphological grounds for diversification and evolutionary change in the amphibian sound-conducting apparatus. *Anatomischer Anzeiger* **166**, 317–322.

Swartz, B. (2012) A marine stem-tetrapod from the Devonian of Western North America. *PloS ONE* **7**, e33683.

Thomson, K.S. (1967). Mechanisms of intracranial kinesis in fossil rhipidistian fishes (Crossopterygii) and their relatives. *Zoological Journal of the Linnean Society* **46**, 223–253.

Vallin, G. & Laurin, M. (2004) Cranial morphology and affinities of *Microbrachis*, and a reappraisal of the phylogeny and lifestyle of the first amphibians. *Journal of Vertebrate Paleontology* **24**, 56–72.

Voigt, S. (2012) Tetrapodenfährten im Rotliegend. In: H. Lützner & G. Kowalczyk (eds.), Stratigraphie von Deutschland X. Rotliegend. Teil 1: Innervariscische Becken. *Schriftenreihe der deutschen geologischen Gesellschaft* **61**, 161–175.

von Wahlert, G. (1966) Atemwege und Schädelbau der Fische. *Stuttgarter Beiträge zur Naturkunde A* **159**, 1–10.

Wagner, G.P. & Larsson, H.C.E. (2007) Fins and limbs in the study of evolutionary novelties. In: B.K. Hall (ed.), *Fins into Limbs: Evolution, Development, and Transformation*. Chicago: University of Chicago Press, pp. 49–61.

Wagner, G.P. & Schwenk, K. (2000) Evolutionary Stable Configurations: functional integration and the evolution of phenotypic stability. *Evolutionary Biology* **31**, 155–217.

Wake, D.B. (2009) What salamanders have taught us about evolution. *Annual Review of Ecology and Systematics* **40**, 333–352.

Wever, E.G. (1985) *The Amphibian Ear*. Princeton: Princeton University Press.

Witzmann F. (2004) The external gills of Palaeozoic amphibians. *Neues Jahrbuch für Geologie und Paläontologie Abhandlungen* **232**, 375–401.

6 Development and Evolution

The study of ontogeny adds a new dimension to the understanding of morphology and evolution – developmental time. Like functional morphology, ontogeny forms a keystone in understanding organismic diversity. However, it does not always require inference of data from extant to fossil taxa, because developmental stages are sometimes preserved. Ontogeny has a fossil record, and in the case of amphibians this is especially fortunate, because many taxa undergo profound developmental changes. Like few other extant vertebrates, modern amphibians have highly complex ontogenies. They exemplify substantial and taxon-specific morphological change through larval and adult life. *Metamorphosis* exemplifies this developmental transformation. But amphibian ontogenies involve many more diverse events than the morphological change that accompanies the transition from water to land. The last few decades have revealed that life cycles of extinct taxa were complex and often radically different from those of modern amphibians.

The fossil record of ontogeny is much better for amphibians than for other vertebrates. This results from the preferential preservation of habitats in which the young of Paleozoic and Mesozoic amphibians lived: lake and stream deposits. Larvae and juveniles of early amphibians have been reported from a broad range of fossil *Lagerstätten*, and are often better preserved, or available in much larger quantities, than fossils of their adults. This not only provides us with the opportunity to study extinct life cycles, but poses additional taxonomic problems not known in other vertebrate groups. Here I elucidate why developmental data matter in paleobiology, and what specifically they tell us about the life of early amphibians. I shall argue that development holds the potential to change the picture as a whole, rather than simply to add a few peculiar observations on life histories. Viewed from this perspective, it is the life cycle that evolves, not just the morphological trait or the gene.

Amphibian Evolution: The Life of Early Land Vertebrates, First Edition. Rainer R. Schoch.
© 2014 Rainer R. Schoch. Published 2014 by John Wiley & Sons, Ltd.

6.1 Ontogeny in modern amphibians

Despite their morphological differences, salamanders, frogs, and caecilians share basic features of ontogeny (Figure 6.1). Many salamanders and the vast majority of anuran species have a biphasic life cycle, with a short metamorphosis transforming an aquatic larva into a terrestrial adult. Most present-day caecilians are viviparous, but two groups retain the larval stage, and metamorphosis also occurs in these. It is generally concluded that this biphasic life cycle is the primitive condition for the Lissamphibia. According to this view, a predatory aquatic larva, a drastic short-term metamorphosis, and a terrestrial carnivorous adult characterized the ancient lissamphibian life cycle. This hypothesis has been developed almost without reference to fossil data, but is derived from the study of salamander ontogenies in particular. Mounting fossil evidence – especially from Mesozoic salamanders and frogs but also from Paleozoic temnospondyls – supports this conclusion.

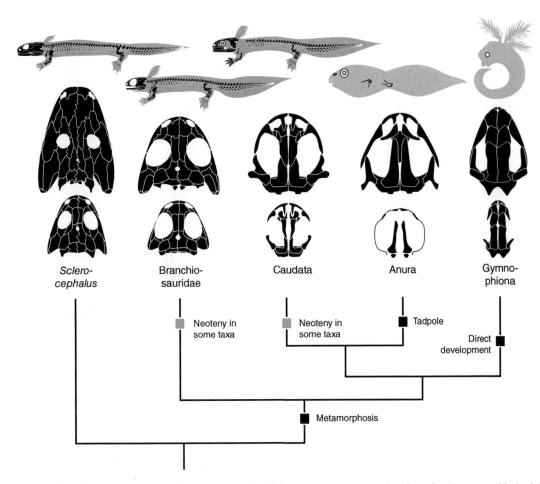

Figure 6.1 Life cycles in Paleozoic and extant amphibians. Primitive temnospondyls had juveniles that resembled adults but had external gills. Branchiosaurids evolved true larvae, similar to those of salamanders. Anurans further modified the larva into the herbivorous tadpole morph. Most caecilians are live-bearing, with embryos passing through a larva-like stage retaining external gills.

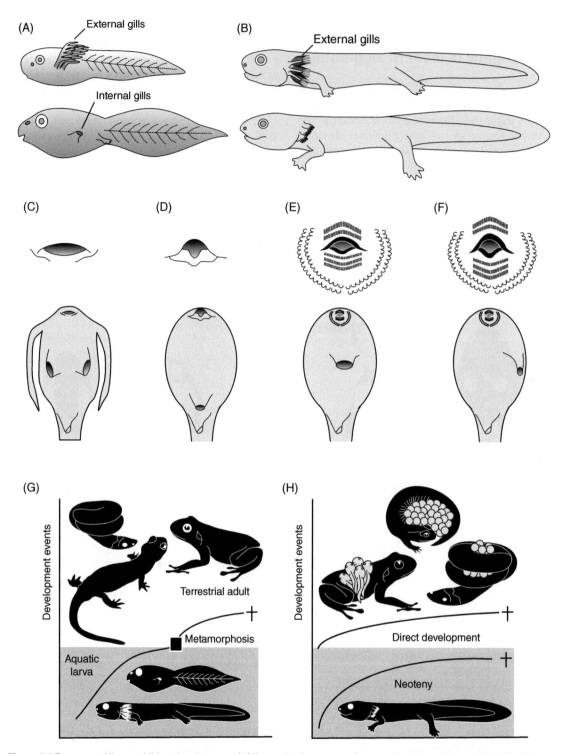

Figure 6.2 Features of lissamphibian development. (A) Two tadpole stages: above, early stage with external gills; below, later stage with secondarily enclosed internal gills (adapted from Haas 2010). (B) Two caudate larval stages (adapted from Lebedkina 2004). (C–F) Four different types of tadpole mouths (above) and bodies (below) (adapted from Mickoleit 2004). (G) Biphasic life cycle, with larval morphs transforming into terrestrial adults. (H) Modified uniphasic life cycles, either by evolving neoteny or by direct development.

The aforementioned points should not distract attention from the fact that many modern amphibian species have diverse life cycles, in which metamorphosis has often been abandoned (Figure 6.2). This may occur in at least two different ways. (1) The larva attains sexual maturity and remains in the larval habitat without transforming – this is neoteny. It means that metamorphosis is skipped and the terrestrial adult does not develop. The classic example is the Mexican axolotl (*Ambystoma mexicanum*), a large salamander that retains most larval features as an adult by failure of the thyroid gland to produce sufficient metamorphosis-inducing hormone. In its natural environment, the axolotl does not transform, but its hormonal system is still sensitive to the hormone thyroxin. If thyroxin is added to the food of axolotls in captivity, they eventually metamorphose into land salamanders. Neoteny may either occur by default (as in the axolotl) or by the evolution of a broader reaction norm that enables the organism to respond to environmental inputs by two different modes: transformation or neoteny. Only when a built-in threshold is reached is metamorphosis initiated in such species. The capacity to become neotenic is only possible when sexual maturity can be reached in the larval period. In modern amphibians, this is only observed in salamanders. In anurans, sexual maturity occurs during or even after metamorphosis, and hence neoteny is not an option. (2) Rather than extend the larval period, many amphibians have abandoned it completely. This is accomplished by retaining the embryos inside the womb for a longer period. The larval phase hence occurs within the mother's body and proceeds at a faster rate, without the need to form larval specializations for active feeding and swimming in the water. Such embryos develop directly into hatchlings that have the adult morphology. This mode of life cycle evolved in various species of salamanders, anurans, and caecilians. In plethodontid salamanders, the most speciose group of the clade, all species are live-bearing. In caecilians, secretions supplied by the mother nourish the embryo, and feeding on these is accomplished by a highly specialized embryonic dentition. It is clear that live-bearing does not evolve in a single step, as exemplified by the European fire salamander

(*Salamandra salamandra*): these salamanders retain embryos in the womb to give birth to advanced larvae, which still complete their development in the water to metamorphose into a fully terrestrial adult. Most important for evolution is the variability of reproductive modes in many amphibian species, again exemplified by the fire salamander. Individuals of *S. salamandra*, sometimes within the same population and season, may either lay eggs or give birth to larvae; its close relative *Salamandra atra* (the alpine salamander) retains embryos for much longer (2–4 years) to give birth to fully developed terrestrial hatchlings.

Amphibian reproduction may be grouped into three different strategies: (1) oviparity is the production of eggs from which aquatic larvae or terrestrial juveniles hatch; (2) ovoviviparity is the retention of eggs inside the mother's womb to give birth to more advanced larvae or juveniles; and (3) viviparity is the retention of larvae in the womb and the provision of nutrients in addition to the yolk of the egg (Wake 1982). Each of these modes has evolved independently in all three clades of lissamphibians, but the most diverse and derived forms of viviparity may be said to occur in anurans.

It is thus important not to underestimate the diversity and complexity of amphibian life cycles. This diversity underlines the enormous evolutionary flexibility of modern amphibians, made possible by a combination of high levels of developmental plasticity, the structure of reaction norms, and divergent modes of reproduction.

Salamanders. Salamanders have two divergent modes of fertilization: external and internal. The external mode is practiced by sirenids, hynobiids, and cryptobranchids, which are regarded as the basalmost caudates. Internal fertilization is performed by the female grasping a spermatophore (sperm capsule) from the male with her cloaca, in order to bring sperm and eggs together inside her body. This is an alternative mode of internal fertilization that does not require a penis, and, among vertebrates, it is only known in derived salamanders. The male produces the spermatophore with secretions provided by cloacal glands, and after uptake of the spermatophore the female stores the semen until the eggs are fertilized.

Salamander larvae usually have three pairs of external gills (Figure 6.2A,B), a thin-layered unprotected skin, small and poorly developed eyes, a lateral-line system, electroreceptors, poorly ossified small limbs, and a long, laterally compressed tail with a continuous fin. As such, they share features with larvae of lungfishes, although some of these, such as the gills, are probably convergent. Among lissamphibians, larval salamanders most closely resemble the aquatic taxa of Paleozoic amphibians. Nevertheless, they have many derived features, highlighting that larvae must have undergone their own evolution within the long-lived clade Caudata. Metamorphosis is usually a short phase during which the animals do not feed and when lungs develop, gills are resorbed, the tail is shortened, and the sense organs develop. The limbs and their musculoskeletal support become larger, and the skin grows additional layers concealed by a keratinized epidermis for protection on land against wear and water loss. The much larger eyes are protected by lids and moistened by a tear duct (nasolacrimal duct). Terrestrial salamanders employ a modified version of the larval gill arch skeleton (hyobranchium) to move the tongue, which plays an important role in feeding. Sticky secretions produced by the intermaxillary gland assist in catching prey. Vision is the most important sense outside the water, replacing the lateral and electric senses, whose receptors have been resorbed during metamorphosis.

Cryptobranchids, sirenids, amphiumids, and proteids have lifelong larvae (neotenes) living as obligatorily aquatic suction-feeders (Figure 6.2H). They inhabit cold freshwater creeks, ponds, or caves, and have a low metabolism and a slowed-down rate of development. Sirenids and amphiumids acquired their eel-like bodies convergently by failure to form or differentiate their limbs, and are extremely neotenic, which means that their adult morphology has early larval features, resembling very early larvae of other salamander families. Cryptobranchids, the giant salamanders, are unique in size and have a wide, parabolic skull with a unique mobility, which permits focused suction of rather small prey into the mouth (Elwood and Cundall 1994). Neotenic species evolved repeatedly in salamanders, and both facultative and obligate neoteny widespread (Reilly 1987; Whiteman 1994). Facultative neoteny

results from a wide reaction norm that is sensitive to different environments, while obligate neoteny arises from fixation of the neotenic state and the failure to transform in any environment.

In contrast, hynobiids, salamandrids, ambystomatids, and dicamptodontids usually have terrestrial adults, but return to the water for reproduction. Most species lay clutches of soft eggs in the water, from which aquatic larvae hatch. Shortly after hatching, they commence feeding on small invertebrates with their fully developed teeth, and swim by using their long tails. In most of these taxa, the larvae transform into terrestrial adults. When returning to the water for mating, some species spend a prolonged period in the water (e.g., European *Triturus* and *Ichthyosaura*). Neoteny has also evolved in some species of salamandrids, ambystomatids, and dicamptodontids, with both obligate (*Ambystoma mexicanum*, *Dicamptodon tenebrosus*) and facultative neotenes (*A. tigrinum*, *D. ensatus*).

Plethodontids are the most speciose clade (220 species), with fully terrestrial adults that lay eggs on land (Figure 6.2H). They are always direct developers without a larval period. Their embryos undergo an abbreviated larval development in the womb, retaining the most important features of their larvae-bearing ancestors. Plethodontids have neither gills nor lungs, thus relying entirely on skin breathing, enabled by their minute size.

Anurans. Modern anurans practice external fertilization, except for *Ascaphus*, which uses an outgrowth of the tail as an intromittent organ. A few other species reproduce by pressing the cloacae together (e.g., *Eleutherodactylus*). Usually, reproduction is initiated by favorable environmental conditions (temperature, day length, rainfall) (Haas 2010). Several males fighting for a single female are often seen during the breeding season in ponds. The successful male clings to the female (amplexus) and fertilizes the eggs after they have been released into the water. Aquatic eggs and larvae hatching in freshwater are the primitive condition, but fossil evidence for tadpoles is restricted to a few Cretaceous and Tertiary anuran taxa. The evolution of tadpoles thus has to be reconstructed largely on the analysis of extant taxa. Some derived taxa deposit eggs outside the water, but tadpoles hatching from these return to the water. Other, more abundant modes are ovoviviparity and viviparity, which evolved numerous times in anurans. In many small ponds, anuran larvae are the largest organisms and the only vertebrates, being able to feed on the only resource available in larger quantity, plant material of all sizes (detritus, algae, plankton). By pumping water through their mouths, the tadpoles extract microscopic food particles. Tadpoles are the only amphibians that managed to move down the trophic chain, thereby invading small, ephemeral water bodies. Some anurans settled in arid environments, where they survive in the soil or other moist places, with shed epidermal layers forming an extra "skin" and the bladder acting as a water reservoir.

Much more than salamander and caecilian larvae, tadpoles evolved their own body plan, which differs radically from that of the adult frog (Figure 6.2C–F). This includes numerous apomorphies: cartilaginous jaws bearing keratinous larval "teeth" and "beaks," forelimbs developing much later than hindlimbs, head and body forming a single rounded structure for most of the tadpole's life. Phylogenetic analyses of tadpoles have been conducted, revealing a world of their own (Haas 2003). Four types of tadpoles are distinguished (Orton 1953) on the basis of their jaws and the structure of the excurrent opening ("spiracle"). Type I (pipids and rhinotrematids) has paired "spiracles," Type II (microhylids) a median unpaired "spiracle" – both groups lack keratinized jaws and are obligate filter-feeders. Types III and IV have mouths with numerous papillae, keratinized "teeth," and an internal horny "beak," differing in the position of the unpaired "spiracle," which is medial in Type III (ascaphids, discoglossids) and located on one side in Type IV (pelobatoids, neobatrachians). The tadpole mouth is supported by unique cartilaginous structures. The larvae use their jaws and keratinized "teeth" to rasp food from surfaces and chop it into small particles that fit their small mouth opening. Food particles small enough to pass the mouth are sorted mechanically: large pieces go directly to the oesophagus, whereas smaller particles are first sieved by branchial filters and covered with mucus (Wassersug 1980). Water is pumped through the buccal cavity and the internally located gills by

coordinated movements of mouth and "spiracle." In this buccal pumping system, three valves control water flow: the mouth, choanae, and the ventral velum. For instance, increased pressure in the buccal cavity, resulting from the uptake of water, closes the choanae and prevents a backflow of water (Gradwell 1969). As in other aquatic vertebrates, the hyobranchium mediates pressure changes in the buccal cavity (Duellman and Trueb 1994). Although the majority of tadpoles feed on plant material, there are many carnivorous species, notably cannibalistic types that live under crowded conditions or with limited food supply, or in ephemeral ponds located in arid regions (Duellman and Trueb 1994).

Anurans have evolved diverse modes of reproduction and parental care. Numerous species are direct developers, with young kept inside the vocal sac, borne on the mother's back, contained in pouches inside the mother's skin, carried on the hindlimbs, or at least transported to the water (Duellman and Trueb 1994). Anurans may lay clutches of eggs in soil, within leaves, or on trees.

Caecilians. All caecilians practice internal insemination by use of a penis-like organ (phallodeum). Their young usually hatch at a much more advanced stage than larvae of anurans or salamanders (Duellman and Trueb 1994). Caecilian hatchlings thus resemble adults more closely than salamander larvae or tadpoles do, with larval features restricted to gill slits and a tail fin in those species that have aquatic larvae.

Primitively, the caecilian life cycle is biphasic, with an aquatic larva possessing three external gills. The basal Ichthyophiidae and Rhinatrematidae lay eggs close to the water, from which aquatic larvae hatch (Duellman and Trueb 1994). Some species of the Caeciliidae and the aquatic Typhlonectidae also have aquatic larvae, but the latter are viviparous. The larvae orientate themselves with a lateral sense organ similar to those of other lissamphibian young. About 25% of caecilian species are oviparous, and in *Ichthyophis* maternal care has evolved, with the female guarding the eggs.

However, in most species larval development takes place within the eggs, and the majority of species are live-bearing. The delay of hatching may be regarded as the essential initial step towards viviparity: first the larva develops inside the egg and hatches as a miniature terrestrial adult, then (in more derived taxa) the eggs are retained inside the womb and the terrestrial hatchlings are born when the larval phase is completed. Live-bearing thus evolved within the clade, clearly convergently to anurans and caudates. Caecilian embryos evolved specialized multicuspid teeth, either feeding on nutritious secretions supplied by the mother or practicing cannibalism on siblings.

6.2 Fossil ontogenies

The fossil record of amphibian development is rich, but confined to certain groups and often biased by preservation and ecological factors (habitat change). The main hurdle in the study of extinct ontogenies is the identification of size classes of specimens as belonging to the same taxon. Suppose for a moment that lissamphibians had become extinct and we did not know about metamorphosis – would we ever consider that a tadpole and an adult frog found in the same fossil beds belonged to the same species? Even if we thought about this possibility, would it not be more parsimonious to conclude that the two belonged to separate clades, as long as we did not have a large sample with a continuous series of specimens spanning metamorphosis?

Sampling specimens from one locality and (optimally) the same horizon is therefore a major requirement for identifying ontogenetic series in fossils. A second criterion is the recognition of shared apomorphic characters in the larva and adult. Both criteria should be applied together in order to draw a conclusion, and there will undoubtedly remain cases that cannot be resolved. A further difficulty is the preservation of fossils. This not only varies between deposits or even within a single bedding plane, but is also selective: thin bones or bone primordia are rarely preserved, and finding and identifying small amphibian larvae can be a difficult task. Fossil growth stages are therefore successive samples of specimens hypothesized to form an ontogenetic series – a hypothesis rather than a fact. This notwithstanding, there are many Paleozoic, Mesozoic, and Cenozoic amphibians from which growth stages have been reported (Boy 1974; Klembara 1995; Steyer 2000; Anderson *et al.* 2003; Schoch 2009a; Fröbisch *et al.* 2010) (Figure 6.3).

Fish-like stem-tetrapods. The best-studied sample of fish-like tetrapodomorphs was reported from *Eusthenopteron* (Schultze 1984; Cote *et al.* 2002). This sample is important because it shows that small juveniles of *Eusthenopteron* had fully formed skulls and opercular bones, indicating that there were no external larval gills of the type found in some dipnoans or lissamphibian

Figure 6.3 Ontogeny has been studied in several groups of early tetrapods, especially temnospondyls, seymouriamorphs, and lepospondyls. In other groups, larvae or juveniles are virtually unknown. Temnospondyls, seymouriamorphs, and nectrideans are known to have had larval morphs with external gills, whereas the primitive condition was to have internal gills, covered by opercular bones (black).

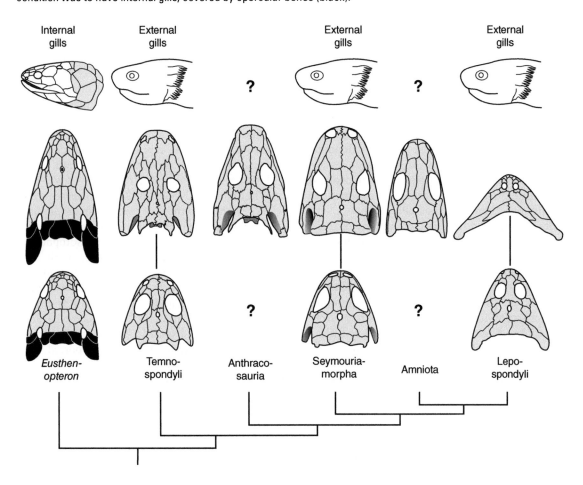

larvae (Figure 6.3). The skull was essentially complete, but had disproportionately large orbits and a short snout, as in most other vertebrate juveniles. The interesting pattern is found in the vertebral column: the centra ossified at a much later stage than did the dermal scales or the bones of the skull (Cote *et al.* 2002). This pattern is consistent with findings in temnospondyls (see below).

Limbed stem-tetrapods. Unfortunately, there are no ontogenetic data of *Acanthostega* or *Ichthyostega* available. Among those Carboniferous taxa that

fall outside crown tetrapods, there are only two clades that preserve growth series: the colosteids *Greererpeton* (Godfrey 1989) and the baphetid *Baphetes orientalis* (Milner *et al.* 2009). In both cases, only moderate morphological changes were reported, and true larval forms are unknown. *Baphetes* is interesting because its juvenile skull is very similar to that of adults, suggesting either that morphological change was confined to very early stages, or that the skull grew isometrically (Milner *et al.* 2009). The potentially most informative sample is that for *Whatcheeria*, of which several hundred specimens were excavated in a single locality and horizon (Bolt and Lombard 2000).

Temnospondyls. The most detailed samples of ontogenetically informative fossils are reported from temnospondyls. In some cases they provide insight into early larval phases of bone formation (Boy 1974; Schoch 2010), whereas others span conspicuous size ranges (Boy 1988, 1990). The fossil record of ontogeny is extraordinarily detailed in some temnospondyls. The classical example is the "branchiosaurs," originally a collective name for larvae of various Paleozoic temnospondyls (Boy 1972). In most of these taxa, development progressed slowly without major change, except for skull proportions (elongation of snout, decrease in size of orbits). As most genera were aquatic or amphibious, there was no habitat change, and adults remained in the same environment as larvae (Schoch 2009a). Both larvae and adults were usually predators, as indicated by the conical dentition and intestine fillings (mostly fish: Boy and Sues 2000). This changed only in the small dissorophoids, in which larvae underwent a profound transformation to become heavily ossified terrestrial adults. Their larvae were specialized filter-feeders, ecologically resembling frog tadpoles, but morphologically very different (Schoch and Milner 2008). A key to the evolutionary success of temnospondyls appears to have been developmental plasticity: taxa from various clades had extremely flexible ontogenies that varied between different environments. If the temnospondyl hypothesis of lissamphibian origin is correct, then lissamphibian plasticity and the ability to metamorphose were both inherited from temnospondyls, notably the dissorophoid clade.

- Basal temnospondyls. Larval or juvenile specimens have been described from the edopoid *Cochleosaurus* (Sequeira 2004) and the small temnospondyl *Balanerpeton* (Milner and Sequeira 1994). A well-sampled growth series of *Cochleosaurus* from Nýřany revealed that the postcranium developed slowly, with many girdle and limb elements unossified in larvae (Sequeira 2009). Although gills are not preserved, the similarity of baby *Cochleosaurus* to other temnospondyl larvae suggests that they were present.

- Dvinosaurians. This aquatic clade is represented by large samples of *Trimerorhachis* from numerous deposits and horizons, most of which are late juvenile or adult. The smaller dvinosaurian *Isodectes* is known from larval specimens that preserve external gills in addition to other skin impressions, gut contents, and poorly ossified bones (Milner 1982). The skull bones formed early, whereas the vertebral centra appeared much later and the limb elements and ribs were short and poorly ossified rods throughout the larval period.

- Zatracheids. This small clade of terrestrial carnivores is best represented by *Acanthostomatops*, which preserves a large part of its larval and metamorphic ontogeny (Boy 1989; Witzmann and Schoch 2006). Larvae were more similar to those of eryopids in developing an elongated snout early and ossifying the limbs at a slower rate than dissorophoids. Metamorphosis was pronounced, with the trunk foreshortening proportionately and the skull becoming very large and much wider than in larvae. The hyobranchium was apparently remodeled from the typical larval pattern into a zatracheid-specific set of numerous tiny ossified rods that probably supported the tongue during feeding on land.

- Dissorophoids. The largest samples of fossil ontogenetic series and the most diverse types of development were found in dissorophoids. Extended ontogenetic series were reported for micromelerpetids (Credner 1881; Werneburg 1994; Boy 1995; Witzmann and Pfretzschner 2003), *Apateon*, and *Melanerpeton* (Boy 1974; Schoch 1992). Smaller, but very intriguing samples were reported from *Amphibamus* (Milner 1982), *Platyrhinops* (Clack and Milner 2010), *Branchiosaurus* (Fritsch 1879), and *Mordex* (Milner 2007). In *Micromelerpeton*, small larvae had fully formed skulls but incomplete limbs without finger bones and rudimentary ribs (Witzmann and Pfretzschner 2003). This genus did not fully metamorphose, but one population (or species) is known with relatively large specimens that are more similar to adults of other dissorophoids (Boy 1995). In the branchiosaurid *Apateon*, nearly complete ossification sequences are known from several species (Schoch 2010). These are unique among temnospondyls in revealing much of the cranial ontogeny, which showed that jaw and palate bones were the first to form, followed by medial skull roof bones, cheek elements, until finally the bones surrounding the eyes appeared (Boy 1974; Schoch 1992). In the postcranium, the neural arches and humerus came first, followed by hindlimb elements and ribs, and only gradually did the limbs, girdles, and tail vertebrae form (Schoch 1992; Fröbisch *et al.* 2007). Branchiosaurids closely resembled modern neotenic salamanders, and individual age data gathered from thin sections indicate that they became sexually mature at the larval state (Sanchez *et al.* 2010a). Most branchiosaurid species apparently remained in the water as true neotenes, but metamorphosed specimens have been reported from at least one other species, *Apateon gracilis* (Schoch and Fröbisch 2006). This taxon provides the most convincing evidence of a drastic metamorphosis like that in extant amphibians – a feature probably shared with most other dissorophoids except for the micromelerpetids (Schoch 2009a). The miniature amphibamids appear to have had larvae similar to branchiosaurids, whereas the larger, armored dissorophids and trematopids are still known largely from adult specimens.

- Eryopids. The Early Permian genus *Onchiodon*, a close relative of the amphibious *Eryops*, is known from a complete series of growth stages (Boy 1990; Werneburg 1993). Development of the postcranium was slow, with girdles and limbs developing only just before the transition to land. Instead of a brief metamorphosis, *Onchiodon* acquired its adult morphology gradually, including the heavy and wide skull, the fully ossified pectoral and pelvic girdle, and the carpals and tarsals.
- Stereospondylomorphs. In the Saar–Nahe basin of Germany, three genera have been found in large quantities that permit recognition of growth stages. Phylogenetically, these form a grade at the base of the Stereospondyli, the dominant Mesozoic clade of temnospondyls. This grade also forms a cline from more heavily built, probably amphibious *Sclerocephalus* to

more gracile and fully aquatic genera such as *Glanochthon* and *Archegosaurus*. Ontogeny was generally similar to *Onchiodon*, but with decreasing levels of ossification from *Sclerocephalus* to *Archegosaurus* (Witzmann 2006). This is most apparent when ontogenetic trajectories are compared (see below). The ontogenetic size range of *Sclerocephalus* is most remarkable, spanning a range of 5–180 cm body length. Stereospondylomorph larvae had rather short external gills, a short snout, but a fully formed skull with large fangs. In contrast to the filter-feeding branchiosaurids, they probably fed on large invertebrates or larval fish.
- Stereospondyls. The largest temnospondyl clade is known mostly from adult material. Some Late Permian rhinesuchids are known from deposits that also yielded small skulls which have short snouts but otherwise resemble the adults; nothing is known about their postcranial ontogeny. Notable exceptions form the capitosaurs *Watsonisuchus* and *Mastodonsaurus*, the metoposaurid *Callistomordax*, the trematosaurid *Trematolestes*, and the plagiosaurid *Gerrothorax* (Early–Late Triassic). These taxa all highlight that ontogenetic changes were minor and came at a slow pace. Even tiny juveniles resembled adults closely: the skulls of *Gerrothorax*, *Trematolestes*, and *Callistomordax* were of remarkably adult appearance, both proportionally and in dermal ornament. In contrast, *Sclerocephalus* larvae of the same size are much more larval in appearance (Boy 1988).

Anthracosaurs. This important stem-amniote clade is represented almost entirely by adult specimens. The only exception is the partial skull of *Calligenethlon*, a probable anthracosaur juvenile, which shows little to distinguish it from adult eogyrinids (Panchen 1970).

Seymouriamorphs. The ontogenetic record of seymouriamorphs is substantial and has been studied in great depth by Jozef Klembara and colleagues in the last 20 years (Klembara 1995; Klembara *et al.* 2006). These studies are extremely important because they highlight that temnospondyl ontogenies are not necessarily representative of all early tetrapods. Indeed, the discosauriscid seymouriamorphs reveal rather distinct ontogenetic patterns. Superficially resembling large branchiosaurids, these stem-amniotes had wide skulls, external gills, and rather poorly ossified axial and appendicular skeletons (Klembara and Bartík 2000; Klembara and Ruta 2005). They also had lateral-line sulci and additional pits, which might have housed electroreceptors by analogy to extant salamander larvae (Klembara 1994). As in branchiosaurids, ossification proceeded at a slow pace, and small larvae lacked vertebral centra, endoskeletal girdle elements, and carpals and tarsals. In contrast to branchiosaurids, they eventually formed all these bones and also the braincase elements (Klembara and Bartík 2000). Although Klembara often refers to "metamorphosis," this was not a brief and drastic transformation as in dissorophoids, but much more similar to that of the temnospondyl *Onchiodon*, requiring up to two years (Sanchez *et al.* 2008). In contrast to branchiosaurids, bones appeared and were completed much more slowly, as did the changes in the proportions of the skull (Klembara *et al.* 2006).

Lepospondyls. This diverse group includes mostly small species that fall in the same size classes as larvae of temnospondyls and seymouriamorphs. In contrast to these, lepospondyls are almost always known from adults and were well ossified. However, a few taxa reveal some developmental data suggesting rather diverse ontogenies (Carroll and Gaskill 1978; Milner 1996; Fröbisch *et al.* 2010). They have in common that larval morphologies were not really established, but rather the juveniles were small adults. Lepospondyl development appears to have been similar to that of amniotes, with little change after hatching and the development of the skeleton confined to embryonic stages that are not preserved.

- A remarkable feature among lepospondyls is that their vertebrae (which normally include neural arches fused to cylindrical centra) were completely ossified even in the smallest individuals (Fröbisch *et al.* 2010). Unlike in temnospondyls and seymouriamorphs, centra formed as a single block and probably at about the same time as the neural arches.
- Microsauria. The aquatic *Microbrachis* is known from growth stages that appear

throughout to be larval by their possession of lateral-line sulci and branchial denticles (or ossicles) in the gill region (Carroll and Gaskill 1978), but gills themselves are never preserved. Still, the postcranial skeleton was fully ossified, in contrast to those of temnospondyl larvae. However, Milner (2008) showed that the tail was successively elongated by the addition of vertebrae with age. All other microsaurs appear to have been terrestrial, and there is not a single taxon for which aquatic larvae or juveniles can be made plausible. For instance, the minute brachystelechids appear so fully formed even at the smallest stages that a larval phase seems improbable. In these microsaurs, juveniles were essentially small adults, probably hatching as such from eggs. If proved, this would parallel the situation in many extant amphibians that lay eggs on land rather than in the water. It is equally plausible, however, that brachystelechids were live-bearing.

- Lysorophians. Like many long-bodied microsaurs, lysorophians show little ontogenetic change. Interestingly, they have robust hyobranchial skeletons composed of numerous elements, resembling those of neotenic salamanders. It remains unclear whether they supported gills or were involved in tongue movements. Lysorophians were preserved in burrows that were interpreted as mud cocoons, not unlike those of extant lungfishes. There are often dozens, possibly hundreds of burrows in a single bed, and they look very similar to the burrows of the lungfish *Gnathorhiza* from the same formations. This could indicate that they led an aquatic, gill-breathing life during the wet season, but there is little more to support this hypothesis. In microsaurs and lysorophians, which probably form a clade within leopospondyls, neural arches and centra formed as separate units and fused only later in development (Carroll 1989).
- Aïstopods. This small group of aquatic forms stands out by an amazing series of growth stages in *Phlegethontia* (Anderson 2003). As in branchiosaurids, the smaller size classes are characterized by fewer ossifications in the skull, which permits recognition of an

ossification sequence (Anderson 2003, 2007). In stark contrast to all temnospondyls, seymouriamorphs, and extant amphibians, the dermal skull bones formed later than braincase elements. Vertebrae were already fully ossified in the smallest specimens, as well as the maxilla, mandible, frontal, prefrontal, and the endochondral palatoquadrate; the parietal and cheek bones formed during successive later stages (Anderson 2003).

- Nectridea. Most nectrideans were aquatic at all stages of development, as indicated by their feebly ossified small limbs and the disproportionately long and deep swimming tails (Bossy and Milner 1998). The smaller, long-headed nectrideans appear like juveniles of the large Napoleon-hat taxa *Diplocaulus* and *Diploceraspis*, and the discovery of small specimens has confirmed that baby *Diplocaulus* had narrow skulls (Olson 1951; Milner 1996). Apart from its different head shape, a 4 cm long baby *Diplocaulus* was essentially like a 1 m long adult, and all bones were ossified at this extremely early stage. Rinehart and Lucas (2001) showed that by 40 mm skull length, *Diplocaulus* underwent a phase of drastic proportional change in skull shape. Unlike the lissamphibian or dissorophoid metamorphosis, this change was not accompanied by a change of habitat. Despite exquisite preservation of details and skin impressions in a tiny specimen from Texas, there were no remains of external gills found. Therefore, external gills like those in branchiosaurids and seymouriamorphs appear to have been absent. In nectrideans and aïstopods, the vertebrae formed as a single unit probably from the perichordal tube, paralleling the condition in modern teleosts and tetrapods (Carroll 1989).

6.3 Ontogeny as a sequence: developmental trajectories

Development can be viewed as a sequence of countless events. Cell division, differentiation, signaling interactions between cell populations, migration of stem cells and their communication with bypassing cells, formation and resorption of skeletal matrix, and controlled cell death are just a few examples. In the long run, averaged over a large number of cases across the body, these events pattern body regions, build up tissues, and form organs. It is impossible to enumerate them all, let alone study the totality of their causal relations. Therefore, developmental biologists focus on single cascades of events, concentrating on the formation of a particular body part. For instance, they study the time window in which genes are expressed in a particular cell population, what the gene products do, and how some genes regulate others to start or stop protein supply: this is the field of gene regulation. They may also study the differentiation of tissue types and the formation of new body parts, as in the classical field of morphogenesis.

On a very gross scale, aspects of development can be summarized by linear coordinate plots of developmental events (*y*-axis) versus ontogenetic time (*x*-axis). The resulting diagram is a developmental trajectory (Figure 6.4). Alberch *et al.* (1979) provided examples of such trajectories and their significance for the study of developmental evolution. (Instead of developmental events, morphological parameters are also often plotted on the *y*-axis – but this is shape change, the topic of section 6.5.) Diagrams of event sequences are not just summaries of ontogeny, but aimed at comparing ontogenetic data on various scales of taxonomy and evolution. In amphibians, trajectories provide easy-to-grasp overviews of major developmental phases, highlighting metamorphosis and other changes at a glimpse. Reilly *et al.* (1997) have described numerous types of trajectories, and that comparison can be made both within and between species. Classic examples are the different populations of neotenic salamanders such as the axolotl (Semlitsch *et al.* 1990). Here, I focus on the formation of the skeleton, particularly the skull, providing a platform for the comparison of extant and fossil developmental data. Such data are referred to as ossification sequences.

Salamanders. Cranial ossification sequences have been studied in numerous salamanders, both neotenic and metamorphosing populations and species (Figure 6.4). Among these, Lebedkina (2004) conducted the most extensive survey,

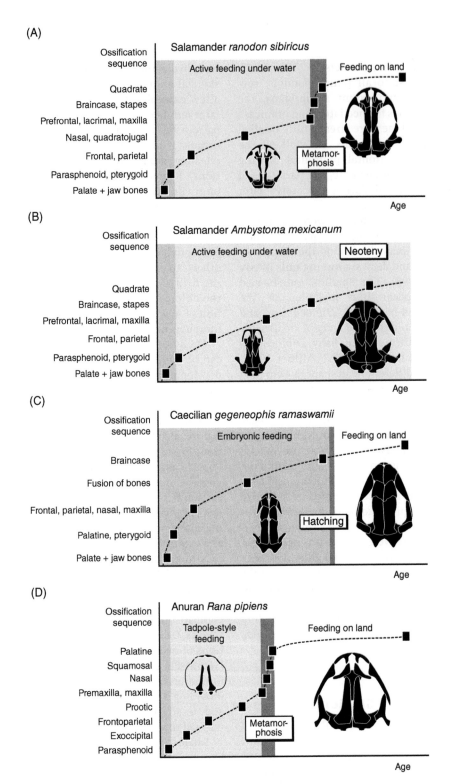

Figure 6.4 Sequences of bone formation (ossification) have been examined in many extant amphibians. (A, D) In transforming taxa, metamorphosis is usually identified by sudden leaps in the trajectory. (B) In neotenes (non-transforming taxa), the trajectory is flat. (C) Like some anurans and caudates, many caecilians are direct developers. Sequences adapted from Lebedkina (2004), Trueb (1985), and Müller *et al.* (2005).

focusing on basal taxa (hynobiids) as well as salamandrids and the axolotl. Recently, Germain and Laurin (2009) analyzed event sequences for 23 caudate taxa and mapped them on recent phylogenetic trees obtained from cladistic analyses. They found relatively high variation, but also confirmed the finding of Lebedkina that jaw and tooth-bearing palate bones were the first to ossify in most taxa, followed by elements of the dermal skull roof, the braincase, and some bones surrounding the eye (prefrontal, lacrimal). In plethodontids with larval development (*Eurycea*, *Hemidactylium*), the braincase bones form much earlier than in other salamanders, but most plethodontids give birth to fully developed terrestrial morphs (Wake *et al.* 1983; Rose 2003). In most salamanders, metamorphosis involves a saltation in the trajectory, with numerous events occurring in rapid succession. In addition to the appearance of new bones, larval elements (bone, teeth, cartilage) are resorbed during metamorphosis (Wintrebert 1922; Lebedkina 2004).

Anurans. Ontogenetic trajectories of tadpoles are radically different from those of other vertebrates. Trueb (1985) gave an overview of the quite diverse ossification sequences in anurans. Despite this variation, many species share the early ossification of parasphenoid and frontoparietal, exoccipital and prootic bones, all of which lie along the main axis of the skull (Kemp and Hoyt 1969). Palatal and jaw elements form at much later stages than in salamanders, related to the derived cartilage-dominated feeding apparatus in tadpoles. Metamorphosis is still more pronounced than in salamanders (Figure 6.4), with most bones forming during that short period. Hence, the tadpole body exemplifies larval adaptation, which relies on an almost completely cartilaginous skeleton; unlike in other amphibians, the bony skeleton of the frog is really only established during metamorphosis. In a more inclusive trajectory of anuran development (e.g., including cartilage), numerous additional events show up in the tadpole period which are not known in other vertebrates.

Caecilians. So far, ossification sequences have been studied in only a few gymnophionans (Wake and Hanken 1982; Müller *et al.* 2005; Müller 2006). As in basal salamanders, the palate

and jaw elements are the first to form, followed by the median bones of the skull roof (frontal, parietal), cheek, nasal, and braincase bones (Müller 2006). Most of these events occur before hatching in direct-developing species. In adults, elements of the palate, skull roof, and jaws fuse to form large compound bones (Wake and Hanken 1982). In caecilians, metamorphosis occurs only in species with free aquatic larvae, involving a widening of the palatal windows (Reiss 2002).

Trajectories of extinct amphibians. The study of size classes and recognition of growth stages in fossil taxa have paved the way for the analysis of extinct ontogenetic trajectories. However, there is a major problem in assessing such data. How can ontogenetic time be measured, when each specimen is but a snapshot of development? A fossil does not carry a label with its age on it. Hence, size has been used as a proxy for individual age, and although this may be generally sound, size has often been shown to be too variable to correlate reliably with age. There is a way out: absolute age data are now within reach, for the new discipline of skeletochronology has identified lines of arrested growth (LAGs) in microscopic bone analysis. These LAGs correlate with seasons and provide a reliable measure of ontogenetic age (see section 6.4).

Branchiosaurids were the first fossil amphibians in which developmental sequences were identified. First reported by Fritsch (1879) and Credner (1881), growth stages were studied by Watson (1963), and then Boy (1972, 1974) provided a detailed analysis. This framework permitted me to fit in new finds from a single locality and lake deposit (Schoch 1992). A sample of some 600 specimens encompassed a wide range of size classes for two different species of *Apateon*, each revealing detailed sequences of bone formation throughout the skeleton, among numerous other changes. This was a platform to study ontogenetic trajectories (Figure 6.5) and compare them to extant amphibians (Schoch 2002).

General features of temnospondyl trajectories. All better-known temnospondyl trajectories share the following chronology of larval phases: (1) an early period in which an aquatic predator was

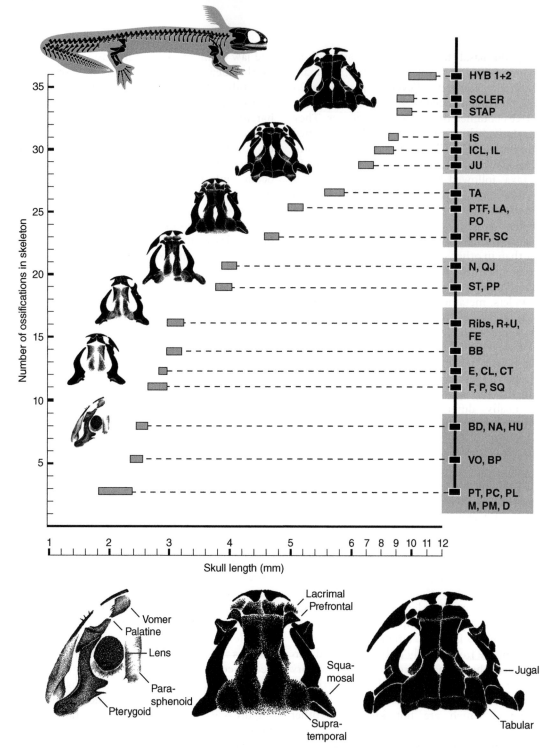

Figure 6.5 Sequence of ossification in the best-sampled branchiosaurid, *Apateon caducus*. From Schoch (1992). BB, basibranchial; BD, branchial denticles; BP, basal plate; CL, clavicle; CT, cleithrum; D, dentary; E, ectopterygoid; F, frontal; FE, femur; HU, humerus; HYB, hypobranchials; ICL, interclavicle; IL, ilium; IS, ischium; JU, jugal; LA, lacrimal; M, maxilla; N, nasal; NA, neural arch; P, parietal; PC, cultriform process; PL, palatine; PM, premaxilla; PO, postorbital; PP, postparietal; PRF, prefrontal; PT, pterygoid; PTF, postfrontal; QJ, quadratojugal; R+U, radius and ulna; SC, scapula; SCLER, scleral ossicles; SQ, squamosal; ST, supratemporal; STAP, stapes; TA, tabular; VO, vomer.

established (jaw elements with teeth, median skull bones, and cheek elements), (2) an intermediate period in which the axial skeleton was strengthened (neural arches, ribs) and the limbs started to form, and (3) a final period during which the jaw joint, braincase, and limbs were fully ossified, giving an adult capable of terrestrial locomotion if ossification was completed (Schoch 2010).

- **Neotenic trajectories**. The large samples for branchiosaurids are unparalleled among Paleozoic tetrapods and permit the reconstruction of more complete developmental sequences than in any other extinct amphibian. The main feature of branchiosaurid larvae is that even tiny, poorly ossified larvae were preserved that revealed early stages of bone formation. The smallest specimens have a skull length of 2 mm and a body length of 3 cm. In these, only a few elements are preserved in the head (jaw and palate bones with primordial teeth), and a few neural arches in the vertebral column; limbs were still entirely absent, as evidenced by skin preservation that shows no appendages at all (Schoch 1992). In successive stages, the skull roof formed, and the axial skeleton, dermal girdle bones, and limbs developed. The last bones to form in branchiosaurids were the plates of the scleral ring in the orbit and the hyobranchium. In most species of *Apateon*, the trajectories were less steep than in other temnospondyls, because the formation of skull bones required a longer time than, for instance, in *Micromelerpeton* or *Sclerocephalus*. After the early larval period was completed (defined by the consolidation of the dermal skull roof and the appearance of the last finger and toe bones), no additional ossifications appeared. *Apateon* thus had a flat, stagnating trajectory. These trajectories are interpreted as neotenic, because the crucial bones correlating with a life on land are all absent: the articular facets for the limbs, the carpals and tarsals, and the vertebral centra. Neoteny appears to have been obligate in all branchiosaurids that show it.
- **Metamorphosing trajectories**. Only one species, *Apateon gracilis*, is known from large, metamorphosed specimens (Schoch and Fröbisch 2006). Although not substantially larger than larval specimens of the same species and locality, metamorphs have numerous additional bones in the skeleton, a pronounced skull ornament, and ossified facets for the articulation of limbs. These bones must have formed during a brief period of drastic change, which plots as a leap in the trajectory of *A. gracilis*. The amphibamid dissorophoids probably had a similar trajectory, as indicated by the (much less numerous) finds (Milner 1982).

Conclusions. The study of ontogenetic trajectories reveals phases of slow progression and phases of drastic change (Figure 6.6, Figure 6.7). Metamorphosis is a short ontogenetic phase that contains many events occurring at a fast pace, and, in lissamphibians, it is usually of great ecological importance. Metamorphosing animals repattern their feeding apparatus, resorb the gills, modify the hyobranchium from gill support to tongue support, and grow lungs for air breathing. In the trajectory, this phase is marked by a leap. In contrast, directly developing or neotenic life cycles often have no such leap, but show a slower progression of development. The ontogenies of most temnospondyls and other early tetrapods were more like the latter, without a drastic metamorphosis (Figure 6.6). The only exceptions are the Paleozoic dissorophoids, in which metamorphosis and neoteny were first identified by the study of trajectories.

6.4 Histology: the skeleton as archive

In recent times, examination of bone microstructure has not only supplemented data on growth, but has become a powerful tool for studies of evolution and development in vertebrates (de Ricqlès 1975; Scheyer *et al.* 2010; Sanchez 2012). The study of bone microstructure requires thin sections (20–30 µm) to be examined under a polarization microscope, and it needs much experience to interpret the various fine structures. This

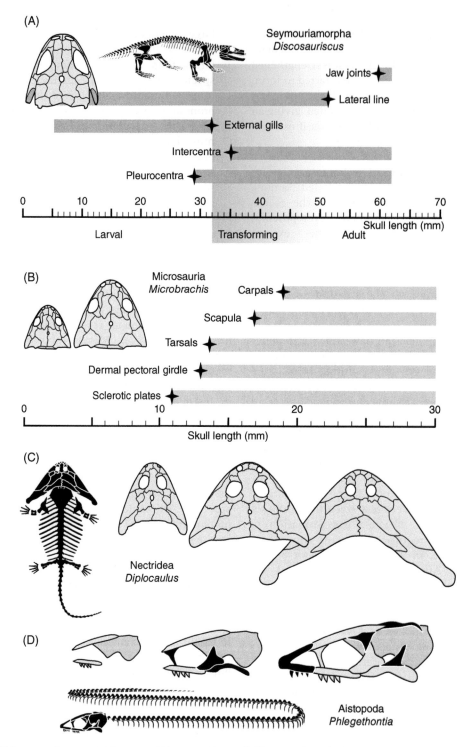

Figure 6.6 Ontogeny in stem-amniote taxa: (A) seymouriamorph *Discosauriscus* (adapted from Klembara and Bartík 2000); (B) lepospondyl *Microbrachis* (adapted from Olori 2008); (C) nectridean *Diplocaulus* (adapted from Olson 1951); (D) aïstopod *Phlegethontia* (adapted from Anderson *et al.* 2003, reproduced with permission of Canadian Science Publishing), new bones for each stage added in black.

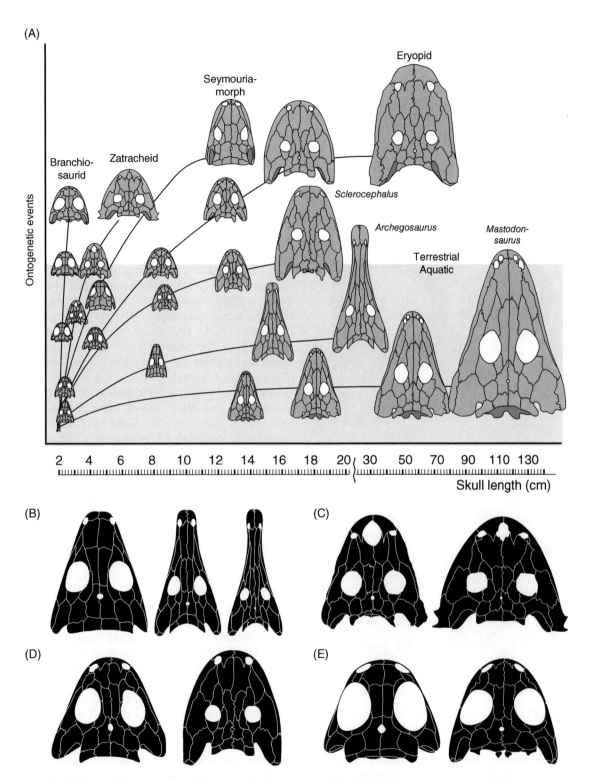

Figure 6.7 (A) In all major clades of temnospondyls, larval to adult development is preserved in at least some taxa. Despite many differences, ontogenetic events were generally similar, in that skeletal correlates of terrestrial life formed late in ontogeny; aquatic taxa thus had flat, stretched-out trajectories. Adapted from Schoch (2009a). (B–E) Growth series of temnospondyl skulls: (B) *Archegosaurus* (adapted from Witzmann 2006b, reproduced with permission of Taylor & Francis); (C) *Acanthostomatops* (adapted from Witzmann and Schoch 2006); (D) *Onchiodon* (adapted from Boy 1990); (E) *Apateon* (adapted from Schoch and Fröbisch 2006, reproduced with permission from John Wiley & Sons).

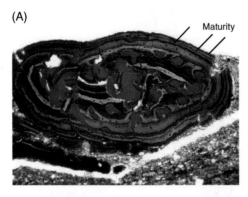

(A)

Maturity

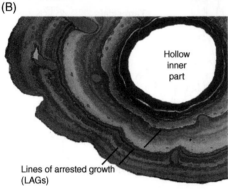

(B)

Hollow
inner
part

Lines of arrested growth
(LAGs)

(C)

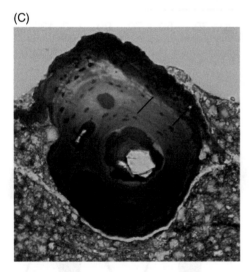

Figure 6.8 Paleohistology, exemplified by long bones of the branchiosaurid temnospondyl *Apateon pedestris*. (A) Lines of arrested growth (LAGs) are more closely set in older specimens, indicating sexual maturity has been reached. (B) LAGs are not always as well defined as in the present example. (C) Cell sizes and volumes may be assessed by measuring osteocyte caves (lacunae).

exercise is worth the effort, because the histology of long bones in particular has revealed age data for individual specimens that would otherwise have been inaccessible to paleontology. These insights have opened a new avenue of research, called skeletochronology (Francillon-Vieillot *et al.* 1990). In modern amphibians, this field has been applied in various species (Castanet *et al.* 2003).

In poikilotherms ("cold-blooded" vertebrates with low metabolism), thin lines of arrested growth (LAGs) are formed by annual cessation of bone apposition (Castanet and Smirina 1990). The correlation of these LAGs with annual cycles has been confirmed in a wide range of extant species (Castanet *et al.* 2003). Seasonal changes in temperature, light intensity, and rainfall are factors that influence cessation of bone deposition (Peabody 1961). LAGs are most completely preserved in the central parts of long bones, especially stylopodials (humerus, femur, tibia) (Figure 6.8). Although theoretically simple, the recognition of LAGs is not always easy, and bone remodeling has often erased older LAGs, for instance through expansion of the marrow cavity. At any rate, counting LAGs requires a sample of several specimens of different sizes in order to span as much of the size range (and LAGs) as possible. Even when retrocalculated, the total available number of LAGs only provides a minimum number of years.

Recently, LAGs have been studied in early tetrapods: the seymouriamorph *Discosauriscus* (Sanchez *et al.* 2008), the metoposaurid *Dutuitosaurus* (Steyer *et al.* 2004), and the branchiosaurid *Apateon* (Sanchez *et al.* 2010a). Sanchez *et al.* (2010b) also developed a new method for ontogenetic staging based on bone histology, exemplified by the branchiosaurid *Apateon*. This might help overcome the ambiguities in assessing age on purely morphological grounds. Numerous research directions are conceivable once histological sections are available: (1) bone density sheds light on the terrestriality of a taxon (Laurin *et al.* 2004); (2) LAGs permit the assessment of the absolute age of individual specimens, which allows the measurement of rates of bone deposition and the determination of the age when sexual maturity was reached (Steyer *et al.* 2004; Sanchez *et al.* 2008, 2010a, 2010c); (3) the study of calcified cartilage, excessive bone deposition (pachyostosis), and bone resorption (osteoporosis) yields data on adaptations to aquatic life (de Ricqlès 1975, 1979); (4) the size of bone lacunae, the cavities in which the bone cells were located, permits the measurement of cell volumes, which in turn shed light on the size of the genome (Organ et al. 2010) and properties of the metabolism (Wake 2009).

6.5 Changing shape: allometry

Ontogeny can be reduced to a purely temporal sequence of events, but is more often described as change in shape (allometry) (Figure 6.9). When an organism grows, the rates at which different parts expand will inevitably diverge (Thompson 1941; Alexander 1990). This has purely functional reasons, because not all tissues and organs work the same way: some properties are defined by area (n^2) and others are by volume (n^3). Consider a mouse growing to the size of an elephant: doubling its length requires squaring its foot area and cubing its lung volume in order to keep the animal viable – the associated parts must change their shape dramatically.

There are several types of allometry (Klingenberg 1998): (1) *static allometry* is measured in different individuals of the same species at the same stage of ontogeny; (2) *ontogenetic allometry* is assessed in the same species across different stages; (3) *evolutionary allometry* is measured in different species at the same ontogenetic stage; and finally (4) *plastic allometry* is analyzed in the same species in different environments (Klingenberg 1998; Schlichting and Pigliucci 1998). Static allometry often accompanies sexual dimorphism, and need not be related to ontogeny at all. In contrast, ontogenetic allometry is the shape change observed during growth. Evolutionary and plastic allometry will be discussed under heterochrony (section 6.6) and plasticity (Chapter 8, section 8.1), respectively. The present section therefore exclusively deals with ontogenetic allometry.

When focusing on a compound unit such as the skull, allometries soon become more complicated.

Emerson and Bramble (1993) calculated ratios required to maintain geometric similarity (i.e., conserving shape through growth) for certain organs and components. For instance, the volume of the ear capsule and braincase scales as n^1, muscle force as $n^{2/3}$, and jaw length as $n^{1/3}$, where n is the mass. Only in the unlikely case that these organs scale by the given ratios is shape conserved (isometry); any departure from this growth trajectory results in allometry. The listed ratios are thus required to maintain geometry rather than function. Consequently, change in size can lead to relative loss of function, unless there are compensatory adjustments in shape (Emerson and Bramble 1993). For instance, scaling up a muscle by $n^{2/3}$ means that its force decreases. Accordingly, the muscle has to increase its cross-sectional area by a higher rate than $n^{2/3}$ if functional properties are to be maintained. This is why the jaw muscles usually grow with positive allometry, in contrast to the negative allometry at which brain and sense organs grow. These examples reveal the existence of constraints on shape and function, and they exemplify how size evolution is limited or directed by geometric factors.

In amphibians, allometry can be quite impressive. If the smallest salamander (*Urspelerpes brucei*, 2.6 cm) is compared with the largest (*Andrias davidianus*, 1.8 m), many discrepancies are immediately apparent. In the tiny species, the skull and body are roundish in cross-section and the sense organs and brain are huge, accounting for most of the head volume. In the giant species, the skull and body are flattened, with minute eyes and a proportionately small brain, but voluminous muscles and numerous skin folds along the flank of the body and limbs. This comparison is especially interesting because both species breathe through the skin, without contributions from the gills or lungs (Ultsch 2012). In the tiny salamander, the body surface is proportionately much larger than in the giant species. In larger skin breathers, the body surface area is not sufficient to supply the much larger body volume with oxygen; the required increase in respiratory area is accomplished by the folds along the flanks. Likewise, the musculature requires more space, for the reasons outlined above.

The skeletons of early amphibians were much more complex than those of their extant relatives,

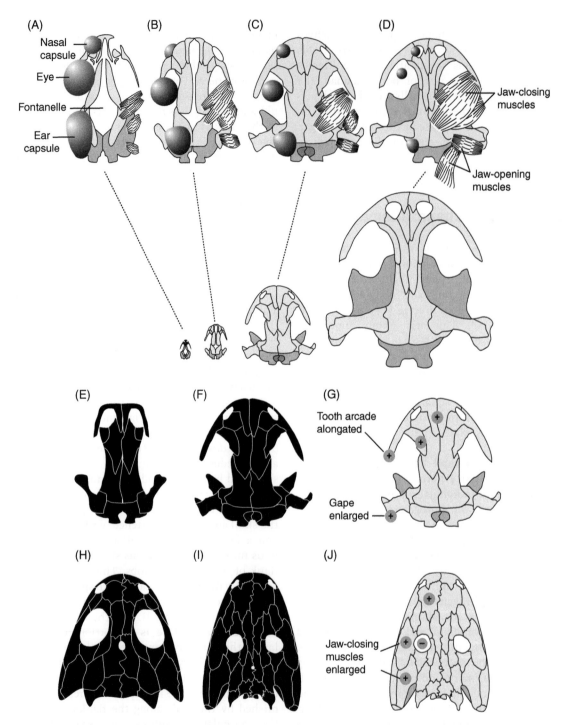

Figure 6.9 Scaling and allometry. (A–D) Evolutionary allometry in salamanders of different body size, with important organ systems mapped to show scaling effects: (A) *Thorius* (adapted from Hanken 1983, reproduced with permission from John Wiley & Sons); (B) *Plethodon*; (C) *Ambystoma*; (D) *Andrias*. (E–J) Ontogenetic allometry: (E–G) *Ambystoma*. (H–J) *Sclerocephalus* (adapted from Boy 1988).

and thus allometry is a rich field of study in these extinct groups. Some of the Triassic temnospondyls attained three times the length of the giant salamander. Allometric patterns are widespread there, as exemplified by the following:

- **Vision and brain**. The most trivial case is that the sensory capsules, eyes, and brain scale with negative allometry in relation to body length or mass (Emerson and Bramble 1993; Ivanović *et al.* 2007). This is evident in the decrease of orbit diameter and braincase volume, a pattern known from temnospondyls, lissamphibians, seymouriamorphs, microsaurs, and amniotes. However, inferring the size of the eye from that of the orbit is problematic, as revealed by the scleral ring in some taxa. For instance, in the dissorophoid *Cacops*, the proportionately large orbits housed a small scleral ring and eye. As for allometry, this means that the proportionate decrease of the eye may have been even stronger. Contrasting the general trend (negative allometry), the eyes of terrestrial temnospondyls grew with positive allometry during metamorphosis (Witzmann and Pfretzschner 2003), a trend also common in anurans and salamanders.
- **Eye muscles**. In many temnospondyls, the size of the palatal openings grew with slightly negative allometry, in contrast to the positive allometry in lissamphibians (Reiss 2002). In salamander metamorphosis, the positive allometry is exceptionally strong, caused by the resorption of pterygoid and palatine, which are remodeled to anchor enlarged eye muscles (Wintrebert 1922).
- **Jaw muscles**. Usually, the cross-sectional areas of jaw muscles and their attachment areas increase by $n^{2/3}$ or more. Accordingly, large salamanders and frogs have proportionately larger areas occupied by jaw muscles, and in the giant Triassic temnospondyls, adductor muscles not only filled the voluminous cheek chamber but also expanded anteriorly to attach in front of the eye. (Part of this phenomenon is also caused by the need to reorient muscles in the strongly flattened skull.) The length and volume of the cheek and adductor chamber in the mandible scaled with positive allometry in temnospondyl growth. Likewise, the subtemporal openings in the palate (which outline the cross-section of the adductor musculature) increased markedly during ontogeny in all early tetrapods.
- **Breathing**. The width of the skull scaled with positive allometry in terrestrial temnospondyls (eryopids, dissorophoids, zatracheids), but not in anthracosaurs or seymouriamorphs, which had rather slender adult skulls. This probably reflects the two divergent air-breathing pumps: buccal in stem-amphibians and rib aspiration in stem-amniotes. Buccal pumpers such as lissamphibians increase the volume of inspired air by broadening the skull (Schmalhausen 1968), in contrast to rib-basket aspiration in stem-amniotes.
- **Gape**. In early tetrapods, gape length usually increased with growth from larvae to adults, indicating the focus on increasingly larger prey by carnivorous species (Witzmann and Pfretzschner 2003). This also appears to hold for frogs and caecilians, whereas salamanders usually have very short upper jaws when they practice suction feeding in the water, and jaw length may increase markedly during metamorphosis.
- **Bite force**. In anurans, skull size generally correlates with prey size, as exemplified by the voracious horned frog (*Ceratophrys*), which has a huge head. However, other factors come into play as well, sometimes differentiated at an intraspecific level. In *Rana ingeri*, males have larger skulls and more powerful jaw muscles than females (Emerson and Bramble 1993). Both sexes feed on crabs as adults, but males start to handle these hard-shelled prey items earlier. This relates to allometries in head shape and muscle area: males attain the critical muscle force required to crack crab shells earlier than females. This is an example of static allometry, in which males and females of the same size and age differ in head shape and size. Witzmann and Scholz (2007) analyzed skull growth in the long-snouted Permian temnospondyl *Archegosaurus*, where they found evidence of increasing capacity for lateral strikes employed in capturing acanthodian fishes, which were preserved as gut contents. They recognized a negative allometry of jaw muscle attachment area in the cheek, in contrast to other temnospondyls, and concluded that bite force was smaller than in other taxa.

- **Locomotion**. In larvae of dissorophoids, the forelimb started to ossify earlier than the hindlimb, a feature shared with salamanders (Boy 1974). During the early larval phase, when the skull roof bones formed, the hindlimb eventually outgrew the forelimb (Schoch 1992). In frogs, the hindlimbs form earlier and are always much longer than the forelimbs; Emerson (1986) showed that the hindlimb and body length scale allometrically in frogs through ontogeny, reflecting the functional demands of jumping on size. In early tetrapods, the tail was a propulsion organ essential for swimming, and therefore usually longer than the trunk. However, even in aquatic forms, adults had proportionately shorter tails than their larvae. In the zatracheid temnospondyl *Acanthostomatops*, both trunk and tail were shortened relative to the length of skull and limbs (Witzmann and Schoch 2006).
- **Metamorphosis**. Allometries are profound in lissamphibian skeletal development, involving changes in head size, width, and the hyobranchial apparatus. Reilly and Lauder (1990) conducted morphometric analyses of *Ambystoma tigrinum* through larval and adult periods. They found that head length decreases by 9%, which largely results from negative allometry of the frontal region (eye size), whereas the width of the preorbital region scales with positive allometry by 18%. In transforming salamanders and frogs, the eyes grow with strong positive allometry during metamorphosis, requiring rearrangement of eye and jaw muscles, as well as resorption of palate bones in salamanders, which results in manifold allometries.

6.6 Heterochrony: the evolution of development

Even 150 years after the scientific discovery of *Ambystoma*, the story of this neotenic species is still fascinating: a salamander that stays a larva for its entire life, remains in the water, and breeds there (Figure 6.10). Although not quite the rule, this story has become a stereotyped tale of amphibian life histories. Neoteny often occurs as an "anomaly" in metamorphosing populations, but thereafter evolves into an adaptive strategy. Development thus forms an important hinge in amphibian evolution, and thoughts about the conceptual relationship between ontogeny and

(A)

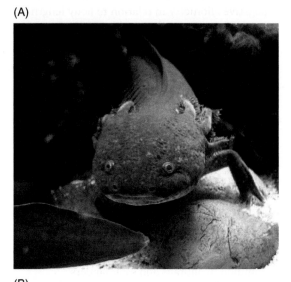

(B)

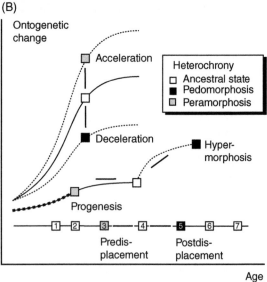

Figure 6.10 (A) The Mexican axolotl has often been emphasized in studies of development and evolution. (B) Heterochrony describes patterns of developmental evolution, such as pedomorphosis and peramorphosis.

phylogeny often begin with a survey of amphibian exemplars. This is the domain of heterochrony, the problem of how development and evolution relate to each other. How does development evolve, and, in turn, do developmental properties impose constraints on evolution? How is ontogeny reorganized or repatterned by evolution?

Considering the relationship between ontogeny and phylogeny opens the door to a field rich in great insights, but also major historical fallacies. Ever since Ernst Haeckel's assertive and finally doomed "biogenetic law," this topic was considered to be a minefield, and as a student I was warned not to get too close. The situation resembles those medieval world maps where monsters were drawn in dangerous regions. However, recent decades have seen renewed interest in the relation of development and evolution, leading to the emergence of a new field known as evo-devo or *evolutionary developmental biology* (Raff 1996; Hall 1999a).

Ontogenies evolve. The problem is how can the study of development shed light on developmental evolution. Observing a bone form and grow provides a dynamic perspective on morphology. Ontogeny is organismic change that can be observed, in contrast to evolutionary transformations that we infer from a given phylogenetic pattern. At the same time, knowledge of ontogeny does not, and cannot, by itself reveal how evolutionary change proceeds. There is no simple parallelism between ontogeny and phylogeny.

A troubled history. Heterochrony holds that evolution proceeds by altering development: rate, timing, and duration of life differ between species, and evolution proceeds along these lines. How do these differences come into being, and what are the rules behind them, if there are any? Ever since organisms have been compared, researchers had ideas about shared features of development. For instance, von Baer (1828) observed that embryos of different vertebrate groups were more similar to each other than adults. Later this pattern became known as "von Baer's law," and up to the present day it is a topic for statistical approaches in embryology (Poe 2006). After Darwin (1859) had argued that similarity between organisms is caused by common ancestry, Haeckel and others proceeded to reformulate von Baer's observation in an evolutionary framework. Haeckel's answer to von Baer's

question was simple (in fact, too simple): embryos are more similar because ontogenies start from the same point, and taxonomic differences develop only during later phases of ontogeny. Haeckel (1866) further proposed that "higher" vertebrates went through developmental stages of "lower" taxa in their embryonic period. In Haeckel's concept, humans were thought to recapitulate (repeat) the ontogeny of fishes, amphibians, and basal amniotes in their early development. This was viewed as a mechanical necessity, almost like a clockwork mechanism, but Haeckel failed to name any *adaptation* explaining why ontogeny should be recapitulated. The idea was that for ontogeny to evolve in such a mechanism, new features were added at the end of the ontogenetic trajectory, and the phylogenetically older early stages were believed to be increasingly condensed. An important tenet, treated like an axiom by its advocates, was that *ontogeny paralleled phylogeny*. In other words, the evolutionary history of groups went through phases of youth, adulthood, and senescence (Haeckel 1866). This is called cyclism, and had an enormous impact on paleontology in the first half of the twentieth century (Schindewolf 1950; Reif 1986). Although long abandoned, these ideas are still implicit in many recent paleontological studies, and often their consequences are not fully explored.

By the early twentieth century, it became increasingly clear that recapitulation had failed (Garstang 1920). Too many exceptions to the "rule" were discovered: (1) ontogenies were much less conservative than previously thought, (2) new features may appear at any stage of development, and (3) there was no simple recapitulation of embryonic (or even adult) stages of fishes in mammals. Developmental sequences do not necessarily form causal chains (Alberch 1985). Haeckel's mechanical model of ontogeny was simply wrong. Cyclism had failed as well – evolution does not parallel ontogeny. However, it is probably fair to say that, despite the shortcomings of their ideas, Haeckel and his fellow evolutionists had the right motive: the search for a proper causal explanation in morphology.

After the advent of the Modern Synthesis of evolutionary biology, the focus of interest shifted away from the relationship between ontogeny and phylogeny (Wake and Roth 1989; Raff 1996; Hall 1999a; West-Eberhard 2003; Sanchez 2012), although a few now-esteemed evolutionary biologists continued to work on the relation of development and evolution (Waddington 1942; Schmalhausen 1949). Gould's (1977) book *Ontogeny and Phylogeny* was the first monograph dealing with heterochrony thereafter, and this author signaled his discontent with the lack of development in the Modern Evolutionary Synthesis. Having studied allometry in detail (Gould 1966), his basic question was, once again: how do the dynamics of ontogeny relate to the dynamics of evolution? More recent decades have shone much light on this field, with both the map and the monsters more clearly visible now. It is essential to consider the following questions: (1) is heterochrony a pattern or a process; (2) what are the requirements to analyze heterochrony; (3) how can heterochrony be studied in paleontology; and finally (4) what explanatory power does heterochrony have by itself, and how does it fit into evolutionary explanations?

Heterochrony: pattern *and* process. Heterochrony is an evolutionary *pattern*, resulting from changes in the timing, rate, or duration of development. There are two principal types of heterochrony: pedomorphosis and peramorphosis. Pedomorphosis is the case when a given adult feature resembles a juvenile trait in the ancestral species. An extreme case of pedomorphosis is the axolotl: many larval features are retained in the adult, and often the species *Ambystoma mexicanum* as a whole is referred to as "pedomorphic." However, like phylogenetic statements about character states (apomorphic, plesiomorphic, homoplastic), heterochrony should be restricted to the feature itself, not its bearer. It makes no sense to say that salamanders are plesiomorphic compared to frogs without naming specific features. Peramorphosis is the opposite pattern, when a trait extends beyond the ontogeny of the ancestral species in rate or timing.

There are six principally different *processes* that may result in heterochrony (Figure 6.10B): (1) *slow-down* of developmental rate (deceleration), (2) *truncation* of the trajectory (progenesis), and (3) *delaying* the start (post-displacement, a shift on the trajectory) all result in pedomorphosis, whereas (4) *speed-up* of development (acceleration), (5) *extension* of the trajectory (hypermorphosis), and (6) *pre-dating* an event (pre-displacement)

all produce peramorphosis (Alberch *et al.* 1979). The term neoteny, a common life history strategy in salamanders, was reformulated by Gould (1977) to refer to a slow-down of developmental rate, but this is now termed deceleration (Reilly *et al.* 1997); neoteny is therefore used in his original sense here (Kollmann 1885).

Heterochrony requires a phylogenetic hypothesis. Analysis of heterochrony is impossible without a phylogenetic hypothesis (Wake 2009). Species whose ontogenies are compared need to be placed in a cladogram, and ontogeny must be known in at least one outgroup, as well. Only within such a framework does heterochrony make sense, and only with the help of the outgroup can polarity be determined for heterochronic change.

Heterochrony and paleontology. In the fossil record, examples of heterochrony have been much highlighted (McNamara 1995). Most spectacular and intuitive are cases of scaling, such as gigantism in dinosaurs and mammals, or miniaturization in salamanders. The shape changes that accompany evolutionary scaling are often tremendous, and trajectories describing them have come to be viewed as a powerful tool to analyze evolutionary trends (Thompson 1941; Gould 1977). However, it must not be forgotten that these are patterns. Without absolute age data and the possibility to study development directly, there is no way to interpret these phenomena. As has been shown above, a single case of heterochrony can be referred to different heterochronic processes. The application of skeletochronology has changed the situation in recent times (Sanchez *et al.* 2008, 2010a). In those taxa that preserve LAGs in their long bones, reasonably sized samples permit the study of growth rate, sexual maturity, and life span. For instance, the seymouriamorph *Discosauriscus* and the branchiosaurid temnospondyl *Apateon* have been shown to have spent a long time in the water (6–10 years). Whereas *Discosauriscus* slowly transformed into a land dweller, most species of *Apateon* remained in the water as adults. *Discosauriscus* reached sexual maturity in the post-aquatic period (Sanchez *et al.* 2008), but *Apateon* was a true neotene because it became mature in the larval state (Sanchez *et al.* 2010a). This forms the first unequivocal evidence of neoteny as a life history strategy in early tetrapods (Fröbisch and Schoch 2009; Sanchez *et al.* 2010a).

Evolutionary allometry: shape heterochrony. An elegant aspect of allometry is that it can be measured in both development and evolution. Ontogenetic and evolutionary allometries are often similar (Schlichting and Pigliucci 1998). In recent years, evolutionary allometries have been studied in a range of Paleozoic and Triassic tetrapods (Stayton and Ruta 2006; Witzmann *et al.* 2009). In a multivariate analysis of landmarks in the skull roof, Witzmann *et al.* (2009) identified two common modes of skull growth in temnospondyls: (1) gradual development with extensive allometries in long-parabolic skulls, and (2) isometric growth in taxa with broad and short crania. Fortunately, growth series are known from a wide range of temnospondyls, including basal Carboniferous taxa (e.g., *Cochleosaurus*) which served as outgroups for the heterochronic analysis. Peramorphic skull evolution occurred in stereospondyls, where the positive allometry of the snout and the negative allometry of the orbits went beyond the adult values in the ancestral growth trajectory. In contrast, a pedomorphic pattern is found in dissorophoids, where the snout was foreshortened and the skull widened. In dissorophoids, evolutionary trajectories diverge for larvae and adults, which is consistent with the recognition of a drastic metamorphosis in amphibamids and branchiosaurids. More subtle are patterns shared by eryopids and zatracheids, both of which have wide skulls with short postorbital tables.

Evolution of ontogenetic trajectories: sequence heterochrony. In contrast to allometry, changes in the sequence of developmental events provide more precise insight into heterochrony. Irrespective of individual age, the trajectory automatically carries a measure of developmental time with it, namely the sequence itself. Any shift of an event along the trajectory is a heterochrony. In this case, pedomorphosis is caused by post-displacement, peramorphosis by pre-displacement. Recent analyses of ontogenetic trajectories have yielded rich data on sequence heterochronies (Schoch 2009a). Despite considerable variation in timing and other parameters, developmental sequences of temnospondyls have numerous features in common (Schoch 2010). The trajectories include (1) an early period in which a larval aquatic predator developed, (2) an intermediate phase in which the axial skeleton was strengthened, and

(3) a final period during which endoskeletal elements such as the jaw joint, the braincase, and the limb and girdle bones were ossified, resulting in a terrestrial adult if completed. Basically, the temnospondyl trajectories are consistent with those of salamanders with biphasic life cycles: traits associated with a terrestrial life appeared long after the larval characters but before the transition to land (Lebedkina 2004). Crucial differences between temnospondyls and lissamphibians are that most temnospondyls did not metamorphose and that lissamphibians generally have a reduced complement of bones in the skeleton, resulting in a much shorter trajectory of skeletal development.

Evolutionary changes of temnospondyl trajectories could be analyzed on systematic scales that range from family to intraspecific levels (Schoch 2009a, 2009b). The plesiomorphic trajectory (parts of which are known from *Cochleosaurus*) was probably similar to that of the Carboniferous–Permian genus *Sclerocephalus*, which was aquatic as an adult but left the water occasionally, based on various lines of evidence (Schoch 2009b). The evolutionary changes to the trajectory involved several modes: (1) truncation of the developmental sequence, which produced various kinds of more aquatic taxa in which the features associated with a terrestrial life failed to developed or were not completed; (2) shifts of events within the developmental sequence, sometimes resulting in more substantial morphological changes; (3) condensation of developmental events, producing phases of drastic change, such as the rapid build-up of the skull in small larvae or the lissamphibian metamorphosis at the end of the larval period; and finally (4) the opposite process: unpacking of condensed periods of development, such as the much slower progression of skull development in branchiosaurids and lissamphibians as compared to temnospondyls and most other Paleozoic tetrapods.

The origin of lissamphibians has also been viewed in the light of heterochrony (Milner 1988). The consistent loss of various skull bones forms an apparent feature of all lissamphibians. Interestingly, some of these bones were the last to ossify in branchiosaurids, a clade close to the sister taxa of Lissamphibia in the temnospondyl hypothesis (Schoch 2002): the postfrontal, postorbital, tabular, and jugal. This suggests that truncation of the dissorophoid trajectory or a further slow-down of skull development resulted in the simplified lissamphibian skull, in either case a pedomorphic pattern. Analogous cases of trajectory evolution have been reported in salamanders (Wake 2009).

Heterotopy. More rarely analyzed are cases of spatial rearrangement in evolution. These so-called heterotopic changes are apparently not affected by timing or rate of developmental processes, but result from altered modes of embryonic patterning. The fusion of two skeletal elements, or the novel arrangement of bones following the loss of an element, exemplify heterotopy. In Paleozoic tetrapods, heterotopies involve the origin and loss of the intertemporal and tabular bones in the skull table, the restructuring of the palate in temnospondyls, or the repatterning of tarsal bones in salamanders (Hanken 1983). In lissamphibians, substantial heterotopies have occurred in the skull: the evolutionary loss of bones surrounding the eye correlates with the novel attachment of jaw musculature on the dorsal surface of the skull table and cheek. In the palate of caecilians, the tooth row on the vomer has been rearranged, to run lateral to the choana rather than medial as in all other lower tetrapods. In batrachians, the pterygoid, palatine, and vomer are substantially altered compared with both larvae and the ancestral state, and the choana has a different morphology. Although heterotopy and heterochrony appear to be distinct modes of developmental evolution, they really are two sides of the same coin. Typical heterotopies – such as the loss of a skeletal element or the different arrangement of body parts – may result from alterations in the *developmental rate* of cell aggregations and tissues. For instance, the failure of a bone to form results from the failure of cell condensations to reach a certain size, which in turn results from a lower rate of cell division. Heterotopy on the morphological level may thus result from heterochrony at the cellular scale. However, there are numerous heterotopies identified at the cellular and condensation level, and this appears to be a more decisive level than morphology (Köntges and Lumsden 2000).

Heterochrony needs adaptive explanation. The study of ontogeny adds a dynamic perspective to morphology, and tracing phylogenetic modifications of development provides a richer picture of evolutionary change. The impressive Cartesian transformations figured by Thompson (1941) might suggest that growth trajectories alone pave the way for evolutionary change. Phylogeny appears like an extrapolation of ontogeny, following allometric curves (Gould 1977). This view misses an important point: adaptation. Growth trajectories and developmental sequences are subject to selection, irrespective of whether they change rapidly or are highly conserved in evolution (a problem to which I shall return below). Heterochrony is often considered an "explanation" for evolutionary change, but it is simply an analytical description of the evolutionary process that altered development. True evolutionary explanation requires an adaptive scenario: what was the focus of selection within the frame of the studied features, and what kind of selection pressure was active?

The above argument does not imply that each and every feature is adaptive (see Wake 2009 for discussion). We know many examples where a trait does not in itself form an adaptation. Pan-adaptationism is as problematic as pan-heterochronism (structuralism). Yet for evolutionary change to occur, selection has to exert influence on some related trait. Heterochrony might not always be the direct focus of selection, but might result as a by-product of changes imposed by selection. For instance, an evolutionary decrease in body size has often resulted in extreme dwarfism (miniaturization: see Chapter 10). Especially in lissamphibians, such dwarfs have a truncated ontogenetic trajectory, and thus (adaptive) miniaturization has produced the pattern of pedomorphosis. In the following section, the embryonic mechanisms of skeleton formation will be discussed, to shed some light on how bones may appear or disappear due to minor genetic changes. The study of adaptive strategies driving developmental evolution opens yet another avenue of research: *ecological evolutionary developmental biology* (eco-evo-devo). Despite the new name, this field has a long tradition: the numerous studies summarized by Woltereck (1909), Schmalhausen (1949), and Waddington (1957) already provided a platform for this kind of research, and recently paleontology has started to contribute data in this field (see Chapter 8).

6.7 Body plans: gene regulation and morphogenesis

What determines organismic form? How are morphological traits produced in the first place? And what contribution do gene products make to morphological features? These are central questions that have only recently been approached by developmental biology and genetics. The causation of form in animals and plants puzzled the founding father of morphology, the German naturalist, writer and politician Goethe, who developed a theory of the origin of the skull long before natural selection was proposed as the cause of evolution. The French anatomist Geoffroy St.-Hilaire, Goethe's fellow spirit, studied embryology in great detail to counter Cuvier's idea of four separate types of design for animals. Geoffroy and Cuvier were at the center of a debate that continues to the present day: is organismic form determined purely by functional necessity (functionalism) or by structural properties of development (structuralism) (Rieppel 1990)?

Ever since these early attempts, the search has intensified for the organismic building blocks, the "architects," and their "construction plans." Yet only through the advent of genetics and its combination with modern techniques of developmental biology have answers to this question come within reach. Amphibians have played a pivotal role in this debate from the early days of embryology onward (Spemann and Mangold 1924; Hörstadius and Sellman 1946), and they are still among the preferred study organisms today (Olsson and Hanken 1996; Hall 2003; Olsson *et al.* 2005; Ericsson *et al.* 2009).

Builders and organizers in development. The molecular building blocks of life are so numerous that they will always provide ground for further analysis. Proteins rank among the most important molecules, because they are not only essential but extremely diverse, and they are the sources supplied by genes. Genes thus do not "code for" a morphological feature, but they provide raw material (proteins) that contributes to its formation (Nijhout 1990; Hall and Olson 2003). In addition to providing proteins as building blocks, some genes code for proteins (transcription factors) that regulate other genes, which sets up a hierarchy within the genome, the gene regulatory network (GRN: Davidson and Erwin 2006). This highly complicated network consists of multiple circuits between genes and other parts of DNA that activate, enhance, silence, or block gene expression. In sum, the GRN forms a highly buffered, evolutionarily conservative network that controls the timing and rate of developmental processes. "Buffered" means here that mutations that would change the network are usually lethal, and thus there is a strong stabilizing selection for network conservation (Galis *et al.* 2001). To ask the question in metaphors: If proteins are the building blocks, genes are their suppliers, and the GRN provides the construction plan – who are the builders?

In recent years, evo-devo has brought together several avenues of biological research (cell biology, histology, embryology, genetics) to come closer to an answer. The builders of morphology are small populations of cells called *condensations* (Hall 2005). They form the level at which genes exert their influence on morphology – by supplying proteins for specific cell properties and for signaling (inductive) communication between cells. Condensations and other primordial cell agglomerates (also called fields, germ layers, *anlagen*) are usually not present in adult organisms. Rather, they are the precursors of adult organs and tissues, existing only during well-defined embryonic periods. In working with salamanders, Spemann and Mangold (1924) discovered an early embryonic cell population they named "organizer." This primordial tissue deserves its name, because without its inductive action the notochord, gills slits, and dorsal nerve cord fail to develop (Gerhart 2001). *Induction* is the important word here, referring to signaling interactions between different cell populations. Only by induction do cell populations receive the positional information they need to migrate, settle, form condensations, and eventually start transforming into specialized cells (differentiation).

In the patterning of the cranial and branchial skeleton, the neural crest plays a pivotal role. This ectodermal tissue layer is yet another embryonic cell population, and forms a crucial autapomorphy of chordates (Gans and Northcutt 1983). It is so quintessential that it has sometimes been called a "fourth germ layer" (Hall 1999a). The neural crest cells originate in the dorsal part of the neural tube and later migrate into various regions, producing an amazing range of adult structures. Among others, it gives rise to nine cell types and twelve organs or tissue types, such as the brain, teeth, cranial cartilage and bone, connective tissue, eye, heart, thyroid and adrenal glands (Hall 1999b). The anterior or cranial neural crest is of particular importance for skeletal development and evolution, because it patterns most of the skull and branchial skeleton. In the developing embryo, the cells follow gradients of signal molecules, which guide their migration to the point where they will settle. Once that is reached, they will form condensations and start differentiation into a cell type such as cartilage-producing (chondroblast) or bone-producing (osteoblast), which then form the skeletal elements.

Skeleton formation. Building the skeleton is a multi-step process, with each step requiring the ones before (Hall 2005). Bone or cartilage is only formed when a specific threshold is reached, measured by the size of the cell condensation (Atchley and Hall 1991). Condensation size results from the number of constituent cells, and is regulated through signaling pathways, for instance the bone morphogenetic proteins (Bmp). Homeobox genes (*Hox*) further modulate the proliferation of cells within condensations. This is the level at which mutations exert their influence, and which can be "seen" by selection (Hall 2003). This is also the level where morphology is determined: slight variation in the size of a condensation may lead to a smaller or larger bone primordium and, eventually, a smaller or larger skeletal element (Hall 2005). More severe mutations may slow down proliferation or otherwise fail to produce condensations of sufficient size; the complete absence of the bone will be the result (Atchley and Hall 1991). It has long been known that both toxic substances and mutations of the genes involved will disrupt cell condensation processes. They do so by disturbing the environment within which the condensations develop, or to which they respond. *Hox* genes are crucially involved in determining the timing, position, and shape of condensations, and mutations of these genes teach interesting

lessons about buffering against mutational perturbations (Hall 2005).

Ossification sequences in the skull. The modification of the skull and branchial arches has played a major role both during the fish–tetrapod transition and in the origin of the modern amphibians. A proper understanding of these changes requires research along two lines of reasoning: functional and developmental. The functional aspects have been discussed above, by integration of fossil data and experimental functional evidence in an extant phylogenetic bracket (EPB). Developmental data can be integrated in a similar way, as both primary fossil evidence of cranial ontogeny and developmental-genetic data of extant taxa are now available. However, this field is still at an early stage, and will require much more work on the developmental patterning and formation of skeletal tissues.

Recently, comparative data on head formation have been supplemented by ossification sequences of fossil taxa (Boy 1974; Schoch 1992, 2004; Anderson 2007). Branchiosaurids and aïstopods have provided particularly detailed insights into ossification sequences of skull bones (Schoch 1992; Anderson 2002). These include two principal types of information: (1) temporal sequences of bone formation, and (2) spatial information on the initial shape of bone primordia, their growth and allometries (Schoch 2002). A rich sample of the branchiosaurid *Apateon caducus* has given the most detailed set of ontogenetic data, which parallel that of basal salamanders to a remarkable degree (Schoch 2002; 2006). The shared ossification sequence includes (1) the early formation of jaw and palatal elements including teeth, (2) the formation of elements along the long axis of the skull roof (frontal, parietal) and cheek (squamosal), (3) the succeeding ossification of supratemporal, nasal, and quadratojugal, and (4) the final appearance of prefrontal, postfrontal, postorbital, tabular, and jugal (Schoch 2002). Interestingly, it is not only this temporal sequence that appears to have been conserved, but also the shape of most bone primordia and their mode of growth (e.g., frontal, parietal, squamosal, premaxilla, and the palate bones). An exception is represented by the nasal, which develops from two separate centers in most salamanders and grows at a much slower rate than it did in *Apateon*. The parasphenoid reveals some most interesting patterns of heterochrony in the two genera, rooted by outgroup comparison with the temnospondyls *Micromelerpeton* and *Sclerocephalus*: in both branchiosaurids and salamanders, the growth trajectory of the bone is similar (separate *anlage* of plate and process, longitudinal growth of process, widening of plate, formation of a channel for the carotid arteries). However, *Apateon* went through these phases more quickly, and the trajectory of the bone ends with a very wide, differentiated basal plate, whereas in salamanders the trajectory ends at about two-thirds that of *Apateon*, a clear case of heterochrony and evolutionary allometry (Schoch 2002). This shows that both time and space are important in patterning morphology.

Studies on a broader taxonomic scale have revealed that the ossification sequence shared between branchiosaurids and larval salamanders also resembles that of bony fishes (*Amia*, *Polypterus*) and amniotes (turtles, *Sphenodon*, crocodilians, and mammals) (Schoch 2006). The early formation of jaws and palate bones is especially puzzling from a functional point of view. Although it makes sense for fishes and larval amphibians that have to feed shortly after hatching in the water, in amniotes, too, early embryonic ossification of jaws (and often teeth) occurs long before hatching or birth. This suggests that in amniotes the ossification sequence is conserved for some unknown epigenetic reason rather than because of functional demands (Schoch 2006). This adds to the common picture of "developmental constraints," a topic to which I return below.

Integrating data on skull development. Understanding the developmental basis for skull evolution is a major task, still requiring numerous steps to tackle experimental difficulties, but also promising great insights. Big questions emerge: Why are some units of the skull so conservative in evolution, whereas others are continuously modified? What gives identity to the cartilages and bones? How are bones changed, duplicated, and lost in evolution? For instance, ossification sequences and primordial growth patterns are remarkably similar between salamanders, caecilians, temnospondyls, and actinopterygians, but radically different in frogs (Trueb 1985; Schoch

2006). The answer must lie, time and again, at the level of condensations and the primordia they pattern. Studies on salamanders and lungfishes have shown how cranial neural crest cells migrate in separate streams towards the head to form different structures (Olsson and Hanken 1996): a *mandibular stream* contributes to the nasal capsule, anterior braincase, and jaws (cartilages and bones), a *hyoid stream* forms the hyoid arch, and a *branchial stream* the branchial arches. Ongoing research on the axolotl will shed more light on the skeletal elements formed by these streams (N. Piekarski, personal communication 2012). These streams form domains that are further parceled to generate more specific structures, as revealed by studies on the maxillary primordium in chicks (Bogardi *et al.* 2000). This primordium, deriving from the mandibular stream, gives rise to an integrated set of bones: the quadratojugal, jugal, maxilla, pterygoid, and palatine. Experiments with this primordium suggested that the skeletal elements form a series in which bone is patterned sequentially (Bogardi *et al.* 2000). Homeobox genes (*Dlx*, *Msx*) and growth factors (*Bmp*, *Fgf*) are involved in patterning these bones, and mutations lead to duplicated elements, enlarged bones, and novel arrangements. A similar parcellation was found by Cassin and Capuron (1979) in the palate of the salamandrid *Pleurodeles*, where vomer, palatine, and pterygoid are formed only when cartilage has formed in the anterior braincase. Signaling interactions are important on several levels, and in bird embryos, tissues that induce skeletal elements have been identified: the notochord for the parasphenoid and basisphenoid, the mesencephalic brain portion for the squamosal and parietal, the prosencephalon for the frontal, etc. (Hall 1999b). Schmalhausen (1968) reported that the nasolacrimal duct is required for the formation of septomaxilla, nasal, and lacrimal in salamanders. In addition to induction by surrounding tissues, mechanical forces also contribute to the formation and growth of skeletal elements (Hall 1999b).

Future research will have to address evolutionary questions by fitting the manifold experimental results into an EPB. At the present stage, only pieces of information on induction, involved *Hox* genes, gene regulation, and patterning gradients are available. It is unclear how universal or clade-specific a particular pattern is. The comparative approach applied by Olsson and Hanken (1996) in anurans is a promising line of research. This is necessary, because the contribution of the neural crest to skull bones has been found to be quite labile, indicating taxonomic diversity of development (Hanken and Gross 2005). It is likely that various developmental modules exist that pattern the skull, which are defined by signaling interactions and neural crest streams. How do sequences of chondrification and ossifications relate to these modules? How are sequence heterochronies and skeletal heterotopies generated? Answering these questions will require a lot more work (Gross and Hanken 2008), but they promise an unprecedented

depth of understanding. A truly causal morphology, as sought by Goethe, Oken, and Haeckel and laid out by Schmalhausen (1949) and Riedl (1978), is finally showing its contours (Köntges and Lumsden 2000; Hall 2005).

Formation of limbs. A second, fascinating line of research has focused on the developmental origin of limbs (Clack 2009). This topic is an excellent example of fruitful cooperation between developmental biology, genetics, and paleontology within the framework of evo-devo (Figure 6.11). From a molecular and cellular point of view, fins and limbs develop in similar ways (Hall 1999a). Signaling interaction between mesenchyme and ectoderm regulates the proliferation and patterning along the three spatial axes. In contrast

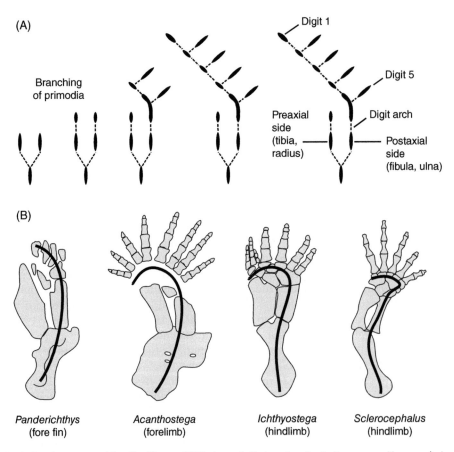

Figure 6.11 Limb development and fossil evidence. (A) Embyronic limb patterning in the anuran *Xenopus* (adapted from Shubin and Alberch 1986): the digits arise from the post-axial side of the main limb axis, contrasting the pre-axial fin rays of fishes. (B) A series of tetrapodomorph limbs with the main axis mapped, showing the curved end in the hand or foot.

to the neural crest-derived neurocranium, the endoskeleton of limbs is of mesodermal origin, while the dermal rays (lepidotrichia), which are lost in tetrapods, develop in the ectodermal layer.

Shubin and Alberch (1986) and Oster *et al.* (1988) identified two mechanisms crucial for the patterning of limbs: *branching* and *segmentation* of the limb primordium. As in the skull, skeletal elements are preformed as condensations which form in a sequential fashion: the main limb axis develops by successive branching (bifurcation) and subsequent parcellation of primordia, resulting in a hierarchically nested set of skeletal primordia. Rather than a symmetrical bush, the resulting pattern is a main axis with successive side branches. The main axis is formed by the sequence humerus – radius – carpal – digit 1. The ulna originates by branching at the first node, the carpals at the second, the first digit at the third, and so forth. In such a nested primordium, digits may be added or lost readily.

The main axis is shared by fishes and tetrapods, as is the sequentially nested pattern (Figure 6.11. However, Shubin and Alberch (1986) found that digits – the almost quintessential autapomorphy of tetrapods – develop in a completely different way from the radials in fishes. They branch off along the posterior side of the main axis (*post-axial*), whereas the radials of fishes branch off along the anterior side (*pre-axial*). In Shubin and Alberch's model, the nested pattern is shared, but mirror-imaged. This is a case of heterotopy, and at the morphological level the hand (and foot) is a new structure. In this concept, there is no phylogenetic precursor of the hand in bony fishes. A major advantage of the Shubin and Alberch model is that the number of digits need not be confined to five, which was later found to be consistent with fossil discoveries (*Acanthostega*: 8; *Ichthyostega*: 7; *Tulerpeton*: 6 digits). Later it was found that digits which derive from the primary axis appear to be the most stable in evolution.

Recent evidence has revealed that the story may be more complicated (Wagner and Larsson 2007; Clack 2009). The patterns are less universal than thought, and although a primary axis has been confirmed, it appears not to be the same as the metapterygoid axis (Wagner and Larsson 2007). Substantial progress has been made in understanding the molecular causation of limb development, and again deep homology has been found. The same (= orthologous) cluster of *Hox* genes is involved in the patterning of both fin and limb structures. For instance, *Hox* A and D are the key genes for appendage development in the zebrafish, chicken, and mouse, involving FGF signal molecules and *Tbx* transcription factor. The very important *Shh* molecule patterns the distal end of the limbs in all gnathostomes, and without it no digits are formed in tetrapods. All these data indicate that the interplay between different *Hox* genes is decisive for the formation of limbs, and that the autopodium evolved by slight changes in the regulation and expression domains of particular genes in the *Hox* clusters (Wagner and Larsson 2007).

Developmental and fossil data also yielded evidence on a crucial evolutionary change within amphibians (Fröbisch 2008). The digits of anurans and amniotes form in a consistent fashion, starting with those holding the highest number (post-axial) and progressing sequentially to the lowest (pre-axial). In salamanders, the opposite pattern is observed, called pre-axial dominance. Based on a detailed fossil ossification sequence of *Apateon caducus*, Fröbisch (2008) showed that the pre-axial pattern of ossification is also found in that branchiosaurid. Future finds of very small larval specimens might reveal whether pre-axial dominance was a dissorophoid apomorphy, or whether it reaches back further within temnospondyls.

Causation of development and evolution. Three major problems emerge from the new field of evo-devo: (1) the origin of order and organization; (2) the remarkable conservation of body plans; and, most importantly, (3) the origin of novelties. The discovery of *Hox* genes was the big surprise of the 1980s and 1990s; even distantly related metazoans were shown to share the same genes and, more stunningly still, the same regulatory networks (Hall 1996; Davidson and Erwin 2006). The sequential patterning of body regions and the regulated behavior of cell condensations lie at the heart of biological organization. This reveals a first major principle of organismic order: *modularity* (Riedl 1978; Schlosser and Wagner 2004). The conservation of body plans, exemplified by *Hox* genes as well as invariant anatomical features, highlights the tight *integration* of mechanisms and morphological units. Integrated features are often referred to "developmental constraints," "genetic constraints," or "burdens," but they are mostly caused by pleiotropy, which is in turn upheld by stabilizing selection (Schwenk and Wagner 2004; Galis *et al.* 2001). Of course, modularity and integration are two sides of one coin, but they also have antagonistic properties.

The third problem is the least understood: How do new features originate in evolution? Here, a consensus has not yet been reached. Some argue that adaptation (an ultimate-level explanation) is sufficient to explain novelty, whereas others hold that it also needs a new property of development (a proximate-level explanation) in order to create a novelty. The main question appears to be: Is evolution only a tinkerer, using and exapting pre-existing structures, or does it also *create entirely new structures*? "There is nothing new under the sun," goes an old saying. Wagner and Larsson (2007) counter this view by defining novelty as a *new morphogenetic option*, suggesting that changes in gene regulation are a likely cause. Davidson and Erwin (2006) showed that such "genetic rewiring" does occur and has profound effects in evolution. New body parts, such as the autopodia of tetrapods, are obvious candidates for novelties that evolved by changes to the regulatory control of the limb field. Recruitment of existing *Hox* genes permitted the control of a new morphogenetic module. This highlights again how important modularity can be for evolution.

References

Alberch, P. (1985) Problems with the interpretation of developmental sequences. *Systematic Zoology* **34**, 46–58.

Alberch, P., Gould, S.J., Oster, G.F., & Wake, D.B. (1979) Size and shape in ontogeny and phylogeny. *Paleobiology* **5**, 296–317.

Alexander, R.M. (1990) *Animals*. Cambridge: Cambridge University Press.

Anderson, J.S. (2002) Revision of the aïstopod genus *Phlegethontia* (Tetrapoda: Lepospondyli). *Journal of Paleontology* **76**, 1029–1046.

Anderson, J.S. (2003) Cranial anatomy of *Coloraderpeton brilli*, postcranial anatomy of *Oestocephalus amphiuminus*, and reconsideration of Ophiderpetontidae (Tetrapoda: Lepospondyli: Aïstopoda). *Journal of Vertebrate Paleontology* **23**, 532–543.

Anderson, J.S. (2007) Incorporating ontogeny into the matrix: a phylogenetic evaluation of developmental evidence for the origin of modern amphibians. In: J.S. Anderson & H.-D. Sues (eds.), *Major Transitions in Vertebrate Evolution*. Bloomington: Indiana University Press, pp. 182–212.

Anderson, J.S., Carroll, R.L., & Rowe, T.B. (2003) New information on *Lethiscus stocki* (Tetrapoda: Lepospondyli) from high-resolution computed tomography and a phylogenetic analysis of Aïstopoda. *Canadian Journal of Earth Sciences* **40**, 1071–1083.

Atchley, W.R. & Hall, B.K. (1991) A model for development and evolution of complex morphological structures. *Biological Reviews* **66**, 101–157.

Bogardi, J.P., Barlow, A., & Francis-West, P. (2000) The role of FGF-8 and BMP-4 in the outgrowth and patterning of the chick embryonic maxillary primordium. In: L. Olsson & C.O. Jacobson (eds.), *Regulatory Processes in Development*. London: Portland Press, pp. 173–177.

Bolt, J.R. & Lombard, R.E. (2000) Palaeobiology of *Whatcheeria deltae*, a primitive Mississippian tetrapod. In: H. Heatwole & R.L. Carroll (eds.), *Amphibian Biology. Volume 4. Palaeontology*. Chipping Norton, NSW: Surrey Beatty, pp. 1044–1052.

Bossy, K.A. & Milner, A.C. (1998) Order Nectridea. In: P. Wellnhofer (ed.), *Handbuch der Paläontologie*. Munich: Pfeil, Vol. 1, pp. 73–131.

Boy, J.A. (1972) Die Branchiosaurier (Amphibia) des saarpfälzischen Rotliegenden (Perm, SW-Deutschland). *Abhandlungen des Hessischen Landesamts für Bodenforschung* **65**, 1–137.

Boy, J.A. (1974) Die Larven der rhachitomen Amphibien (Amphibia: Temnospondyli, Karbon-Trias). *Paläontologische Zeitschrift* **48**, 236–268.

Boy, J.A. (1988) Über einige Vertreter der Eryopoidea (Amphibia: Temnospondyli) aus dem europäischen Rotliegend (? höchstes Karbon – Perm). 1. *Sclerocephalus. Paläontologische Zeitschrift* **62**, 107–132.

Boy, J. A. (1989) Über einige Vertreter der Eryopoidea (Amphibia: Temnospondyli) aus dem europäischen Rotliegend (? höchstes Karbon –

Perm). 2. Acanthostomatops. *Paläontologische Zeitschrift* **63**, 133–151.

Boy, J.A. (1990) Über einige Vertreter der Eryopoidea (Amphibia: Temnospondyli) aus dem europäischen Rotliegend (? höchstes Karbon – Perm). 3. *Onchiodon*. Paläontologische Zeitschrift **64**, 287–312.

Boy, J.A. (1995) Über die Micromelerpetontidae (Amphibia: Temnospondyli). 1. Morphologie und Paläoökologie des *Micromelerpeton credneri* (Unter-Perm; SW-Deutschland). *Paläontologische Zeitschrift* **69**, 429–457.

Boy, J.A. & Sues, H.-D. (2000) Branchiosaurs: larvae, metamorphosis and heterochrony in temnospondyls and seymouriamorphs. In: H. Heatwole & R.L. Carroll (eds.), *Amphibian Biology. Volume 4. Palaeontology*. Chipping Norton, NSW: Surrey Beatty, pp. 973–1496.

Carroll, R.L. (1989) Developmental aspects of lepospondyl vertebrae in Paleozoic tetrapods. *Historical Biology* **3**, 1–25.

Carroll, R.L. & Gaskill, P. (1978) The order Microsauria. *Memoirs of the American Philospohical Society* **126**, 1–211.

Cassin, C. & Capuron, A. (1979) Buccal organogenesis in *Pleurodeles waltli* Michah (urodele, amphibian). Study by intrablastocoelic transplantation and in vitro culture. *Journal du Biologie buccale* **7**, 61–76.

Castanet, J., Francillon-Vielleiot, H., de Ricqlès, A., & Zylberberg, L. (2003) The skeletal histology of the Amphibia. In: H. Heatwole & R.L. Carroll (eds.), *Amphibian Biology. Volume 5. Osteology*. Chipping Norton, NSW: Surrey Beatty, pp. 1597–1683.

Castanet, J. & Smirina, E. (1990) Introduction to the skeletochronological method in amphibians and reptiles. *Annales des Sciences Naturelles – Zoologie et Biologie animale* **11**, 191–196.

Clack, J. A. (2009) The fin to limb transition: new data, intepretations, and hypotheses from paleontology and developmental biology. *Annual Reviews of Earth and Planetary Sciences* **37**, 163–179.

Clack, J.A. & Milner, A.R. (2010) *Platyrhinops* from the Upper Carboniferous of Linton and Nýřany and the family Amphibamidae (Amphibia: Temnospondyli). *Transactions of the Royal Society of Edinburgh: Earth and Environmental Sciences* **100**, 275–295.

Cote, S., Carroll, R.L., Cloutier, R., & Bar-Sagi, L. (2002) Vertebral development in the Devonian sarcopterygian fish *Eusthenopteron foordi* and the polarity of vertebral evolution in non-amniote tetrapods. *Journal of Vertebrate Paleontology* **22**, 487–502.

Credner, H. (1881) Die Stegocephalen (Labyrinthodonten) aus dem Rothliegenden des Plauenschen Grundes. 2. Theil. *Zeitschrift der deutschen geologischen Gesellschaft* **33**, 298–330.

Darwin, C. (1859) *On the Origin of Species by Means of Natural Selection*. London: John Murray.

Davidson, E.H. & Erwin, D.H. (2006) Gene regulatory networks and the evolution of animal body plans. *Science* **311**, 796–800.

de Ricqlès, A. (1975) Quelques remarques paléohistologiques sur le problème de la néotenie chez les stégocéphales. *CNRS Colloquium International* **218**, 351–363.

de Ricqlès, A. (1979) Relations entre structures histologiques, ontogenese, strategies demographiques et modalites évolutives: le cas des reptiles captorhinomorphes et des stégocéphales temnospondyles. *Comptes Rendus Académie des Sciences Paris* **228**, 1147–1150.

Duellman, W.E. & Trueb, L. (1994) *Biology of Amphibians*. Baltimore: Johns Hopkins University Press.

Elwood, J.R.L. & Cundall, D. (1994) Morphology and behavior of the feeding apparatus in *Cryptobranchus alleganiensis* (Amphibia: Caudata). *Journal of Morphology* **220**, 47–70.

Emerson, S.B. (1986) Heterochrony and frogs: the relationship of a life history trait to morphological form. *American Naturalist* **127**, 167–183.

Emerson, S.B. & Bramble, D.M. (1993) Scaling, allometry, and skull design. In: J. Hanken & B.K. Hall (eds.), *The Skull*. Chicago: University of Chicago Press, Part 3, pp. 384–421.

Ericsson, R., Ziermann, J.M., Piekarski, N., Schubert, G., Joss, J., & Olsson, L. (2009) Cell fate and timing in the evolution of neural crest and mesoderm development in the head region of amphibians and lungfishes. *Acta Zoologica* **90** (Suppl.), 264–272.

Francillon-Vieillot, H., de Buffrenil, V., Castanet, J., *et al.* (1990) Microstructure and mineralization of vertebrate skeletal tissues. In: J.G. Carter (ed.), *Skeletal Biomineralization: Processes and Evolutionary Trends.* New York: Van Nostrand Reonhold, pp. 471–533.

Fritsch, A. (1879–1901) *Fauna der Gaskohle und der Kalksteine der Permformations Böhmens,* Vols. 1–4. Prague: self-published.

Fröbisch, N.B. (2008) Ossification patterns in the tetrapod limb: conservation and divergence from morphogenetic events. *Biological Reviews* **83**, 571–600.

Fröbisch, N.B. & Schoch, R.R. (2009) Testing the impact of miniaturization on phylogeny: Paleozoic dissorophoid amphibians. *Systematic Biology* **58**, 312–327.

Fröbisch, N.B, Carroll, R.L., & Schoch, R.R. (2007) Limb ossification in the Paleozoic branchiosaurid *Apateon* (Temnospondyli) and the early evolution of preaxial dominance in tetrapod limb development. *Evolution and Development* **9**, 69–75.

Fröbisch, N.B., Olori, J., Schoch, R.R, & Witzmann, F. (2010) Amphibian development in the fossil record. *Seminars in Cell and Developmental Biology* **21**, 424–431.

Galis, F., van Alphen, J.J.M., & Metz, J.A.J. (2001) Why five fingers? Evolutionary constraints on digit numbers. *Trends in Ecology and Evolution* **16**, 637–646.

Gans, C. & Northcutt, R.G. (1983) Neural crest and the origin of vertebrates. *Science* **220**, 268–274.

Garstang, W. (1920) The theory of recapitulation: a critical restatement of the biogenetic law. *Zoological Journal of the Linnean Society* **35**, 81–101.

Gerhart, J. (2001) Evolution of the Organizer and the chordate body plan. *International Journal of Developmental Biology* **45**, 133–153.

Germain, D. & Laurin, M. (2009) Evolution of ossification sequences in salamanders and urodele origins assessed through event-pairing and new methods. *Evolution and Development* **11**, 170–190.

Godfrey, S.J. (1989) Ontogenetic changes in the skull of the Carboniferous tetrapod *Greererpeton burkemorani* Romer 1969. *Philosophical Transactions of the Royal Society London B* **323**, 135–153.

Gould, S.J. (1966) Allometry and size in ontogeny and phylogeny. *Biological Reviews* **41**, 587–638.

Gould, S.J. (1977) *Ontogeny and Phylogeny.* Cambridge, MA: Belknap Press.

Gradwell, N. (1969) The function of the internal nares of the bullfrog tadpole. *Herpetologica* **25**, 120–121.

Gross, J.B. & Hanken, J. (2008) Review of fate-mapping studies of osteogenic cranial neural crest in vertebrates. *Developmental Biology* **317**, 389–400.

Haas, A. (2003) Phylogeny of frogs as inferred from primarily larval characters (Amphibia: Anura). *Cladistics* **19**, 23–89.

Haas, A. (2010) Lissamphibia. In: W. Westheide & R. Rieger (eds.), *Spezielle Zoologie. Wirbel- oder Schädeltiere.* Heidelberg: Spektrum Akademischer Verlag, pp. 330–359.

Haeckel, E. (1866) *Generelle Morphologie der Organismen.* Berlin: Reimer.

Hall, B.K. (1996) Baupläne, phylotypic stages, and constraint. Why are there so few types of animals? *Evolutionary Biology* **29**, 215–261.

Hall, B.K. (1999a) *Evolutionary Developmental Biology,* 2nd edition. Dordrecht: Kluwer.

Hall, B.K. (1999b) *The Neural Crest in Development and Evolution.* Heidelberg: Springer.

Hall, B.K. (2003) Developmental and cellular origins of the amphibian skeleton. In: H. Heatwole & R.L. Carroll (eds.), *Amphibian Biology. Volume 5. Osteology.* Chipping Norton, NSW: Surrey Beatty, pp. 1551–1597.

Hall, B.K. (2005) *Bone and Cartilage. Evolutionary Skeletal Biology.* Amsterdam: Elsevier.

Hall, B.K. & Olson, W.M. (2003) Introduction: evolutionary developmental mechanisms. In: B.K. Hall & W.M. Olson (eds.), *Keywords and Concepts in Evolutionary Developmental Biology.* Cambridge, MA: Harvard University Press, pp. xii–xvi.

Hanken, J. (1983) Miniaturization and its effects on cranial morphology in plethodontid salamanders, genus *Thorius* (Amphibia: Plethodontidae). II. The fate of the brain and sense organs and their role in skull morphogenesis and evolution. *Journal of Morphology* **177**, 155–268.

Hanken, J. & Gross, J.B. (2005) Evolution of cranial development and the role of neural crest: insights from amphibians. *Journal of Anatomy* **207**, 437–446.

Hörstadius, S. & Sellman, S. (1946) Experimentelle Untersuchungen über die Determination des knorpeligen Kopfskeletts bei Urodelen. *Novae*

Acta Regiae Societatis Scientiarum Uppsala, Serie IV, **13**, 1–170.

Ivanović, A., Vukov, T.T., Dzukic, G., Tomasevic, N., & Kalezic, M. (2007) Ontogeny of skull size and shape changes within a framework of biphasic lifestyle: a case study in six *Triturus* species (Amphibia, Salamandridae). *Zoomorphology* **126**, 173–183.

Kemp, N.E. & Hoyt, J.A. (1969) Sequence of ossification in the skeleton of growing and metamorphosing tadpoles of *Rana pipiens*. *Journal of Morphology* **129**, 415–443.

Klembara J. (1994) Electroreceptors in the Lower Permian Discosauriscus austriacus. *Palaeontology* **37**, 609–626.

Klembara, J. (1995) The external gills and ornamentation of skull-roof bones of the Lower Permian Discosauriscus (Kuhn 1933) with remarks to its ontogeny. *Paläontologische Zeitschrift* **69**, 265–281.

Klembara, J. & Bartík, I. (2000) The postcranial skeleton of *Discosauriscus* Kuhn, a seymouriamorph tetrapod from the Lower Permian of the Boskovice Furrow (Czech Republic). *Transactions of the Royal Society of Edinburgh: Earth Sciences* **90**, 287–316

Klembara, J. & Ruta, M. (2005) The seymouriamorph tetrapod *Ariekanerpeton sigalovi* from the Lower Permian of Tadzhikistan. Part I: cranial anatomy and ontogeny. *Transactions of the Royal Society of Edinburgh: Earth Sciences* **96**, 43–70.

Klembara, J., Berman, D.S., Henrici, A.C., Cernansky, A., & Werneburg, R. (2006) Comparison of cranial anatomy and proportions of similarly sized Seymouria sanjuanensis and Discosauriscus austriacus. *Annals of the Carnegie Museum* **75**, 37–49.

Klingenberg, C.P. (1998) Heterochrony and allometry: the analysis of evolutionary change in ontogeny. *Biological Reviews* **73**, 79–123.

Kollmann, J. (1885) Das Überwintern europäischer Frosch- und Tritonlarven und die Umwandlung des mexikanischen Axolotl. *Verhandlungen der Naturforschenden Gesellschaft in Basel* **7**.

Köntges, G. & Lumsden, A. (2000) Hindbrain neural crest segmentation: tracing compartments through development and evolution. In: L. Olsson & C.O. Jacobson (eds.), *Regulatory Processes in Development*. London: Portland Press, pp. 89–100.

Laurin, M., Girondot, M., & Loth, M.L. (2004) The evolution of long bone microstructure and lifestyle of lissamphibians. *Paleobiology* **30**, 589–613.

Lebedkina, N.S. (2004) *Evolution of the Amphibian Skull*. Sofia: Pensoft.

McNamara, K. (1995) *Evolutionary Change and Heterochrony*. Chichester: Wiley.

Mickoleit, G. (2004) *Phylogenetische Systematik der Wirbeltiere*. Munich: Pfeil.

Milner, A.C. (1996) A juvenile diplocaulid nectridean amphibian from the Lower Permian of Texas and Oklahoma. *Special Papers in Palaeontology* **52**, 129–138.

Milner, A.R. (1982) Small temnospondyl amphibians from the Middle Late Carboniferous of Illinois. *Palaeontology* **25**, 635–664.

Milner, A.R. (1988) The relationships and origin of living amphibians. In: M.J. Benton (ed.), *The Phylogeny and Classification of the Tetrapods*. Oxford: Clarendon Press, Vol. 1, pp. 59–102.

Milner, A.R. (2008) The tail of *Microbrachis* (Tetrapoda; Microsauria). *Lethaia* **41**, 257–261.

Müller, H. (2006) Ontogeny of the skull, lower jaw, and hyobranchial skeleton of *Hypogeophis rostratus* (Amphibia: Gymnophiona: Caeciliidae) revisited. *Journal of Morphology* **267**, 968–986.

Müller, H., Oommen, V.O., & Batrsch, P. (2005) Skeletal development of the direct-developing caecilian *Gegeneophis ramaswamii* (Amphibia: Gymnophiona: Caeciliidae). *Zoomorphology* **124**, 171–188.

Nijhout, F. (1990) Metaphors and the role of genes in development. *BioEssays* **12**, 441–446.

Olori J. (2008) Postcranial ossification sequence and morphogenesis of Microbrachis pelikani (Lepospondyli: Tetrapoda). *Journal of Vertebrate Paleontology* **28**, 123A.

Olson, E.C. (1951) *Diplocaulus*, a study in growth and variation. *Fieldiana: Geology* **11**, 57–149.

Olsson, L. & Hanken, J. (1996) Cranial neural crest migration and chondrogenic fate in the oriental fire-bellied toad *Bombina orientalis*: defining the ancestral pattern of head development in anuran amphibians. Journal of Morphology **229**, 105–120.

Olsson, L., Ericsson, R., & Cerny, R. (2005) Vertebrate head development: segmentation, novelties, and homology. *Theory in Bioscience* **124**, 145–163.

Organ, C.L., Canoville, A., Reisz, R.R., & Laurin, M. (2010) Paleogenomic data suggest mammal-like genome size in the ancestral amniote and derived large genome size in amphibians. *Journal of Evolutionary Biology* **24**, 372–380.

Orton, G.L. (1953) The systematics of vertebrate larvae. *Systematic Zoology* **2**, 63–75.

Oster, G.F., Shubin, N., Murray, J.D., & Alberch, P. (1988) Evolution and morphogenetic rules. The Shape of the vertebrate limb in ontogeny and phylogeny. *Evolution* **45**, 862–884.

Panchen, A.L. (1970) Anthracosauria. In: O. Kuhn (ed.), *Encyclopedia of Paleoherpetology*, Vol. 5A. Stuttgart: Fischer.

Peabody, F.E. (1961) Annual growth zones in living and fossil vertebrates. *Journal of Morphology* **108**, 11–62.

Poe, S. (2006) Test of von Baer's law of the conservation of early development. *Evolution* **60**, 2239–2245.

Raff, R.A. (1996) *The Shape of Life: Genes, Development, and the Evolution of Animal Form*. Chicago: University of Chicago Press.

Reif, W.E. (1986) The search for macroevolutionary theory in German paleontology. *Journal of the History of Biology* **19**, 79–130.

Reilly, S.M. (1987) Ontogeny of the hyobranchial apparatus in the salamanders *Ambystoma talpoideum* (Ambystomatidae) and *Notophthalmus viridescens* (Salamandridae): the ecoolgical morphology of two neotenic strategies. *Journal of Morphology* **191**, 205–214.

Reilly, S.M. & Lauder, G.V. (1990) Metamorphosis of cranial design in tiger salamanders *(Ambystoma tigrinum)*: a morphometric analysis of ontogenetic change. *Journal of Morphology* **204**, 121–137.

Reilly, S.M., Wiley, E.O., & Meinhardt, D.J. (1997) An integrative approach to heterochrony: distinguishing intraspecific and interspecific phenomena. *Biological Journal of the Linnean Society* **60**, 119–143.

Reiss, J.O. (2002) The phylogeny of amphibian metamorphosis. *Zoology* **105**, 1–12.

Riedl, R. (1978) *Order in Living Organisms*. New York: Wiley.

Rieppel, O. (1990) Structuralism, functionalism, and the four Aristotelian causes. *Journal of the History of Biology* **23**, 291–320.

Rinehart, L.F. & Lucas, S.G. (2001) A statistical analysis of a growth series of the Permian nectridean *Diplocaulus magnicornis* showing two-stage ontogeny. *Journal of Vertebrate Paleontology* **21**, 803–806.

Rose, C.S. (2003) The developmental morphology of salamander skulls. In: H. Heatwole & R.L. Carroll (eds.), *Amphibian Biology. Volume 5. Osteology*. Chipping Norton, NSW: Surrey Beatty, pp. 1684–1781.

Sanchez, M. (2012) *Embryos in Deep Time: the Rock Record of Biological Development*. Berkeley: University of California Press.

Sanchez, S., Klembara, J., Castanet, J., & Steyer, J.S. (2008) Salamander-like development in a seymouriamorph revealed by palaeohistology. *Biology Letters* **4**, 411–414.

Sanchez, S., de Ricqlès, A., Schoch, R.R., & Steyer, J.S. (2010a) Developmental plasticity of limb bone microstructural organization in *Apateon*: histological evidence of paedomorphic conditions in branchiosaurs. *Evolution and Development* **12**, 315–328.

Sanchez, S., Ploeg, G, Clement, G., & Ahlberg, P.E. (2010b) A new tool for determining degrees of mineralization in fossil amphibian skeletons: The example of the late Palaeozoic branchiosaurid Apateon from the Autun Basin, France. *Comptes Rendus Palevol* **9**, 311–317.

Sanchez, S., Steyer, J.S., Schoch, R.R., & de Ricqlès, A. (2010c) Palaeoecological and palaeoenvironmental influences revealed by long-bone palaeohistology: the example of the Permian branchiosaurid *Apateon*. *Geological Society Special Publication* **339**, 139–150.

Scheyer, T., Klein, N., & Sander, P.M. (2010) Developmental palaeontology of Reptilia as revealed by histological studies. *Seminars in Cell and Developmental Biology* **21**, 462–470.

Schindewolf, O.H. (1950) *Grundfragen der Paläontologie*. Stuttgart: Schweizerbart.

Schlichting, C. & Pigliucci, M. (1998) *Phenotypic Plasticity: a Reaction Norm Perspective*. Sunderland: Sinauer.

Schlosser, G. & Wagner, G.P. (2004) *Modularity in Development and Evolution*. Chicago: University of Chicago Press.

Schmalhausen, I.I. (1949) *Factors of Evolution*. Chicago: University of Chicago Press.

Schmalhausen, I.I. (1968) *The Origin of Terrestrial Vertebrates.* London and New York: Academic Press.

Schoch, R.R. (1992) Comparative ontogeny of Early Permian branchiosaurid amphibians from southwestern Germany. Developmental stages. *Palaeontographica A* **222**, 43–83.

Schoch, R.R. (2002) The early formation of the skull in extant and Paleozoic amphibians. *Paleobiology* **28**, 378–396.

Schoch, R.R. (2004) Skeleton formation in the Branchiosauridae as a case study in comparing ontogenetic trajectories. *Journal of Vertebrate Paleontology* **24**, 309–319.

Schoch, R.R. (2006) Skull ontogeny: developmental patterns of fishes conserved across major tetrapod clades. *Evolution and Development* **8**, 524–536.

Schoch, R.R. (2009a) The evolution of life cycles in early amphibians. *Annual Review of Earth and Planetary Sciences* **37**, 135–162.

Schoch, R.R. (2009b) Developmental evolution as a response to diverse lake habitats in Paleozoic amphibians. *Evolution* **63**, 2738–2749.

Schoch, R.R. (2010) Heterochrony: the interplay between development and ecology in an extinct amphibian clade. *Paleobiology* **36**, 318–334.

Schoch, R. R. & Fröbisch, N.B. (2006) Metamorphosis and neoteny: alternative developmental pathways in an extinct amphibian clade. *Evolution* **60**, 1467–1475.

Schoch, R.R. & Milner, A.R. (2008) The intrarelationships and evolutionary history of the temnospondyl family Branchiosauridae. *Journal of Systematic Palaeontology* **6**, 409–431.

Schultze, H.-P. (1984) Juvenile specimens of *Eusthenopteron foordi* Whiteaves, 1881 (osteolepiform rhipidistian, Pisces) from the Upper Devonian of Miguasha, Quebec, Canada. *Journal of Vertebrate Paleontology* **4**, 1–16.

Schwenk, K. & Wagner, G.P. (2004) The relativism of constraints on phenotypic evolution. In: M. Pigliucci & K. Preston (eds.), *Phenotypic Integration: Studying the Ecology and Evolution of Complex Phenotypes.* Oxford: Oxford University Press, pp. 390–408.

Semlitsch, R.D., Harris, R.N., & Wilbur, H.M (1990) Paedomorphosis in Ambystoma talpoideum: maintenance of population variation and alternative life-history pathways. *Evolution* **44**, 1604–1613.

Sequeira, S.E.K. (2004) The skull of *Cochleosaurus bohemicus* Fric, a temnospondyl from the Czech Republic (Upper Carboniferous) and cochleosaurid interrelationships. *Transactions of the Royal Society of Edinburgh: Earth Sciences* **94**, 21–43.

Sequeira, S.E.K. (2009) The postcranium of *Cochleosaurus bohemicus* Frič, a primitive Upper Carboniferous temnospondyl from the Czech Republic. *Special Papers in Palaeontology* **81**, 137–153.

Shubin, N.H. & Alberch, P. (1986) A morphogenetic approach to the origin and basic organisation of the tetrapod limb. *Evolutionary Biology* **20**, 319–387.

Spemann, H. & Mangold, H. (1924) Über Induktion von Embryonalanlagen durch Implantation artfremder Organisatoren. *Archiv für mikroskopische Anatomie und Entwicklungsmechanik* **100**, 599–638.

Stayton, T. & Ruta, M. (2006) Geometric morphometrics of the skull roof of stereospondyls (Amphibia: Temnospondyli). *Palaeontology* **49**, 307–337.

Steyer, J.S. (2000) Ontogeny and phylogeny in temnospondyls: a new method of analysis. *Zoological Journal of the Linnean Society* **130**, 449–467.

Steyer, J.S., Laurin, M., Castanet, J., & de Ricqlès, A. (2004) First histological and skeletochronological data on temnospondyl growth; palaeoecological and palaeoclimatological implications. *Palaeogeography, Palaeoecology, Palaeoclimatology* **206**, 193–201.

Thompson, D'A.W. (1941) *On Growth and Form.* Cambridge: Cambridge University Press.

Trueb, L. (1985) A summary of osteocranial development in anurans with notes on the sequence of cranial ossification in *Rhinophrynus dorsalis* (Anura: Pipoidea: Rhinophrynidae). *South African Journal of Science* **81**, 181–185.

Ultsch, G.R. (2012) Metabolism, gas exchange, and acid–base balance of giant salamanders. *Biological Reviews* **87**, 583–601.

von Baer, K.E. (1828) *Über Entwickelungsgeschichte der Thiere.* Königsberg: Weidmann.

Waddington, C.H. (1942) Canalization of development and the inheritance of acquired characters. *Nature* **150**, 563–565.

Waddington, C.H. (1957) *The Strategy of the Genes*. London: Allen and Unwin.

Wagner, G.P. & Larsson, H.C.E. (2007) Fins and limbs in the study of evolutionary novelties. In: B.K. Hall (ed.), *Fins into Limbs: Evolution, Development, and Transformation*. Chicago: University of Chicago Press, pp. 49–61.

Wake, D.B. (1982) Functional and developmental constraints and opportunities in the evolution of feeding systems in urodeles. In: D. Mossakowski & G. Roth (eds.), *Environmental Adaptation and Evolution*. Stuttgart: Gustav Fischer, pp. 51–66.

Wake, D.B. (2009) What salamanders have taught us about evolution. *Annual Review of Ecology and Systematics* **40**, 333–352.

Wake, D.B. & Roth, G. (1989) The linkage between ontogeny and phylogeny in the evolution of complex systems. In: D.B. Wake & G. Roth (eds.), *Complex Organismal Function: Integration and Evolution in Vertebrates*. Chichester: Wiley, pp. 361–377.

Wake, M.H. & Hanken, J. (1982) Development of the skull of *Dermophis mexicanus* (Amphibia: Gymnophiona), with comments on skull kinesis and amphibian relationships. *Journal of Morphology* **173**, 203–223.

Wake, T.A., Wake, D.B, & Wake, M.H. (1983) The ossification sequence of *Aneides lugubris*, with comments on heterochrony. *Journal of Herpetology* **17**, 10–22.

Wassersug, R.J. (1980) Internal oral features of larvae from eight anuran families: functional, systematic, evolutionary and ecological considerations. *Miscellaneous Publications of the Museum of Natural History University of Kansas* **68**, 1–146.

Watson, D.M.S. (1963) On growth stages in branchiosaurs. *Palaeontology* **6**, 540–553.

Werneburg, R. (1993) *Onchiodon* (Eryopidae, Amphibia) aus dem Rotliegend des Innersudetischen Beckens (Böhmen). *Paläontologische Zeitschrift* **67**, 343–355.

Werneburg, R. (1994) Dissorophoiden (Amphibia, Rhachitomi) aus dem Westfal D (Oberkarbon) von Böhmen – *Limnogyrinus elegans* (Fritsch 1881). *Zeitschrift für geologische Wissenschaften* **22**, 457–467.

West-Eberhard, M.J. (2003) *Developmental Plasticity and Evolution*. New York: Oxford University Press.

Whiteman, H.H. (1994) Evolution of facultative paedomorphosis in salamanders. *Quarterly Review of Biology* **69**, 205–221.

Wintrebert, P. (1922) La voute palatine des Salamandridae. *Bulletin Biologique de la France et de la Belgique* **56**, 275–426.

Witzmann F. (2006). Developmental patterns and ossification sequence in the Permo-Carboniferous temnospondyl *Archegosaurus decheni* (Saar-Nahe Basin, Germany). *Journal of Vertebrate Paleontology* **26**, 7–17.

Witzmann, F. & Pfretzschner, H.U. (2003) Larval ontogeny of *Micromelerpeton credneri* (Temnosposndyli, Dissorophoidea). *Journal of Vertebrate Paleontology* **23**, 750–768.

Witzmann, F. & Schoch, R.R. (2006) Skeletal development of *Acanthostomatops vorax* from the Döhlen Basin of Saxony. *Transactions of the Royal Society of Edinburgh: Earth Sciences* **96**, 365–385.

Witzmann, F. & Scholz, H. (2007) Morphometric study of allometric growth in the temnospondyl *Archegosaurus decheni* from the Permin/Carboniferous of Germany. *Geobios* **40**, 541–554.

Witzmann, F., Scholz, H., & Ruta, M. (2009) Morphospace occupation of temnospondyl growth series: a geometric morphometric approach. *Alcheringa* **33**, 237–255.

Woltereck, R. (1909) Weitere experimentelle Untersuchungen über Artveränderung, speziell über das Wesen quantitativer Artunterschiede der Daphniden. *Verhandlungsbericht der deutschen zoologischen Gesellschaft* **1909**, 110–172.

7 Paleoecology

Species do not exist in isolation, and they cannot evolve independently from others. Predation, competition, symbiosis, and many other interactions between species form highly integrated networks that are referred to as ecosystems. These networks form hierarchical systems that span a wide range of levels, from cellular to planetary scale. Ecosystems and their properties have profound influence on evolution but are themselves subject to evolutionary changes. In the twentieth century, ecology developed into a theory with its own set of tools, models, and approaches (Molles 2009).

Considering the enormous progress in ecological studies of extant organisms, analysis of ecological factors in fossil taxa forms a separate field of research (paleoecology). Most extinct species, notably vertebrates, are not sufficiently well preserved or abundant to permit ecological analyses on the same scale as for extant species. The impact of environmental parameters on extinct species is also often difficult to trace, because the paleoenvironments themselves are not sufficiently known. Reconstructions of ancient sedimentary basins, landscapes, and habitats are often plagued by ambiguous data that permit more than one reconstruction. In addition, numerous essential ecological data can only be gathered in living organisms, such as the quantification of predator–prey relationships, competition between species that perform similar ecological functions, or the identification of indispensable keystone species within a trophic chain. Thus, environmental parameters as well as biotic factors are only fragmentarily known. Still, paleoecology is a fascinating field that holds great potential, and fossil amphibians are well placed as a case study among vertebrates, because their fossil record offers a rich field for paleoecological studies (Boy 1998, 2003; Boy and Sues 2000).

Ecology and paleoecology have a number of research topics in common, as follows: (1) studies of food webs as an example of the relationships between species in a given ecosystem, (2) analysis of habitats and key

Amphibian Evolution: The Life of Early Land Vertebrates, First Edition. Rainer R. Schoch.
© 2014 Rainer R. Schoch. Published 2014 by John Wiley & Sons, Ltd.

environmental parameters that are of importance to the existence and evolution of a given species, (3) studies of life histories and their dependence on environmental factors. In Paleozoic and Mesozoic amphibians, all three fields provide interesting data for paleoecological research, often giving insight into extinct ecosystems that were radically different from those of today. Most interesting are data that suggest the total absence of trophic guilds known from today – for instance those occupied by tadpoles. In turn, the presence of many large predators indicates that early amphibians and basal tetrapods covered a wider range of positions in the food web than modern lissamphibians.

7.1 Lissamphibian ecology

Despite their divergent morphology and the wide range of habitats they occupy, modern amphibians have many ecological properties in common: they are ectotherms, they are highly dependent on water, they are small mid-level consumers of invertebrates, and they themselves form prey for various higher-level predators. In many ecosystems, amphibians make a major contribution to biomass, sometimes as much as all small bird and mammal species combined, or even more (Duellman and Trueb 1994; Davic and Welsh 2004). With their enormous range of plasticity and temperature tolerance, some lissamphibians are able to exist in habitats inaccessible to many other vertebrates. For instance, salamanders can be the dominant predators in ephemeral ponds (Trenham et al. 2001), and tadpoles populate numerous tiny and short-lived pools. At the same time, amphibians are absent from most desert and glacial regions.

Amphibians perform many ecological roles and are thus regarded as of central importance to the structure of many ecosystems. As a result of their biphasic life cycle, many lissamphibians trigger the flow of energy and matter between aquatic and terrestrial landscapes (Davic and Welsh 2004). This is further enhanced by the migratory lifestyle of many terrestrial salamanders and frogs, which includes migration of adults to breeding sites. Many salamanders undertake extensive excursions through forests, especially during wet seasons or at night (Hairston 1987). Some amphibians carry eggs of arthropods or mollusks from pool to

pool, contributing to the dispersal of non-migratory species (Bohonak and Whiteman 1999). Other salamanders move into headwater stream or pond habitats, where waterfalls and insufficient water depth keep fishes out (Vannote et al. 1980).

Predation forms a major means of control by which amphibians exert influence on invertebrates and primary producers. For instance, woodland-dwelling salamanders regulate the density of forest-floor invertebrates, and newts impose similar control in aquatic habitats. Manipulation of salamander densities in such ecosystems has shown how crucial their contributions are. Salamanders indirectly enhance the abundance of collembolans by regulating (lowering) the density of their predators (Rooney et al. 2000). Parker (1992) removed larvae of the marbled salamander *Dicamptodon tenebrosus* from a stream pool, revealing the impact on benthic invertebrates. Such influences can be both direct (by predation) and indirect (by competition with other predators). Species with similar prey preference may form redundant regulators, such as *Notophthalmus viridescens* and *Ambystoma opacum*, which feed on tadpoles of the same anuran species (Morin 1995). Reduction of tadpole density increases the biomass of phytoplankton in a water body. Salamander predation in ponds cascades down through several hierarchical levels to increase the production of algae (Morin 1995), or mosquitoes (Brodman et al. 2003), as well as to affect the population structure of molluscs in the case of snail-eating *Siren* species (Petranka 1998). Finally, some amphibians form so-called keystone species, which prevent prey taxa from monopolizing limited resources and thereby allow the coexistence of additional species (Paine 1969). These contribute disproportionately to biotic regulation – often their status is only identified when they become extinct, which results in major ecosystem reshufflings because there are no redundant species that could take over their ecological roles.

In turn, amphibians form regular prey items for many predators (large arthropods, snakes, turtles, crocodiles, birds, and mammals), thus providing stores of energy and nutrients for tertiary consumers. Burton and Likens (1975) concluded that salamanders in mature forests "represent a higher quality source of energy and nutrients than birds, mice, and shrews."

Adult amphibians are good ecological indicators, because they are highly sensitive, responding to even slight changes in the environment. The dramatic rate at which amphibian species are disappearing, particularly in the tropics, is therefore alarming (Duellman 1999; Wake 2009). Anurans play more divergent roles, because of the radically different lives of tadpoles and adult frogs. Tadpoles are microvores, exploiting nutrients not otherwise used by amphibians and in habitats hardly accessible to others; they make a significant contribution in recycling nutrients. With their fossorial mode of life, caecilians make an essential contribution to soil dynamics in tropical ecosystems. Many adult salamanders also use underground retreats in order to minimize desiccation and hide from predators (Semlitsch 1983). Unlike caecilians, burrowing salamanders do not dig their own burrows, and are thus limited by the availability of burrows constructed by small mammals (Faccio 2003). Finally, metamorphosis is not only an ecological factor that transforms large quantities of aquatic animals into terrestrial ones, but may itself be regulated by predator abundance and the costs of metabolism and water economy (Downie *et al.* 2004).

7.2 Paleoecology: problems and perspectives

Paleoecology, the reconstruction of ecological relations between extinct taxa, requires a completely different approach from that of modern ecology. Much more important than for taxonomy and morphology, the paleoecologist needs to understand the nature of the deposits in which the study objects are found (Figure 7.1). Excavations are therefore not only required to secure fossil material in the first place, but are essential to collect a wide range of data. In this view, single fossils are only pieces of a large puzzle, and the connections between the pieces are what matters.

Two aspects are particularly important here: first, how did the deposit form, and second, which filters have prevented essential paleoecological

---

BOX 7.1: ECOLOGY AND PALEOECOLOGY – KEY TERMS

Taphonomy: Study of processes between the death of an organism and its final burial in a sediment. Specifically, taphonomy studies the filters that prevent parts of organisms or complete taxa from being preserved.

Autochthonous: Living at the site of final burial (habitat = deposit).

Allochthonous: Living in a different place from the site of burial.

Community: A group of species interacting in a given environment, based on the same resources.

Life assemblage: The full set of species occurring in the same habitat at the same time.

Death assemblage: All species of a life assemblage preserved in the same deposit, excluding all those taxa that are not preserved for taphonomic reasons (thanatocenosis).

Grave assemblage: All species of a death assemblage, excluding all taxa whose remains are destroyed by diagenesis, plus allochthonous taxa transported to the deposit (taphocenosis).

Time-averaging: Preservation of organisms from different time slices in the same bed.

Diagenesis: Processes leading to the formation of a sedimentary rock (pressure, temperature, water loss and circulation, mineral growth, recrystallization). Often results in the loss or alteration of fossils.

Figure 7.1 The preservation of fossil communities goes through several successive filters, which result in the loss of taxa, the inclusion of foreign taxa, and other biases. Conversely, the reconstruction of paleocommunities requires a reverse process of identifying the taphonomic filters and assessing the amount of taxa missing from the preserved grave community.

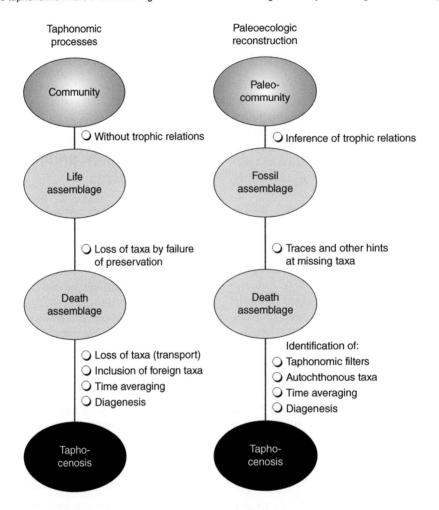

data from being preserved? For the morphologist, the skeleton forms only a small fraction of the once-complete organism, and the paleontologist has just a few pieces of a large puzzle to work with. Likewise, the paleoecologist has to assess the dimensions of all the missing data (Figure 7.2). Naturally, the further one goes back in Earth's history, the wider the gaps and the greater the amount of missing data.

Paleoecology has two foci, which are largely complementary: (1) the ecological features of single extinct species (their living conditions, feeding strategies, preferred habitats), which is referred to as *autecology*, and (2) the interaction between different species within an ecosystem (trophic relations, competition, regulation, etc.), which is called *synecology*.

Autecology had long formed the main interest of paleoecologists, aimed at elucidating environmental parameters and their influence on particular taxa (Brenchley 1990; Jablonski and Sepkoski 1996). In such studies, species are ranked in different categories such as "detritus-feeders," "suspension-feeders," and "predators" based on

Figure 7.2 The preservation of a fossil is a relatively unlikely event. Usually, weathering leads to the complete destruction of organic remains. The three most common pathways of preservation in vertebrates are mapped here, with 3D preservation (right) forming the great exception.

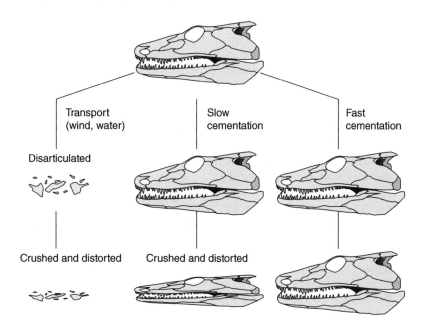

their morphological features. In vertebrates, autecology usually concentrates on the analysis of tooth structure and wear, often complemented by consideration of jaw mechanics and skull mobility. This analysis may help constrain the range of conceivable food items the study taxon might have preyed on, and specify which feeding strikes it probably employed from a purely mechanical point of view.

Increasing interest in extinction events triggered the synecology approach, which emphasizes the interactions within communities (Ricklefs *et al.* 1990). Vertebrates usually occupy the upper levels of the food web, and these are best reconstructed in aquatic paleoecosystems (Maisey 1994; Boy 2003). Synecology in vertebrate paleontology means analysis of food webs in fossil communities, such as by the analysis of stomach contents in predators, bite marks on bones of prey, and coprolites associated with the study taxa. The main aim of synecology is the identification of communities in a former habitat. Communities form a central aspect of paleoecology. They are groups of species that share the same habitat and interact with each other in various ways (predation, competition, symbiosis).

The first step in a synecological analysis of fossil deposits is therefore the identification of paleocommunities. This is not a trivial task and requires the following steps: (1) the formation of the fossil deposit must be understood, especially regarding the question of whether the site of deposition was also a habitat, or which fossils were transported from somewhere else; (2) those taxa that were native to a preserved habitat (autochthonous fauna) must be identified and separated from all others; (3) among the autochthonous fauna, the different trophic guilds must be recognized, i.e., those taxa sharing the same food resources. Only after this procedure can trophic relations between autochthonous species be analyzed.

Thus, an excavation of a fossil-rich horizon does not yield paleocommunities *per se*, but normally produces a set of taxa that are found in the same outcrop. This is called a taphocenosis (grave community). Usually, such taphocenoses represent only a fractional subset of the original fauna and flora, because some taxa were never preserved in the first place or left remains that were destroyed by subsequent taphonomic or diagenetic processes. For instance, many lake and stream deposits include horizons with vertebrate fossils, but plants and invertebrates are often not preserved in the same beds for chemical reasons. In turn, bones are often not preserved in *Lagerstätten* that are rich in plant fossils (Behrensmeyer *et al.* 1992). Important steps in the reconstruction of a paleocommunity are therefore (1) an estimate of how many and which sort of taxa might have been lost by failure to preserve, or taphonomic and diagenetic filters, and (2) an approximation of how large a time interval may be represented by a single bedding plane or any other fossiliferous rock unit (time-averaging). These estimations are necessarily error-laden and require much experience with sedimentary facies and taphonomic processes. Studies of present-day examples are highly useful here.

The resulting set of species forms the thanatocenosis (death assemblage), which, along with trace fossils produced by soft-bodied species, permits a reconstruction of the original life assemblage in the studied habitat. After this long procedure, an understanding of the paleocommunity is eventually within reach, provided that additional evidence of trophic interactions is available.

Extinct food webs are reconstructed on the basis of various different sources of data. First, there can be direct evidence of predation such as stomach and intestine contents in a predator. Fish-eating tetrapods are a good example, and various early tetrapods are preserved in sufficient detail to permit identification of prey species. In a very fortunate case, a three-level trophic chain was identified in a specimen from the Lower Permian of Lebach (Germany). A skeleton of the Permian shark *Triodus* was found to preserve two prey specimens, the aquatic temnospondyls *Archegosaurus* and *Glanochthon*, in its stomach region. The larger of the two prey items, *Glanochthon*, had itself eaten an acanthodian fish whose remains were found in its intestine (Kriwet *et al.* 2008). The Late Paleozoic lake deposits of central Europe contain a rich record of articulated temnospondyls with stomach contents, regurgitated prey, and coprolites (Boy and Sues 2000). In the Lower Keuper deposits of Germany, diagnostic bite marks on large bones and skeletons are common. In an Early Permian red-bed deposit in Texas, a specimen of the amphibamid dissorophoid *Tersomius* sp. was found with deep tooth impressions in the center of the skull, suggesting predation by a large pelycosaur synapsid (A.R. Milner, personal communication). There are also entire localities composed of regurgitated and chopped-up pieces of prey, such as the Carboniferous coal pit near Five Points, Ohio (A.R. Milner, personal communication 2012). A rather common phenomenon is cannibalism, which has been reported from the temnospondyls *Apateon* (Witzmann 2009), *Sclerocephalus* (Schoch 2009), and *Mastodonsaurus* (Schoch, unpublished data).

7.3 Paleozoic and Mesozoic amphibians

In the present section, I shall discuss different and well-studied paleoecosystems from the fossil record in which early tetrapods played a major role. Despite numerous differences, they all appear to have existed for a comparably short time interval (10^2–10^3 years). Evolutionary changes are therefore not to be expected and cannot be identified in such samples. Some *Lagerstätten* preserve changes in paleocommunities between sedimentary beds, suggesting successions in the paleoecosystem (one predator replaced by another, increase or reduction in the number of trophic levels).

These deposits have in common that (1) tetrapods are preserved well and are often articulated, (2) the sediments contain coprolites and other traces of predation. A major difference exists between and within *Lagerstätten* as to whether the single bedding planes are time-averaged or not. Time-averaging is the normal case and does not permit identification of mortality patterns of real populations, but some deposits (Nýřany, Odernheim) appear to preserve populations killed in a single event. These deposits also differ considerably in their position within climatic belts and altitudes, ranging from tropical rainforest (Nýřany) over mountain lakes (Variscan deposits) to a semiarid delta setting in subtropical lowlands (Keuper).

Nýřany as a Pennsylvanian peat lake. Milner (1980) has summarized the knowledge of the fauna and potential paleocommunities in the small peat lake at Nýřany (Czech Republic). Based on a census of the 22 taxa preserved in the gas coal, he separated the common autochthonous taxa from the much rarer terrestrial or riparian dwellers that were only occasionally washed into the lake. The bulk of tetrapod skeletons is fully articulated and embedded in a 30 cm thick sequence of canneloid shales and mudstones. They formed in a small water body, presumably a pond located within a swamp forest of calamite horsetails. This pond lacked a benthic fauna and was characterized by undisturbed sedimentation under stagnating, anaerobic conditions. An occasional inflow of silt carried algae, spores, and plant debris from the surrounding woodland areas. The diversity of aquatic tetrapods indicates that the pond was not too small and sufficiently deep to permit a separation of stagnating and aerated zones. Based on studies of recent analogs, Skoček (1968) suggested a total depositional time of 300–700 years for the fossiliferous sequence at Nýřany. The well-articulated tetrapod skeletons span a wide range of sizes in autochthonous taxa, suggesting that they form census populations which died in single events (e.g., an algal bloom).

The fauna includes a few fishes, the lepospondyls *Oestocephalus*, *Phlegethontia*, *Sauropleura*, and *Microbrachis*, and the aquatic temnospondyls *Branchiosaurus*, *Limnogyrinus*, and *Cochleosaurus*. The lake dwellers are recognized by aquatic adaptations in the skeleton, their much greater frequency in the sample, and the fact that they are preserved in different size classes, including small larvae. In addition to these probably lifelong aquatic taxa, larvae of temnospondyls with biphasic life cycles also lived in the Nýřany lake (*Platyrhinops*, *Mordex*). In this fauna, *Branchiosaurus* was a plankton-feeder (first and second level), *Microbrachis* and the snake-like *Oestocephalus* probably fed on small invertebrates (ostracodes and other crustaceans). *Limnogyrinus*, *Sauropleura*, and especially *Cochleosaurus* were the large predators in the aquatic community, probably all feeding on the smaller tetrapods (Milner 1980). These inferences are based on anatomical data (autecological traits), as gut contents and other direct evidence of trophic relations are lacking.

Variscan intramontane lakes. The next three paleoecosystems are all located within the Variscan mountain belt, fall into the Pennsylvanian–Permian *Rotliegend* facies, and were analyzed in great detail by Jürgen Boy and colleagues over a period of 30 years (Boy 1972, 1977, 1998, 2003). The rich data derived from these deposits include sedimentological and facies analyses, taphonomic studies, anatomical data, and the examination of stomach contents, regurgitated skeletons, and coprolites (Figure 7.3). These lakes have in common that they were short-lived (10^1–10^3 years), were populated through inflowing streams, and the lake ecosystems depended on exchange with river faunas (Boy 1994). The fauna was poorer in species than the preceding Pennsylvanian lakes, and in contrast to Cenozoic ecosystems there were no herbivorous taxa (e.g., snails). The food webs were short and the connections between the different trophic levels not very elaborate, as is evident from the fact that changes on one level had little impact on those of other levels (Boy 1994).

Odernheim. The Odernheim deposits (Figure 7.3A) represent the best-studied Late Paleozoic lake paleoecosystem (Boy 1972, 2003). They formed in a lake that was shallow but relatively large (~40 km long), with a stratified water column (Boy 1972). Its fauna reveals patterns of high environmental stress that was probably caused by seasonal fluctuations (e.g., algal blooms, rhythmic circulation and turnover of water layers). These events led to the sudden death of many fish and branchiosaurids, which accumulated in single layers (census populations, Boy 2003). The less favorable conditions probably formed a limiting factor to growth and taxonomic diversity, and all vertebrates in this lake were substantially smaller than in other lakes (Boy 2003; Schoch 2009). The limestone is finely laminated, with each lamina representing one season; similar sediments are known from recent glacial lakes (varves). Boy (2003) reported that the whole number of varves suggests around 700 years of continued lake deposition. The trophic relations of the Odernheim lake were reconstructed by Boy (1972, 2003). The first and second levels of the food web were probably occupied by phytoplankton and zooplankton, which are not preserved. The third level was held by the small branchiosaurid filter-feeding *Apateon pedestris* and the microvorous actinopterygian fish *Paramblypterus*, which both focused on

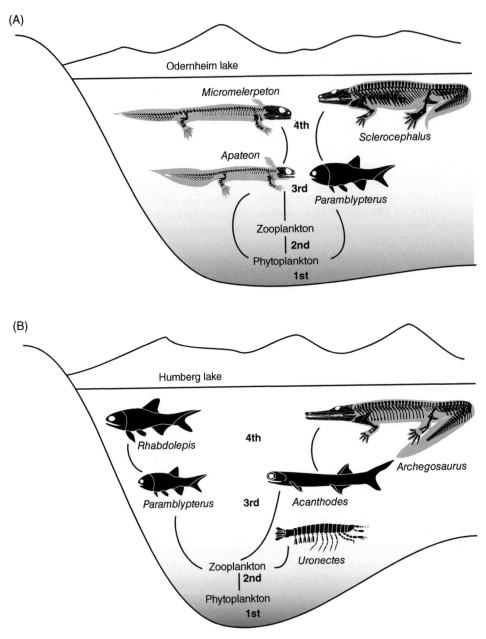

Figure 7.3 (A) The paleocommunity of the 40 km long Odernheim Lake. This intramontane lake was impoverished in vertebrate species and had a simple trophic web. (B) The 80 km long Humberg Lake, where branchiosaurids were absent and *Archegosaurus* formed the top predator; this temnospondyl preyed on acanthodians.

plankton. At the fourth level, the larger temnospondyls *Micromelerpeton* and *Sclerocephalus* formed the main predators, with *Micromelerpeton* focusing on *Apateon* and *Sclerocephalus* on *Paramblypterus*. These trophic relations are all concluded from preserved gut contents in numerous skeletons (Boy 2003).

Lebach. Larger and markedly deeper than the Odernheim example was the Humberg lake (Figure 7.3B), which spanned some 80 km northeast–southwest, also located within the Saar–Nahe basin (Germany). It was deepest in the southwest, where a sequence of ironstone nodules several meters thick preserves fossils of lake dwellers. Boy (1994) was able to distinguish four different phases during the existence of this water body: (1) a seasonally stressed lake (algal blooms), populated by plankton-feeding *Paramblypterus* and *Apateon* and a rather aquatic morph of *Sclerocephalus* (Boy 1994; Schoch 2009); (2) a similar but less stressed lake with *Paramblypterus* and *Acanthodes* as plankton-feeders and two larger sharks as predators (*Triodus*, *Xenacanthus*), which replaced *Sclerocephalus*; (3) a shallower lake, affected by currents, dominated by the crustacean *Uronectes* preying upon conchostracans, with the same vertebrate fauna as in (2) but the temnospondyl *Glanochthon* replacing *Xenacanthus* as top predator; and finally (4) the lake became deeper again, now under the influence of a delta, with a well-aerated bottom zone, *Acanthodes* and *Paramblypterus* as plankton-feeders, the new predatory fishes *Elonichthys* and *Rhabdolepis*, and the large shark *Xenacanthus* and the gharial-like temnospondyl *Archegosaurus*.

Niederkirchen. A much smaller lake was excavated at Niederkirchen, again in the Saar–Nahe basin, but situated in a tectonically isolated depression. This local deposit also preserves a succession of communities (Boy 1995). The first phase was dominated by the filter-feeding branchiosaurid *Apateon pedestris* and the fishes *Acanthodes* and *Aeduella*, and the ~2 m long temnospondyl *Sclerocephalus*. In a second phase, *Sclerocephalus* was replaced by the 2–3 m long *Orthacanthus*, a large shark. Then, after the disappearance of *Orthacanthus* and *Aeduella*, larger branchiosaurids immigrated (*Apateon caducus*, *Melanerpeton*), preying upon *Apateon pedestris*.

During a further short phase, the larger dissorophoid *Micromelerpeton* occurred, forming a larger local morph that preyed upon all other tetrapods and small fishes (Boy 1995).

Middle Triassic subtropical estuaries. The Middle Triassic vertebrate *Lagerstätten* of the Lower Keuper, formed in a deltaic setting, preserve rich aquatic faunas of the subtropical realm. During the Triassic, central Europe was under the influence of a large monsoon triggered by the adjacent Tethys Ocean (Etzold and Schweitzer 2005). In contrast to the *Rotliegend* lake deposits, the paleoecosystems were highly diverse, and the tetrapod faunas included highly disparate temnospondyls, chroniosuchian stem-amniotes, diapsid reptiles, archosauriforms, and peudosuchian archosaurs (Figure 7.4). The problem with these deposits is that they contain mixed faunas, combining skeletons of aquatic, amphibious, and fully terrestrial taxa in the same beds. In addition, articulated skeletons are much rarer than in the Late Paleozoic lake deposits, because the water bodies of the Keuper were usually smaller, shallower, and affected by seasonal hurricanes. These frequent storms probably changed the landscape on a regular basis, leading to the formation of new basins and interconnecting formerly separate water bodies. During Middle Triassic times, a vast system of water bodies covered the south of Germany and adjacent areas, ranging from freshwater to brackish and even hypersaline (Beutler *et al.* 1999). Two lake paleoecosystems have been excavated and studied in detail: Kupferzell and Vellberg. Smaller than the more common brackish water bodies, which often covered dozens of square kilometers, these lakes were populated by freshwater bivalves and ostracodes, as well as numerous bony fishes, temnospondyls, and aquatic reptiles.

Kupferzell. This *Lagerstätte* was excavated in spring 1977, when a road-cut for the construction of a highway crossed mudstones of the Lower Keuper (Wild 1980). The greenish mudstones yield a local mass accumulation of bones from two temnospondyls, *Gerrothorax* and *Mastodonsaurus*, and rarely produce skeletal remains of the large pseudosuchian archosaur *Batrachotomus*. The deposit formed in a shallow, well-aerated freshwater lake (~5–6 km) with a rich fish fauna. Frequent characean algae indicate carbonaceous freshwater

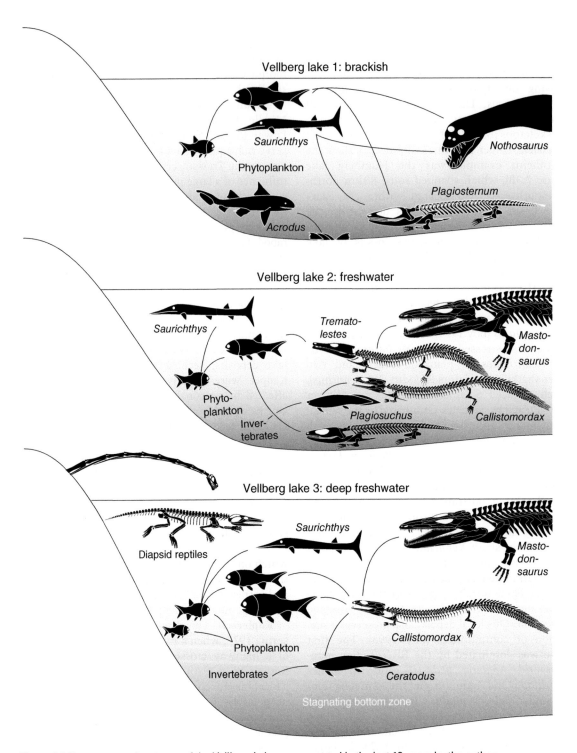

Figure 7.4 Three successive stages of the Vellberg Lake, as excavated in the last 12 years by the author. The rock sequence preserves a shift from a brackish lagoon to a freshwater lake that was rich in fish and temnospondyl species.

conditions, which is confirmed by the rich carbonate content of the beds. The fish-eating bottom-dwelling *Gerrothorax* was abundant. This taxon was unable to leave the water, indeed the lake floor, thus indicating a well-aerated bottom zone, while the 5 m long *Mastodonsaurus* formed the top predator.

Vellberg. The Vellberg locality, a limestone pit in the Middle Triassic Muschelkalk beds, has been quarried for over 60 years. It has yielded large quantities of temnospondyl and diapsid material. The fossiliferous beds include a 20 cm thick sequence of green and brown siltstones and grey mudstones, topped by a massive dolomite. The general trend within this sequence documents a continued withdrawal of the sea, with the siltstones forming the largest water body (probably a relic of a short-term lagoonal basin), and the dolomite a prograding shoreline on a carbonatic mudflat (Figure 7.4). All beds provide evidence for interrupted deposition, when the lakes dried out and soils formed, indicating that several thousand years must have passed during each drying phase. All water bodies were relatively shallow, and thus regularly affected by storms, which led to the frequent disarticulation of skeletons, fragmentation of bivalve shells, and complete destruction of arthropod cuticles. This gives a substantially reduced set of preserved taxa in the death assemblages, which are also enriched by numerous allochthonous taxa washed in by storms. The following three lacustrine paleoecosystems are preserved in the Vellberg sequence (Figure 7.4).

1. The green siltstones contain masses of brackish bivalves, teeth of marine sharks, and bones of the temnospondyl *Plagiosternum*, a taxon usually occurring in marine bone beds. This deposit formed in a brackish lagoon spanning some 25 km, which must have been shallow and affected by heavy storms repeatedly, as bivalve coquinas and microscopic erosional surfaces indicate. In this paleoecosystem, the shell-crushing shark *Acrodus* was feeding on bivalves, the predatory fish *Saurichthys* probably fed on small planton-feeding *Serrolepis*, *Plagiosternum* in turn focused on larger fishes, and the 2–3 m long marine reptile *Nothosaurus* formed the top predator.

2. In the Vellberg region, the green lagoon deposits grade into brown siltstones which formed in a local freshwater lake. This was a shallow, rich habitat, in which skeletons of lungfish babies occur in great quantities, accompanied by the fish-eating temnospondyls *Trematolestes*, *Callistomordax*, and *Plagiosuchus*, which are also present with larval specimens. Apparently this was a protected water body, rich enough in food to permit these three putative piscivores to breed there, and it even harbored two large predators, the 2.5 m long *Kupferzellia* and the 5 m long *Mastodonsaurus*, whose remains are very common and usually heavily affected by predation themselves. During the early phase of this lake, the ecologically more flexible genus *Gerrothorax* appears to have replaced the brackish *Plagiosternum*, subsequently itself replaced by *Plagiosuchus*.

3. After the disappearance of the small freshwater lake, erosion must have changed the landscape to permit a deeper lake to form, again containing freshwater, but this time with a stagnating bottom zone that led to the deposition of dark, pyrite-rich mudstones. Evidence of predation is abundant in this bed, including destruction of skulls, which often contain bite marks, and regurgitated fish and reptile skeletons. As in the Kupferzell deposit, tooth marks of pseudosuchian archosaurs have been identified on limb bones and ribs of *Mastodonsaurus*. Coprolites of all sizes are very common, indicating that the deposit was indeed a habitat. As in the other deposits, there is no direct evidence of plankton in the deposit. Plankton-feeders are found among the 5–10 cm long actinopterygians (*Dipteronotus*, Redfieldiidae). Lungfishes (*Ceratodus* spp.) were abundant but are mainly represented by juveniles; they probably fed on snails and crustaceans. The temnospondyls *Callistomordax* (1 m) and *Mastodonsaurus* (3–5 m) formed the most common large tetrapods, accompanied by 30–50 cm long aquatic reptiles (choristoderes), which fed on the 5 cm long actinopterygian *Dipteronotus*, preserved as stomach contents. Rarer temnospondyls constitute the long-snouted *Trematolestes* and the crocodile-like

Kupferzellia, which probably dwelled along the shoreline or occasionally immigrated from inflowing rivers. Bound to the lake floor, *Gerrothorax* is very rare and must have been confined to the habitable shoreline. Both *Callistomordax* and *Mastodonsaurus* are represented by complete growth series from tiny larvae to adults. *Callistomordax* is likely to have fed on larger actinopterygians and juvenile lungfishes, whose regurgitated skeletons are frequently found. *Mastodonsaurus* was the top predator and left characteristic bite marks on skeletons of smaller temnospondyl species as well as its own juveniles. There are 40 cm long skulls of *Mastodonsaurus* with clear bite marks of much larger specimens, suggesting the animals were killed by larger conspecifics who crushed their skulls. Ribs and limb elements often bear tooth marks as well (Figure 7.5), but sometimes they match the rauisuchian archosaur *Batrachotomus* rather than *Mastodonsaurus* itself (Wild 1980). The abundance of juvenile fishes (coelacanths, lungfishes) and the rarity of adults indicate that the lake was also a breeding habitat for these taxa. Adults apparently used inflowing streams to visit the site during the breeding season. An enigmatic

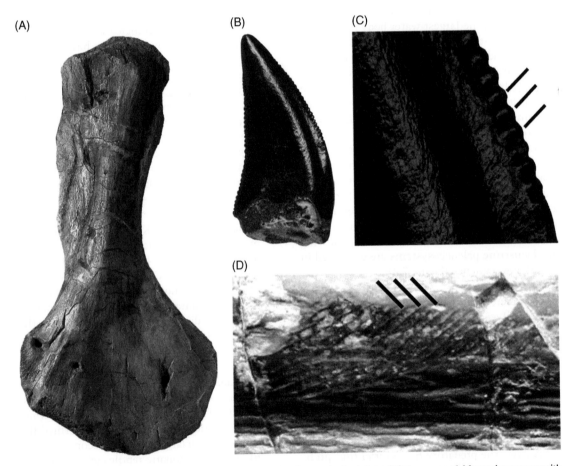

Figure 7.5 Traces of predation form an important source of paleoecological data. (A) Humerus of *Mastodonsaurus* with bite marks. (B) Tooth of the pseudosuchian archosaur *Batrachotomus* from the same deposits. (C) Close-up of B, showing serrated cutting edge. (D) Scratch marks on a rib of *Mastodonsaurus*, matching the shape and size of serrated cutting edges in *Batrachotomus*.

faunal component was the chroniosuchian *Bystrowiella*, whose remains were probably washed in by inflowing streams (Witzmann *et al.* 2008).

Gaildorf. The first Lower Keuper locality to be reported was a coal mine at Gaildorf (southern Germany), which yielded skeletons of *Plagiosuchus* and *Mastodonsaurus*, along with rare remains of *Trematolestes* and *Kupferzellia*. The coal was deposited in an oxbow lake, as indicated by its close connection with a large river sandstone body. As in the lake deposits, the sandstone and coal *Lagerstätten* often harbor mixed faunas, preserving marine, brackish, and freshwater animals. This was probably caused by the repeated flooding of larger river channels by sea water in times of marine incursions, which is a typical phenomenon in modern estuaries. The close resemblance of temnospondyl faunas in these deposits reveals that similar ecosystems were established in a wide range of environments, from stream-dominated smaller ponds to larger lakes and lagoons. Probably most of these subtropical water bodies offered sufficient nutrition for at least two large temnospondyls, once the salinity had declined to a habitable level.

General problems. In contrast to the relatively well-understood Paleozoic lacustrine deposits, the Lower Keuper habitats were populated by diverse faunas whose trophic relations are far from understood. In addition, key guilds such as crustaceans were not preserved in most of these *Lagerstätten*, making these paleoecosystems less completely known. In addition, more diverse communities imply more trophic relations between the different trophic levels, and reconstruction of these is further restricted by incomplete finds, rare occurrence of bite marks and stomach contents, and equivocal evidence such as coprolites, which usually cannot be assigned to particular taxa in such rich paleoecosystems.

A major focus of interest is the potential salinity tolerance of Paleozoic and Triassic taxa (Boy and Sues 2000; Laurin and Soler-Gijón 2001; Schultze 2009). Lissamphibians are largely confined to freshwater, although a few anurans are reported to tolerate brackish or even marine conditions (Haas 2010). Based on a range of criteria, Laurin and Soler-Gijón (2010) suggested that the

(almost) exclusive freshwater dwelling might be an autapomorphy of lissamphibians. It is very possible that many early tetrapods were euryhaline (tolerant of higher water salinities), but this remains difficult to prove without isotope or other geochemical data from most *Lagerstätten*. Some Triassic temnospondyls (trematosaurids, plagiosaurids) occur in demonstrably marine faunas, such as the Posidomya beds of Svalbard (Lindemann 1991) or certain Lower Keuper beds of Germany (Schoch and Milner 2000). Other occurrences are more ambiguous, most particularly the intramontane basin lakes (*Rotliegend*), where "marine influence" (Schultze and Soler-Gijón 2004; Schultze 2009) has been proposed largely on the basis of taxa whose extant members are marine, but Paleozoic relatives may not have been. Strontium isotope and geochemical analysis has meanwhile not confirmed this hypothesis, instead indicating a non-marine origin of most vertebrate-bearing *Rotliegend* sediments in a range of Late Paleozoic European basins (Fischer *et al.* 2013), an interpretation which is in line with studies of basin structure and sedimentation (Lützner and Kowalczyk 2012).

7.4 Amphibian evolution as a walk through trophic levels

The paleoecology of Paleozoic and Mesozoic stem-amphibians differs in many respects from the ecology of lissamphibians. It is tempting to trace the changes in the ecological properties through amphibian evolution, because during their 330 myr long evolutionary history, these taxa walked across various trophic levels. Like modern lissamphibians, temnospondyls and lepospondyls are likely to have contributed substantially to the biomass of forest and freshwater habitats. First confined to aquatic environments, early tetrapods started to contribute to the flow of matter and energy between aquatic and terrestrial ecosystems (Figure 7.6). This probably commenced with a few anthracosaur and dendrerpetid temnospondyl taxa, but was a regular feature of ecosystems by Pennsylvanian times, when seymouriamorphs and dissorophoids spent their early life in lakes and

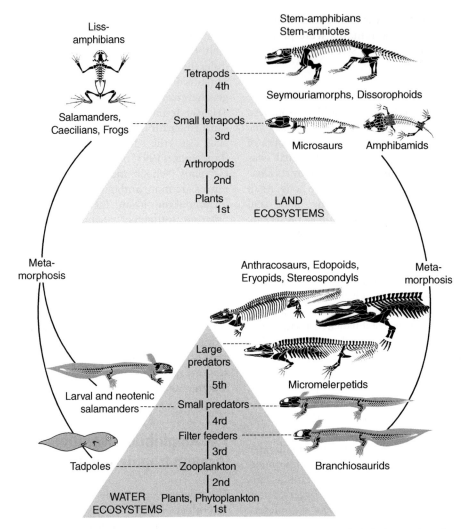

Figure 7.6 Through Earth's history, amphibians have held positions in various trophic levels, in both aquatic and terrestrial ecosystems. The evolution of large aquatic predators reached a climax in the early Mesozoic, with up to 5 m long stereospondyls. Today, lissamphibians hold only lower positions in both ecosystems, with tadpoles having evolved their own herbivorous niches. Amphibians have always made important contributions to energy flow between aquatic and terrestrial habitats.

changed to a terrestrial existence as adults. In the Permian, the ecological disparity appears to have reached a peak, with fully aquatic through to fully terrestrial taxa occupying various habitats. The late-Permian (not necessarily end-Permian!) extinction of many lepospondyls, anthracosaurs, seymouriamorphs, and the larger dissorophoids left wide gaps in the range of trophic levels held by early tetrapods, now increasingly filled by amniotes, especially diapsid reptiles. By the early

Mesozoic, only two large clades had survived, forming ecologically most divergent clusters: the giant stereospondyls and the tiny lissamphibians.

The walk through trophic levels started with tetrapodomorphs, which formed medium-sized aquatic predators, an ecomorph that was conserved in baphetids, anthracosaurs, and basal temnospondyls. As highly diverse aquatic predators, these taxa held niches today occupied by bony fishes (especially teleosts). Only during the Permian and especially in the Triassic did some of these groups evolve larger predators, paralleling the ecological roles of modern crocodiles, gharials, and large salamanders. This occurred in parallel to the evolution of subtropical and tropical ecosystems, whose faunas diversified rapidly, particularly in the Early and Middle Triassic.

Along a separate line, seymouriamorphs and dissorophoid temnospondyls first explored food sources of the terrestrial realm, still entirely focused on carnivory. Among these, seymouriamorphs, dissorophids, and trematopids are likely to have predated small tetrapods, whereas the miniaturized amphibamids must have focused on small invertebrates. The latter probably overlapped with microsaurs, a diverse clade that occupied Pennsylvanian and early Permian forests, but was eventually replaced by lissamphibians. Likewise, the largely aquatic nectrideans were replaced by aquatic temnospondyls, especially stereospondyls, during the Late Permian.

Amphibamids and their close relatives the branchiosaurids appear to have taken new paths, with both larvae and adults focusing on smaller food resources than their ancestors. They continued the descent in trophic levels first explored by dendrerpetids and basal dissorophoids: from fish-eaters through small terrestrial tetrapods towards tiny arthropods. Small invertebrate prey was probably captured by means of the tongue, using a sticky secretion of the intermaxillary gland, and handled by the newly evolved pedicellate teeth. Their larvae became filter-feeders, for the first time in early tetrapods focusing on plankton (microvory), which permitted exploitation of a vast resource available in almost all water bodies. Neotenes, in contrast, were rather conservative in retaining the predatory morphotype. This is retained in salamanders, whose larvae are all predatory as well. A fundamental change happened in frogs, whose tadpoles became herbivores, again opening a new avenue for a vast adaptive radiation. They form the only lissamphibians and indeed basal tetrapods in general to have evolved herbivores. Thus, temnospondyls appear to have shown the most extreme broadening of the range through trophic levels, ranging from the lowest positions (lissamphibians) all the way through to the highest (stereospondyls).

An important paleoecological consequence of the pattern outlined above concerns the origin of land vertebrates (Chapter 10, section 10.6). An orthodox scenario ("central dogma") holds that tetrapods followed arthropods in the conquest of land. This would imply that aquatic fish-eaters became arthropod-eaters with the transition to a terrestrial existence. Late Devonian or Early Mississippian tetrapods should therefore have possessed dentitions suitable to process insect or milliped cuticles rather than fishes or small aquatic tetrapods. Conceivable transitional stages might have involved taxa feeding on crustaceans in the water, with dentitions and jaw mechanics suited to crush arthropod cuticles. However, in the total absence both of such dentitions and of stomach contents, this remains entirely hypothetical. In fact, the evidence for such dentitions and diets is elusive in basal tetrapods of this stratigraphic age. Instead, microsaurs and dissorophoids, the only larger clades to have evolved insectivorous dentitions, became established only during the Pennsylvanian. This means that the evolutionary descent of early tetrapods and stem amphibians down the trophic levels from top predators through first-level carnivores to insectivores required several tens of myr *longer* than the central dogma holds. An alternative explanation, although conceivable, is not very parsimonious: the described pattern might be caused by a preservation (or collection) bias, neglecting small tetrapods. However, it needs to be emphasized that the preservation potential of habitats within rainforests is poor, and unfortunately these environments are the most likely to have housed diverse insect faunas and their putative predators.

References

Behrensmeyer, A.K., Damuth, J.D., DiMichele, W.A., *et al.* (eds.) (1992) *Terrestrial Ecosystems Through Time: Evolutionary Paleoecology of Terrestrial Plants and Animals.* Chicago: University of Chicago Press.

Beutler, G., Hauschke, N., & Nitsch, E. (1999) Fageisentwicklung des Keupers im Germanischen Becken. In: N. Hauschke & V. Wilde (eds.), *Trias. Eine ganz andere Welt.* Munich: Pfeil, pp. 129–174.

Bohonak, A.J. & Whiteman, H.H. (1999) Dispersal of the fairy shrimp *Branchinecta coloradensis* (Anostraca). *Limnology and Oceanography* **44**, 487–493.

Boy, J.A. (1972) Palökologischer Vergleich zweier berühmter Fossillagerstätten des deutschen Rotliegenden (Unter-Perm, Saar-Nahe-Gebiet). *Notizblatt des hessischen Landesamts für Bodenforschung* **100**, 46–59.

Boy, J.A. (1977) Typen und Genese jungpaläozoischer Tetrapoden-Lagerstätten. *Palaeontographica A* **156**, 111–167.

Boy, J.A. (1994) Seen der Rotliegend-Zeit: ein Lebensraum vor rund 300 Millionen Jahren in der Pfalz. In: W. Koenigswald & W. Meyer (eds.), *Erdgeschichte im Rheinland*. Munich: Pfeil, pp. 107–116.

Boy, J.A. (1995) Über die Micromelerpetontidae (Amphibia: Temnospondyli). 1. Morphologie und Paläoökologie des *Micromelerpeton credneri* (Unter-Perm; SW-Deutschland). *Paläontologische Zeitschrift* **69**, 429–457.

Boy, J.A. (1998) Möglichkeiten und Grenzen einer Ökosystem-Rekonstruktion am Beispiel des spätpaläozoischen lakustrinen Paläo-Ökosystems. 1. Theoretische Grundlagen. *Paläontologische Zeitschrift* **72**, 207–240.

Boy, J.A. (2003) Paläoökologische Rekonstruktion von Wirbeltieren: Möglichkeiten und Grenzen. *Paläontologische Zeitschrift* **77**, 123–152.

Boy, J.A. & Sues, H.-D. (2000) Branchiosaurs: larvae, metamorphosis and heterochrony in temnospondyls and seymouriamorphs. In: H. Heatwole & R.L. Carroll (eds.), *Amphibian Biology. Volume 4. Palaeontology*. Chipping Norton, NSW: Surrey Beatty, pp. 973–1496.

Brenchley, P.J. (1990) Biofacies. In: D.E.G. Briggs & P.R. Crowther (eds.), *Palaeobiology: a Synthesis*. Oxford: Blackwell, pp. 395–400.

Brodman, R., Ogger, J., Kolaczk, M., et al. (2003) Mosquito control by pond-breeding salamander larvae. *Herpetological Review* **34**, 116–119.

Burton, T.M. & Likens, G.E. (1975) Salamander populations and biomass in the Hubbard Brook experimental forest, New Hampshire. *Copeia* **1975**, 541–546.

Davic, R.D. & Welsh, H.H. (2004) On the ecological roles of salamanders. *Annual Review of Ecology, Evolution and Systematics* **35**, 405–434.

Downie, J.R., Bryce, R., & Smith, J. (2004) Metamorphic duration: an under-studied variable in frog life histories. *Biological Journal of the Linnean Society* **83**, 261–272.

Duellman, W.E. (1999) Global distribution of amphibians: patterns, conservation, and future challenges. In: W.E. Duellman (ed.), *Patterns of Distribution of Amphibians: a Global Perspective*. Baltimore: Johns Hopkins University Press, pp. 1–30.

Duellman, W.E. & Trueb, L. (1994) *Biology of Amphibians*. Baltimore: Johns Hopkins University Press.

Etzold, A. & Schweizer, V. (2005) Der Keuper in Baden-Württemberg. In: Deutsche Stratigraphische Kommission (Hrsg.), Stratigraphie von Deutschland IV. Keuper. Bearbeitet von der Arbeitsgruppe Keuper der Subkommission Perm-Trias der DSK. *Courier Forschungsinstitut Senckenberg* **253**, 215–258.

Faccio, S.D. (2003) Postbreeding emigration and habitat use by Jefferson and spotted salamanders in Vermont. *Journal of Herpetology* **37**, 479–489.

Fischer, J., Schneider, J.W., Voigt, S., et al. (2013) Oxygen and strontium isotopes from fossil shark teeth: Environmental and ecological implications for Late Palaeozoic European basins. *Chemical Geology* **342**, 44–62.

Haas, A. (2010) Lissamphibia. In: W. Westheide & R. Rieger (eds.), *Spezielle Zoologie. Wirbel- oder Schädeltiere*. Heidelberg: Spektrum Akademischer Verlag, pp. 330–359.

Hairston, N.G. (1987) *Community Ecology and Salamander Guilds*. New York: Cambridge University Press.

Jablonski, D. & Sepkoski, J.J., Jr. (1996) Paleobiology, community ecology, and scales of ecological pattern. *Ecology* **77**, 1367–1378.

Kriwet, J., Witzmann, F., Klug, S., & Heidtke, U.H.J. (2008) First direct evidence of a vertebrate three-level trophic chain in the fossil record. *Proceedings of the Royal Society of London B* **275**, 181–186.

Laurin, M. & Soler-Gijón, R. (2001) The oldest stegocephalian from the Iberian Peninsula: evidence that temnospondyls were euryhaline. *Comptes Rendus Academie des Sciences de la Vie* **324**, 495–501.

Laurin M. & Soler-Gijón, R. (2010) Osmotic tolerance and habitat of early stegocephalians: indirect evidence from parsimony, taphonomy, paleobiogeography, physiology and morphology. In: M. Vecoli & G. Clément (eds.), *The Terrestrialization Process: Modelling Complex Interactions at the Biosphere–Geosphere Interface*. London: Geological Society, pp. 151–179.

Lindemann, F.J. (1991) Temnospondyls and the Lower Triassic paleogeography of Spitsbergen. In: Z. Kielan-Jaworowska, N. Heintz, & H.A. Nakrem (eds.), Fifth Symposium on Mesozoic Terrestrial Ecosystems and Biota. *Contributions from the Paleontological Museum Oslo* **364**, 39–40.

Lützner, H. & Kowalczyk, G. (2012) Paläogeographie und Beckengliederung. In: H. Lützner & G. Kowalczyk (eds.), Stratigraphie von Deutschland X. Rotliegend. Teil 1: Innervariscische Becken. *Schriftenreihe der deutschen geologischen Gesellschaft* **61**, 71–78.

Maisey, J.G. (1994) Predator–prey relationships and trophic level reconstruction in a fossil fish community. *Environmental Biology of Fishes* **40**, 1–22.

Milner, A.R. (1980) The tetrapod assemblage from Nýřany, Czechoslovakia. In: A.L. Panchen (ed.), *The Terrestrial Environment and the Origin of Land Vertebrates*. London and New York: Academic Press, pp. 439–496.

Molles, M.C. (2009) *Ecology: Concepts and Applications*. Columbus, OH: McGraw Hill.

Morin, P.J. (1995) Functional redundancy, non-additive interactions, and supply-side dynamics in experimental pond communities. *Ecology* **76**, 133–149.

Paine, R.T. (1969) A note on trophic complexity and community stability. *American Naturalist* **103**, 91–93.

Parker, M.S. (1992) Feeding ecology of larvae of the Pacific giant salamander (*Dicamptodon tenebrosus*) and their role as top predator in a headwater stream benthic community. PhD dissertation, University of California, Davis.

Petranka, J.W. (1998) *Salamanders of the United States and Canada*. Washington, DC: Smithsonian Institution Press.

Ricklefs, R.E., Buffetaut, E., Hallam, A., et al. (1990) Biotic systems and diversity. *Palaeogeography, Palaeoclimatology, Palaeoecology* **82**, 159–168.

Rooney, T.P., Antolik, C., & Moran, M.D. (2000) The impact of salamander predation on collembolan abundance. *Proceedings of the Entomological Society Washington* **102**, 308–312.

Schoch, R.R. (2009) Developmental evolution as a response to diverse lake habitats in Paleozoic amphibians. *Evolution* **63**, 2738–2749.

Schoch, R.R. & Milner, A.R. (2000) Stereospondyli. In: P. Wellnhofer (ed.), *Handbuch der Paläontologie*. Munich: Pfeil, Vol. 3B.

Schultze, H.-P. (2009) Interpretation of marine and freshwater paleoenvironments in Permo-Carboniferous deposits. *Palaeogeography, Palaeoclimatology, Palaeoecology* **281**, 126–136.

Schultze, H.-P. & Soler-Gijon, R. (2004) A xenacanth clasper from the ?uppermost Carboniferous-Lower Permian of Buxières-les-Mines (Massif Central, France) and the palaeoecology of the European Permo-Carboniferous basins. *Neues Jahrbuch für Geologie und Paläontologie Abhandlungen* **232**, 325–363.

Semlitsch, R.D. (1983) Burrowing ability and behavior of salamanders of the genus Ambystoma. *Canadian Journal of Zoology* **61**, 616–620.

Skoček, V. (1968) Upper Carboniferous varvites in coal basins of central Bohemia. *Vestnik ustred Ust geologicaleski* **43**, 113–121.

Trenham, P.C., Koenig, W.D., & Shaffer, H.B. (2001) Spatially autocorrelated demography and inter-pond dispersal in the salamander *Ambystoma californiense*. *Ecology* **82**, 3519–3530.

Vannote, R.L., Minshall, G.W., Cummins, K.W., Sedell, J.R., & Cushing, C.E. (1980) The river continuum concept. *Canadian Journal of Fish Aquatic Science* **37**, 130–137.

Wake, D.B. (2009) What salamanders have taught us about evolution. *Annual Review of Ecology and Systematics* **40**, 333–352.

Wild, R. (1980) The fossil deposits of Kupferzell, southwest Germany. *Mesozoic Vertebrate Life* **1**, 15–18.

Witzmann, F. (2009) Cannibalism in a small growth stage of the branchiosaurid *Apateon gracilis* (Credner, 1881) from the Early Permian Döhlen Basin, Saxony. *Fossil Record* **12**, 7–11.

Witzmann, F., Schoch, R.R., & Maisch, M.W. (2008) A relic basal tetrapod from the Middle Triassic of Germany. *Naturwissenschaften* **95**, 67–72.

8 Life History Evolution

The study of life histories focuses on the interplay between development and ecology in evolution. Life history theory explores how natural selection shapes key events in the ontogeny of an organism. These events depend on the physical and ecological environment of the particular species, and life histories involve the costs and benefits of growth, reproduction, and survivorship. Life history theory tries to understand the variation and adaptive value of life history traits by analyzing which traits are favored in different environments.

Modern amphibians have diverse and often complex life histories, and extinct taxa were found to have had a wide range of ontogenies correlating with paleoecological features. Ontogenetic trajectories are particularly well suited for the analytical comparison of ontogenies for this purpose. Together with the analysis of fossil *Lagerstätten*, these form the raw material of life history studies in the fossil record. As in paleoecology, one has to deal with numerous unknown traits, caused by inaccessible data in the fossil record. The central question is: What impact did ecology have on the structure of development in extinct amphibians?

Like the developmental and paleoecological data of fossil taxa, extinct life histories are difficult to reconstruct. However, in extinct amphibians, good preservation, large samples, a substantial ontogenetic record, and the potential for paleoecological analysis of deposits form a rich source of data for the study of extinct life history traits. What role did the evolution of development play in extinct amphibians, and are the identified modes any different from extant taxa?

8.1 Plasticity, reaction norm, and canalization

In an unorthodox statement, van Valen (1973) proposed "one could make an argument that evolution is the control of development by ecology." This formulates a central problem in modern evolutionary biology: How can development be influenced by external factors? There are two answers to this question – one obvious and one subtle – both of which are firmly grounded in modern evolutionary biology. First, selection is of course imposed by environmental conditions, preserving the most successful individuals in a given framework of environmental and biotic parameters. Selection favors organisms that grow and develop in a way to optimally use the resources and cope with the challenges of a given environment. Growth rate, timing of developmental events, and life span are important life history traits controlled by genetic factors, and these are constantly under selective pressure.

In a more subtle way, environmental parameters have a direct impact on development, given that the organism is sensitive to these parameters. "Sensitive" here means that the developmental system is sufficiently flexible to time and rate growth according to external influences. In the decades after the New Synthesis, this field was viewed with great caution by evolutionary biologists. Was this Lamarckism, the nightmare of Neo-Darwinism, fighting its way back on stage? Not at all. This concept is called *plasticity*. It is in full agreement with selection theory, and unequivocal cases of plasticity have been found in many different organisms (Stearns 1992; Schlichting and Pigliucci 1998; West-Eberhard 2003). Amphibians provide some of the best examples of plasticity, with respect to both development and morphology. Metamorphosis, for instance, is initiated in response to external factors (temperature, properties of water, nutrition). Although metamorphosis is usually tightly regulated in a given amphibian species, its timing is flexible with respect to external influences. As a result of different metamorphic timing, morphology often differs within populations or even between individuals of the same population in different seasons. These are traits in which the phenotype can be highly plastic.

The study of plasticity adds a new dimension to the understanding of development and evolution (Figure 8.1). A central concept here is the *reaction norm*, which includes all phenotypes that can be produced by an individual genotype (Woltereck 1909; Schmalhausen 1949). Depending on the

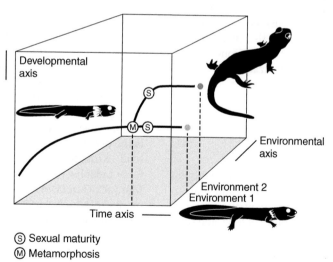

Ⓢ Sexual maturity
Ⓜ Metamorphosis

Figure 8.1 The life histories of modern amphibians described as a function of development and environmental conditions. Two important traits are sexual maturity and metamorphosis, which in combination enable the pathway of neoteny, as realized in numerous salamanders. Variation on the environmental axis may lead to plasticity, a means by which development responds to different external conditions.

environmental conditions to which the genotype is exposed, different phenotypes will be formed as a response or "reaction" to the external parameters. Such plastic responses to the environment can be measured by their amount, pattern, rapidity, and reversibility (Schlichting and Pigliucci 1998). In addition, the developmental system needs to be *competent* to respond to the environment *at a particular time*. All of these parameters are influenced by selection, thus making the genotype sensitive to environmental inputs. In this sense, ecology does control development in that it activates inbuilt developmental switches and mechanisms.

A property related to plasticity but often reciprocal in effect is *canalization*. Phenotypes may be plastic with respect to certain inputs, but buffered against perturbations in other directions. This buffering can result in a pattern opposite to that of the reaction norm: different genotypes may result in similar phenotypes. In describing this aspect of canalization, Waddington (1942) envisioned development as a ball rolling down a slope with a few distinct, well-defined channels. Rather than taking any path, development follows a decision tree of dichotomous pathways, each of which has been shaped by canalization. At a very gross scale, metamorphosis and neoteny are examples of alternative pathways in development. Their timing and the fine-tuning of their boundary conditions form important life history traits. In the following sections, I shall discuss the evolution of life history traits in the context of the concepts of plasticity, reaction norm, and canalization.

8.2 Reaction norms in extant amphibians

Lissamphibians have formed the focus of numerous life history studies (Werner 1986; Leips and Travis 1994). Variations in temperature, quantity and quality of water, seasonality, availability of food, and the pressure exerted by predators rank among the most universal traits. In many cases, trade-offs evolve between important traits, such as size and age: an earlier age at maturity will increase the probability of surviving to reproductive age, but it might do so at the cost of reduced size and fecundity (Stearns 1992). There is widespread empirical evidence showing

that variation in age and body mass at metamorphosis has an impact on survival as well as age of first reproduction; thus conditions of metamorphosis determine reproductive output (Rudolf and Rödel 2007). In general, amphibians are so prone to plasticity because wide reaction norms are a common adaptive response to heterogeneous or fluctuating environments (Doughty and Roberts 2003).

Listed below are some examples of traits for which reaction norms have been identified in extant amphibians, with notes on their evolutionary implications.

- **Temperature**. In salamanders, maximum adult size is often determined by the size at maturation, which in turn is regulated by a negative relationship with temperature on land. In *Plethodon glutinosus*, mean adult size is plastic and increases in cooler climates. This suggests the existence of a reaction norm linking temperature and body size in that species (Camp and Marshall 2000). Joly *et al.* (2005) also found that larval development is very sensitive to temperature in the anuran *Pelodytes punctatus*, whereas size at metamorphosis is highly constrained, apparently buffered against temperature and density fluctuations by canalization. In a further study, Jockusch (1997) found plasticity in the number of trunk vertebrae in the salamander *Batrachoseps*, responding to variations in developmental temperature. However, not all variation in the number of vertebrae was found to be subject to plasticity in *Batrachoseps*; notably, some geographic variation was shown to result from genetic variation instead (Jockusch 1997).
- **Water**. Most critical to the survival of tadpoles is the drying of their habitat ponds. In the warty toad *Rhinella spinulosa*, fast-drying ponds force tadpoles to accelerate their development. Márquez-García *et al.* (2010) reported experiments in which variation in pond desiccation affected the growth rate but not the morphology of metamorphosing tadpoles. However, this reaction norm is not a universal property of amphibians, as shown by another study: in natterjack toads (*Bufo calamita*), some sibships showed the ability to respond to earlier drying by accelerating development, whereas others did not (Reques and Tejedo 1997). This shows

how a reaction norm may evolve in parts of a population. Semlitsch *et al.* (1990) found that pond drying was met by different populations of *Ambystoma talpoideum* using different strategies: whereas some populations have a broad reaction norm responding to desiccation, others have a genetic polymorphism in their propensity to metamorphose as ponds dry.

- **Predation**. In many lissamphibians, timing of metamorphosis is plastic in response to predation risk during the pre-metamorphic stage (Higginson and Ruxton 2010). In populations responding to predation pressure individuals were found to be larger or equal-sized at metamorphosis than if they left the water at a smaller size. Thus, when living conditions are optimal, the anti-predator strategy of these tadpoles is to grow at an accelerated rate to escape predation, with the effect that they metamorphose at a larger size.

- **Maternal effects**. Egg size may exert a critical influence on morphology. Kaplan and Phillips (2006) studied a complex interplay between temperature variation, egg size, morphological development, swimming performance, and survivorship in tadpoles of *Bombina orientalis* from South Korea. In their experimental study, these authors found that higher sprint speeds and a higher rate of development had a significant impact on survival. Sprint speed and development in turn were found to be positively affected by higher temperatures.

- **Density**. Population size and density can be important factors in evoking plastic response. Dense populations of *Ambystoma macrodactylum* provide an example (Wildy *et al.* 2001). In the larvae of these long-toed salamanders, population density and food availability are coupled: high density (= high encounter rate of conspecifics) and/or poor availability of food increase aggressive behavior and cannibalism.

8.3 The biphasic life cycle in lissamphibians

Metamorphosis forms the plesiomorphic condition for the Lissamphibia, and many modern lissamphibians have biphasic life cycles (Wilder 1925;
Fritzsch 1990; Reiss 2002). Once established, biphasic life cycles impose specific constraints on evolution (Figure 8.2). To further evolve such an ontogeny, selection acts in divergent ways in larvae and adults. Larval and adult selection pressures need not be linked, but metamorphosis itself (duration, timing, and amount of change, responsiveness to particular influences) often forms the focus of selection. For instance, the larvae of a given species may be under pressure to mature early in order to escape predation pressure, or to escape predation by growing exceptionally large. Larvae may exploit new niches by becoming planktivores or larger predators, and in both cases the larval period will be modified. Larval specializations add to the importance of metamorphosis, because in these, adult structures depart even more from the larval body plan. In turn, transformed adult morphs usually focus on other sources and employ different feeding strategies from those of their larvae. A central question in amphibian evolution is therefore the maintenance and modification of metamorphosis. How does this transformation structure and constrain the amphibian life cycle, and what options are there to free the life cycle from these constraints?

It is not easy to break out of the tightly regulated system of a biphasic life cycle, but there are three principal options: (1) the adult phase is abandoned and maturity is reached during an extended larval period (neoteny), (2) the larval period is suppressed, and miniature adults hatch from land-laid eggs or are born live (terrestrial oviparity or viviparity), and, finally, (3) the theoretical option that metamorphosis (as a sequence of closely set events) disintegrates and the larva transforms slowly and gradually into a terrestrial adult. In lissamphibians, the former two options have evolved repeatedly, but the third never developed.

Larval morphs evolved in all three clades of lissamphibians, each forming a specific embryonic or larval body plan. Caecilian embryos have specialized multi-cusped teeth employed in scratching nutrients supplied by the mother or in intrauterine cannibalism. Aquatic larvae are known from some caecilian taxa, possessing external gills and lateral-line organs. Salamander larvae and neotenes have conical and pointed larval teeth, a large hyobranchium that is used in creating suction, external gills for aquatic respiration, a powerful swimming

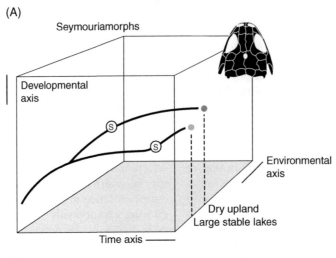

(A)

Seymouriamorphs

Developmental axis

Environmental axis

Dry upland
Large stable lakes

Time axis ——

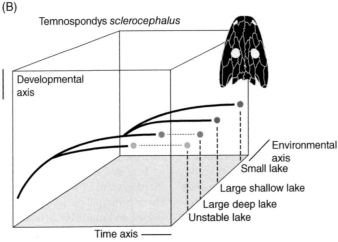

(B)

Temnospondys *sclerocephalus*

Developmental axis

Environmental axis
Small lake

Large shallow lake

Large deep lake

Unstable lake

Time axis ——

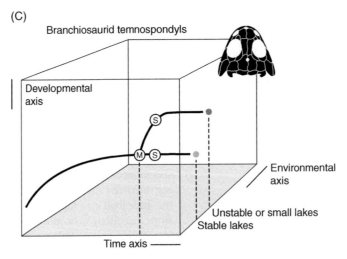

(C)

Branchiosaurid temnospondyls

Developmental axis

Environmental axis

Unstable or small lakes
Stable lakes

Time axis ——

Figure 8.2 Life histories of selected early tetrapods. (A) The most ancient trajectories are exemplified by seymouriamorphs: a biphasic life cycle with long transformation. (B) Some non-metamorphosing temnospondyls were extremely flexible in adjusting to different lake habitats. (C) Dissorophoids evolved a condensed metamorphosis, which opened the new pathway of neoteny, first explored by Late Paleozoic branchiosaurids.

tail, and lateral-line and electroreceptor organs. Anuran tadpoles evolved larval "keratinized teeth," "internal" gills and a remodeled hyobranchium for the processing of plant detritus in small ponds.

Metamorphosis is a short phase in which many physiological and morphological changes occur (Fritzsch 1990). In the ontogenetic trajectory, it plots as a condensation of developmental events (Alberch 1989). In lissamphibians it is usually connected with a transition from an aquatic to a terrestrial existence, and so it probably was in dissorophoid temnospondyls, which also underwent a series of profound morphological changes in a short time window (Schoch and Fröbisch 2006). Metamorphosis may be initiated even after the first year of larval life, and even neotenes may metamorphose in some species (*Notophthalmus viridescens*) (see Reilly 1987). When compared with dissorophoid temnospondyls, transforming salamanders have accumulated various additional metamorphic events, and in anurans metamorphosis is maximally inclusive within a very short period of time. Most importantly, anurans attain sexual maturity only during or after metamorphosis, and therefore cannot evolve neoteny. This results from the fact that the required thyroid hormone level is only reached during metamorphosis (Hayes 1997).

Because transforming amphibians usually do not feed, the duration of metamorphosis is minimized by selection. It is an energetically demanding process, resulting in 30–58% weight loss, and its duration correlates with body size (Downie *et al.* 2004). Predation and desiccation also form important selection pressures. In sum, metamorphosis is favored to occur in small animals in which it requires minimal time. In lung-breathing salamanders, the enhanced growth of lungs usually forces the body to the water surface during metamorphosis, which triggers their migration to land where they hide under small stones or cover in dense vegetation.

8.4 Seymouriamorphs: biphasic life cycles without metamorphosis

Seymouriamorphs had aquatic larvae with bushy external gills and lateral lines and terrestrial adults in the 1–2 m range. They are a good example of how the retention of aquatic larvae paved the way for exploring diverse environments. Seymouriamorph larvae were carnivores, essentially miniature adults with external gills and sensory lines. More specific larval specializations, both morphologically and ecologically, are not apparent. In turn, the terrestrial adults of seymouriamorphs were probably not able to compete with the numerous other terrestrial clades to evolve during the Permian, notably diapsids. Seymouriamorphs shared the principal developmental pattern with basal temnospondyls: they were able to extend or foreshorten their larval period, responding to environmental variation (Figure 8.2). Like most temnospondyls, seymouriamorphs did not undergo a fast transformation. This supports the hypothesis that the temnospondyl developmental trajectory and its ecological implications are primitive for crown tetrapods. Uncertainties remain, however, because anthracosaurs and other stem-amniotes have such poorly known ontogenies. Delayed sexual maturity, found by Sanchez *et al.* (2008) in the largest specimens of *Discosauriscus*, differs from the results of histological studies in branchiosaurids: whereas the latter reached sexual maturity earlier and apparently stayed in the water, *Discosauriscus* might have left the water after maturity was reached, as Klembara (2009) has concluded. This adds to the hypothesis that neoteny as an evolutionary startegy was not an option for seymouriamorphs, or in general for tetrapods outside the Dissorophoidea and Lissamphibia. Seymouriamorphs thus had a biphasic life cycle but no drastic metamorphosis.

8.5 Temnospondyls: flexible uni- and biphasic ontogenies

Despite their overall similarity, temnospondyls show a surprising variation in ontogeny, size, and probable life span, as well as in ecological parameters (Schoch 2009a). The analysis of their ontogenetic trajectories revealed a common theme that departs considerably from the lissamphibian life cycle: most temnospondyls did not metamorphose, and their adults differed much less from larval stages than in any extant amphibian. Slight modification of the temnospondyl trajectory permitted

Figure 8.3 The adaptive background of ontogenetic trajectories has been well studied in temnospondyls. Ancient trajectories were uniphasic and permitted much flexibility on a small scale, in order to respond to environmental gradients or fluctuations. The more advanced pathways of metamorphosis and neoteny form a different type of life history, constrained by the brief period of transformation, but with the option to evolve larval, neotenic, and post-metamorphic pathways separately.

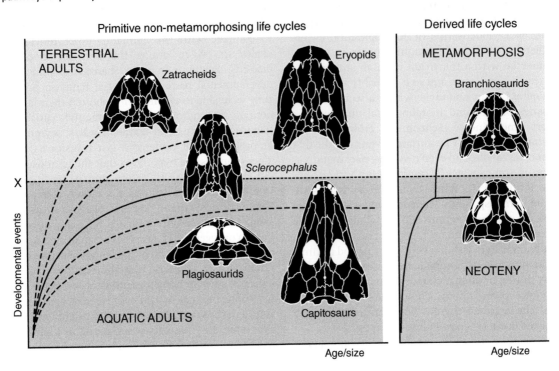

crucial life history traits to evolve into divergent directions (Figure 8.3). Unlike in the biphasic life cycle, temnospondyl ontogeny could be truncated or extended more flexibly. This mode of developmental evolution was recently termed "developmental fine-tuning" (Schoch 2010).

The best-known life cycle is that of *Sclerocephalus haeuseri*, a 1–1.8 m long fish-eating Carboniferous lake dweller (Schoch 2009b). It is known from numerous lake deposits in Germany, some of which have yielded rich samples with a wide range of size classes (Schoch and Witzmann 2009). This species had a stretched-out trajectory without metamorphosis, in which the skull and dermal shoulder girdle formed early (Boy 1988; Schoch 2010), but other postcranial bones formed successively, well into late ontogeny. Notably, the completion of ossification in the limb and girdle elements required a long time, and was not

reached in every population studied (Schoch 2009b). Time-averaged "populations" of *Sclerocephalus* were found to have responded to different environments by variations in development: over a period of 2–3 myr, six different lake habitats were inhabited by this species, which responded to changes by modification of growth rate, adult size, developmental sequence, skeletal features, prey preference, and relative degree of terrestriality. For instance, in the small Kappeln and Niederkirchen lakes, adults were huge with fully ossified limbs, probably able to leave the water. In the large but unstable Odernheim and Humberg lakes, adults were much smaller and less completely ossified, with aquatic morphologies such as long swimming tails and lateral-line grooves in the skull (Schoch 2009b). Despite all these differences, and apart from occasional cannibalism, *Sclerocephalus* was a fish-eater that focused on the basal actinopterygian

Paramblypterus, based on well-preserved gut contents. Evidently, the size and abundance of its prey had an impact on its own size and frequency (Schoch 2009b).

It is far from clear to what extent these evolutionary patterns include plasticity and reaction norms *within* species, or "simple" developmental evolution of fixed traits *between* populations and species. However, these data show how important the fine-tuning of development was in Paleozoic temnospondyls, with ecological parameters such as size and depth of the lake, size of prey items, aeration and other water properties, and seasonality playing important roles.

The major evolutionary directions that temnospondyl life histories underwent are illsutrated by two Permian genera. In *Archegosaurus*, which led a fully aquatic adult life, the trajectory was similar to that of *Sclerocephalus* but truncated, without ossification of carpals, tarsals, and the pelvic girdle (Witzmann 2006). *Archegosaurus* inhabited large and deep lakes and consumed acanthodian fishes (Boy and Sues 2000). In contrast, *Onchiodon* had a longer trajectory with additional events not occurring in *Sclerocephalus*, and this genus had a more terrestrial or amphibious adult life (Boy 1990; Schoch 2009a).

The predominantly large Mesozoic stereospondyls had flat ontogenetic trajectories that usually did not reach the phase in which a terrestrial adult developed. Their ontogenies were much more constrained than those of Paleozoic temnospondyls, with small juveniles closely resembling large adults. True larval morphs apparently did not exist. This indicates that the evolutionary divergence of larvae and adults was not a one-way street. Instead, the 5 m long *Mastodonsaurus* had a much more conservative ontogeny and morphology than the 1.5 m long *Archegosaurus*.

8.6 Lepospondyls: dwarfism and uniphasic life cycles

The life histories of the more diverse and speciose lepospondyls are not as easily understood as those of seymouriamorphs and temnospondyls. The most apparent feature is the pronounced dwarfism of most lepospondyls, and the fact that juveniles resembled adults closely. In most microsaurs and lysorophians, an aquatic larval phase did not exist. But even the fully aquatic nectridean *Diplocaulus* had tiny juveniles that were miniature adults in terms of their postcranial anatomy, with only the shape of the skull subject to major ontogenetic changes. It is quite probable that many microsaurs and lysorophians were live-bearing or hatched from land-laid eggs (like some terrestrial salamanders), although this is very difficult to test.

As a whole, lepospondyls appear to have evolved a different evolutionary strategy from that of other early tetrapods. Their dwarfism probably opened new avenues for the evolution of respiratory mechanisms. By analogy with salamanders, skin breathing might have been an important option especially for tiny microsaurs, both in the water and on land. The long ribs, shared with amniotes and their ancestors, indicate that lepospondyls practiced some kind of primitive rib-driven respiration. Internal gills were apparently not retained, and external larval gills are never preserved (except for branchial ossicles in *Microbrachis*: see below). On the other hand, nectrideans, aïstopods, and adelospondyls might well have been gill-bearing, as some taxa preserve hyobranchial skeletons (Carroll 2009). The ontogeny of lepospondyls is known from several taxa among the microsaurs (Carroll and Gaskill 1978), aïstopods (Anderson *et al.* 2003; Germain 2008), and nectrideans (Rinehart and Lucas 2001; Schoch and Sues in preparation). In general, the skeleton ossified at a faster rate than in temnospondyls or lissamphibians (Fröbisch *et al.* 2010), and the close resemblance of juveniles and adults is a feature shared with amniotes. This means that lepospondyls did not metamorphose and had uniphasic life cycles. Microsaurs and lysorophians were probably terrestrial without a larval morph; their small size and pronounced terrestriality probably permitted them to fill niches inaccessible to temnospondyls. Nectrideans were probably fully aquatic without the option to leave the water as adults. The lifestyles of aïstopods and adelospondyls are poorly known. The only exception to these life history pathways is the enigmatic genus *Microbrachis*, which resembles larvae of temnospondyls and seymouriamorphs (Carroll and Gaskill 1978; Vallin and Laurin 2004).

8.7 The evolution of metamorphosis

Metamorphosis – defined as a short-term phase with drastic morphological change – evolved in the putative ancestors of lissamphibians and is not known from any other basal tetrapod group. The substantial remodeling of the jaws and hyobranchium in all modern clades suggests that feeding might have played a role in the evolutionary origin of metamorphosis. Evidently, the transformation from an aquatic to a terrestrial existence itself does not require drastic morphological change. A range of Paleozoic temnospondyls underwent such changes in a much longer period of time, as exemplified by *Onchiodon*. Despite their capability to cross land, large temnospondyls appear to have been fish-eaters that returned to the water on a regular basis. This diet appears to have been diversified in two clades of relatively small large-headed and short-bodied temnospondyls, the zatracheids and dissorophoids. These not only underrwent more substantial modification in their life cycle, but also had a clearly terrestrial existence as adults (Schoch 2009a). The change in feeding, from fish-dominated to insect-dominated diets, may have triggered the evolution of metamorphosis. Whereas most aquatic temnospondyls were fish-eaters irrespective of their size, amphibamid and zatracheid adults had mostly tiny teeth and modified hyobranchial skeletons suggesting tongue-supported feeding on small terrestrial prey. Some amphibamids even had the derived pedicellate tooth morphology shared by most adult lissamphibians. It has therefore been concluded that dissorophoids and zatracheids first evolved metamorphosis, and amphibamids further intensified it to remodel their aquatic feeding apparatus into a terrestrial one (Schoch 2002, 2009a). As in modern amphibians, the hyobranchium and dentition must have played central roles in this early metamorphosis. For a new feeding apparatus to be built, a condensed phase of rapid remodeling appears to be the easiest solution, even though it means starvation and loss of weight.

Size may also have been an important factor. In extant amphibians, size and duration of metamorphosis are correlated (Downie *et al.* 2004). Thus, for the duration of remodeling to be minimized, body size must be reduced. This could explain the comparably small size of most lissamphibians – and the relatively large size of those salamanders that abandoned metamorphosis (cryptobranchids, sirenids, amphiumids).

The evolution of increasingly specialized aquatic larvae and terrestrial adults appears to have been linked. Indeed, the most apomorphic larval forms are found in the same taxa that also had the most elaborate terrestrial adults: the dissorophoids and zatracheids. This indicates that a biphasic life cycle had already evolved, and well-sampled taxa have confirmed this (*Apateon gracilis*, *Amphibamus grandiceps*). These observations indicate that separate larval and adult selection pressures were already active, at least in dissorophoids. The evolution of a biphasic life cycle from the uniphasic, more flexible trajectory of temnospondyls must have brought substantial changes for the evolutionary toolkit: rather than having unconstrained options to modify any part of the trajectory, metamorphosis now demanded canalized pathways to be chosen (Figure 8.3). Metamorphosis itself was buffered against various kinds of perturbation in order to guarantee a viable terrestrial adult. The high costs of metamorphosis (mortality rates, starvation, predation) are compensated for by the manifold options gained through the dichotomy of larval and terrestrial morphs. In anurans, selection was able to drive larvae so far as to suppress the ancient larval morphology, and evolve the largely new body plan of a tadpole. This permitted the construction of new niches and the exploitation of ephemeral microhabitats. Tadpoles might well have driven anuran evolution by diversifying reaction norms along gradients of ecological parameters. This is a fruitful field that links reaction-norm evolution with speciation, a topic mentioned by Schmalhausen (1949) and explored by West-Eberhard (1989).

8.8 The evolution of neoteny

Neoteny is here considered in its original meaning, as a life history trait (adaptive strategy) that occurs in salamanders, but not in anurans or caecilians. Neoteny does not occur in fishes or amniotes either, and I argue here that an important prerequisite for neoteny is lissamphibian-like metamorphosis. Neoteny can be defined as the failure of an individual to metamorphose before sexual maturity is reached. Some salamanders may transform after a neotenic period (Reilly 1987), but in most cases neoteny extends over the entire adult life. Perennibranchiates are neotenic salamanders that retain their external gills throughout their lives (e.g., proteids, sirenids), but there are many other types of neoteny, depending on how far somatic development proceeds before maturity and adulthood are reached. Neotenic salamanders usually retain lateral-line organs, an unprotected larval skin, and the aquatic feeding technique (suction).

Kollmann (1885) first defined neoteny after having studied life histories of salamandrids and the Mexican axolotl. Originally, it was considered a retardation of somatic development. Later, it became clear that three different modes of neoteny exist: (1) *obligate* in species that are generally insensitive to thyroxin, (2) *facultative* in species that may metamorphose under appropriate environmental conditions, and (3) *inducible* in species that fail to transform in the wild due to a low production of thyroxin, but may metamorphose when the hormone level is increased (Norris 1985).

There are many variants of neoteny known, depending on the mode of larval development, the specific features of metamorphosis, the structure of the salamander population, and the environmental setting. As with metamorphosis, no two cases of neoteny are exactly alike. For instance, in the North American salamandrid *Notophthalmus viridescens*, metamorphosis begins during the first year with three life history options: (1) metamorphosis is completed and terrestrial juveniles form until maturity to return to the water as aquatic adults; (2) metamorphosis is complete, but juveniles remain in the pond; and (3) limited neoteny occurs and variably branchiate juveniles mature in the pond to gradually finish metamorphosis (Reilly 1987). Thus, in salamandrids, the *completion* of metamorphosis is delayed, whereas in ambystomatids the *onset* of metamorphosis is postponed. Neoteny thus is a complex phenomenon that can be produced in different ways: (1) the deceleration of somatic development (pedomorphosis), or (2) the acceleration of gonadal development (peramorphosis) (Reilly 1987). Physiologically,

failure to metamorphose can result either from a lack of thyroxin production or by tissue insensitivity to thyroxin (White and Nicoll 1981). It is therefore likely that neoteny in different salamanders has different genetic bases.

In the fossil record, neoteny has been demonstrated only in branchiosaurid temnospondyls (Boy 1974; Schoch 2009a). Reliable evidence that particular populations of *Apateon pedestris* and *A. caducus* were neotenic comes from the study of histology and skeletochronology, which confirms that larval morphs were already sexually mature (Sanchez *et al.* 2010). The existence of many fully aquatic temnospondyls with larval features (short snout, larval hyobranchial skeleton, poorly ossified limbs) has often been cited as a sign of neoteny in dvinosaurs, brachyopoids, and plagiosaurids. However, these apparently pedomorphic morphologies are combined with many peramorphic traits, and it would be wrong to regard these aquatic adults as sexually mature larvae. None of these taxa underwent metamorphosis, nor did they have metamorphosing ancestors, which is why the concept of neoteny as put forward here is not applicable.

Neoteny has evolved numerous times in many salamander species, and even occurs in small populations of normally metamorphosing taxa. The adaptive value of complete neoteny is probably the retention of effective gill respiration and the advantage of using the unidirectional water flow in suction feeding (Reilly 1987).

8.9 General features of life history evolution

Throughout the history of Earth, amphibian life histories have been diverse, involving manifold cases of iterative microevolution, triggered by the interplay of reaction norms and canalization. Metamorphosis apparently evolved in dissorophoid temnospondyls as a specific response to unstable environments, and forms the bottom line of life history pathways in all extant amphibians. It provides canalized larval and adult pathways on which selection can act separately. Direct development and neoteny are additional essential life history traits that repeatedly evolved from ancestral biphasic life cycles. However, the study of fossil life histories has revealed that metamorphosis is a highly derived version of a biphasic life cycle. Ontogenetic trajectories of temnospondyls and seymouriamorphs exemplify biphasic life cycles with slow and gradual transformation. A common life history in temnospondyls involved adults that occasionally crossed land bridges between lakes, but preferred aquatic prey. By slight modifications of this trajectory, more aquatic or terrestrial taxa evolved rather easily and repeatedly (Schoch 2010). The evolution of these gradual life cycles progressed by "developmental fine-tuning" – a modification of the trajectory at whatever point. Developmental plasticity evidently played important roles in these early tetrapod life histories, although the range of a reaction norm is almost impossible to analyze in a particular fossil taxon. Evolution by "developmental fine-tuning" addressed the same general problems as metamorphosis, responding to drying ponds, unfavorable water conditions, and selection pressures on land.

Metamorphosis not only opened new avenues for adaptation and diversity, but also acted as a constraint for evolution. For instance, larval features often influence adult structures, unless they are entirely remodeled, which, of course, requires energy and time. A means to break such metamorphosis-related constraints has been reported in salamanders, which evolved redundant developmental systems. In some taxa, larval and adult hyobranchial skeletons are formed by different sets of cell populations, such that the adult structures are no longer constrained by larval ones (Alberch 1989). In the fossil record, such redundancy is very difficult to identify; even resorption, a common phenomenon in the palate of metamorphosing salamanders (Larsen 1963; Lebedkina 2004), is not easy to trace. New data on growth series of the miniaturized dissorophoid *Amphibamus* (from the Late Carboniferous of Illinois) could be interpreted as resorption in the marginal palate, but confirmation would require a much larger sample. Given that there was such a resorption of palatine, ectopterygoid, and pterygoid bones, the build-up of adult elements would be likely to have required separate developmental systems for the larval and adult palate. At any rate, developmental redundancy appears to be an alternative way to bypass constraints imposed by development on evolution.

Plasticity may also be important for the evolution of diversity. In a study of closely related spadefoot toads (Pelobatidae), Gomez-Mestre and Buchholz (2006) showed that species within this small group differ substantially in tadpole reaction norms. They further identified post-metamorphic traits (hindlimb, snout length) that were influenced by the duration of the larval period. These differences mirrored within-species plasticity at a higher taxonomic level, thus revealing reaction-norm evolution to accompany cladogenesis. This example also highlights the importance of the extra dimension that plasticity contributes to evolution: new morphological characters do not necessarily evolve in isolation and as static traits, but often result from evolutionary changes in reaction norms. The example of *Sclerocephalus* at least testifies that plasticity and developmental evolution played an important role in early amphibians, and that "developmental fine-tuning" might have paved the way for speciation along environmental gradients such as the diverse intramontane lake habitats of Permo-Carboniferous central Europe.

A final question is how life histories evolve on small scales, especially when neoteny is involved. The heterochronic shift from metamorphosing to neotenic represents a punctuational event even on a small time scale. Harris *et al.* (1990) showed that the propensity to metamorphose has an additive genetic basis in *Ambystoma talpoideum* and that selection uses the available genetic variation to produce an adaptive neotenic phenotype. Morphologically, this pattern may be ranked as a macroevolutionary event, but the responsible process (selection pressure created by variations in pond permanence) acts on the microevolutionary scale (Harris *et al.* 1990). The dichotomy between metamorphosis and neoteny thus forms an adaptation to a variable aquatic environment and an exaptation for rapid evolutionary change to either side; such polymorphisms have been reported to lead to rapid evolution and speciation (West-Eberhard 1989).

Based on numerous first-hand observations, Schmalhausen (1949) held that most morphological features had a physiological component that traveled with them. If viewed from this perspective, one may say that it is reaction norms, rather than characters, that evolve.

References

Alberch, P. (1989) Development and the evolution of amphibian metamorphosis. *Fortschritte der Zoologie* **35**, 163–173.

Anderson, J.S., Carroll, R.L., & Rowe, T.B. (2003) New information on *Lethiscus stocki* (Tetrapoda: Lepospondyli) from high-resolution computed tomography and a phylogenetic analysis of Aïstopoda. *Canadian Journal of Earth Sciences* **40**, 1071–1083.

Boy, J.A. (1974) Die Larven der rhachitomen Amphibien (Amphibia: Temnospondyli, Karbon-Trias). *Paläontologische Zeitschrift* **48**, 236–268.

Boy, J.A. (1988) Über einige Vertreter der Eryopoidea (Amphibia: Temnospondyli) aus dem europäischen Rotliegend (? höchstes Karbon – Perm). 1. *Sclerocephalus. Paläontologische Zeitschrift* **62**, 107–132.

Boy, J.A. (1990) Über einige Vertreter der Eryopoidea (Amphibia: Temnospondyli) aus dem europäischen Rotliegend (? höchstes Karbon – Perm). 3. *Onchiodon. Paläontologische Zeitschrift* **64**, 287–312.

Boy, J.A. & Sues, H.-D. (2000) Branchiosaurs: larvae, metamorphosis and heterochrony in temnospondyls and seymouriamorphs. In: H. Heatwole & R.L. Carroll (eds.), *Amphibian Biology. Volume 4. Palaeontology.* Chipping Norton, NSW: Surrey Beatty, pp. 973–1496.

Camp, C.C. & Marshall, J.L. (2000) The role of thermal environment in determining the life history of a terrestrial salamander. *Canadian Journal of Zoology* **78**, 1702–1711.

Carroll, R.L. (2009) *The Rise of Amphibians.* Baltimore: Johns Hopkins University Press.

Carroll, R.L. & Gaskill, P. (1978) The order Microsauria. *Memoirs of the American Philospohical Society* **126**, 1–211.

Doughty, P. & Roberts, J.D. (2003) Plasticity in age and size at metamorphosis of *Crinia georgiana* tadpoles: responses to variation in food levels and deteriorating conditions during development. *Australian Journal of Zoology* **51**, 271–284.

Downie, J.R., Bryce, R., & Smith, J. (2004) Metamorphic duration: an under-studied variable in frog life histories. *Biological Journal of the Linnean Society* **83**, 261–272.

Fritzsch, B. (1990) The evolution of metamorphosis in amphibians. *Journal of Neurobiology* **41**, 1011–1021.

Fröbisch, N.B., Olori, J., Schoch, R.R, & Witzmann, F. (2010) Amphibian development in the fossil record. *Seminars in Cell and Developmental Biology* **21**, 424–431.

Germain, D. (2008) A new phlegethontiid specimen (Lepospondyli, Aïstopoda) from the Late Carboniferous of Montceau-les-Mines (Saone-et-Loire, France). *Geodiversitas* **30**, 669–680.

Gomez-Mestre, I. & Buchholz, D.R. (2006) Developmental plasticity mirrors differences among taxa in spadefoot toads linking plasticity and diversity. *Proceedings of the National Academy of Sciences* **103**, 19021–19026

Harris, R.N., Semlitsch, R.D., Wilbur, H.M., & Fauth, J.E. (1990) Local variation in the genetic basis of paedomorphosis in the salamander *Ambystoma talpoideum. Evolution* **44**, 1588–1603.

Hayes, T.B. (1997) Hormonal mechanisms as potential constraints on evolution: examples from the Anura. *American Zoologist* **37**, 482–490.

Higginson, A.D. & Ruxton, G.D. (2010) Adaptive changes in size and age at metamorphosis can qualitatively vary with predator type and available defences. *Ecology* **91**, 2756–2768.

Jockusch, E.L. (1997) Geographic variation and phenotypic plasticity of number of trunk vertebrae in slender salamanders, *Batrachoseps* (Caudata: Plethodontidae). *Evolution* **51**, 1966–1982.

Joly, P., Morand, A., Plénet, S., & Grolet, O. (2005) Canalization of size at metamorphosis despite temperature and density variations in *Pelodytes punctatus. Herpetological Journal* **15**, 45–50.

Kaplan, R.H. & Phillips, P.C. (2006) Ecological and developmental context of natural selection: maternal effects and thermally induced plasticity in the frog *Bombina orientalis. Evolution* **60**, 142–156.

Klembara, J. (2009) New cranial and dental features of *Discosauriscus austriacus* (Seymouriamorpha, Discosauriscidae) and the ontogenetic conditions of *Discosauriscus. Special Papers in Palaeontology* **81**, 61–69.

Kollmann, J. (1885) Das Überwintern europäischer Frosch- und Tritonlarven und die Umwandlung des mexikanischen Axolotl. *Verhandlungen der Naturforschenden Gesellschaft in Basel* **7**.

Larsen, J.H. (1963) The cranial osteology of neotenic and transformed salamanders and its bearing on interfamilial relationships. Dissertation, University of Michigan.

Lebedkina, N.S. (2004) *Evolution of the Amphibian Skull*. Sofia: Pensoft.

Leips, J. & Travis, J. (1994) Metamorphic responses to changing food levels in two species of hylid frogs. *Ecology* **75**, 1345–1356.

Márquez-García, M., Correa-Solís, M., & Méndez, M.A. (2010) Life-history trait variation in tadpoles of the warty toad in response to pond-drying. *Journal of Zoology* **281**, 105–111.

Norris, D.O. (1985) *Vertebrate Endocrinology*. Philadelphia: Lea and Feibiger.

Reilly, S.M. (1987) Ontogeny of the hyobranchial apparatus in the salamanders *Ambystoma talpoideum* (Ambystomatidae) and *Notophthalmus viridescens* (Salamandridae): the ecoolgical morphology of two neotenic strategies. *Journal of Morphology* **191**, 205–214.

Reiss, J.O. (2002) The phylogeny of amphibian metamorphosis. *Zoology* **105**, 1–12.

Reques, R. & Tejedo, M. (1997) Reaction norms for metamorphic traits in natterjack toads to larval density and pond duration. *Journal of Evolutionary Biology* **10**, 829–851.

Rinehart, L.F. & Lucas, S.G. (2001) A statistical analysis of a growth series of the Permian nectridean *Diplocaulus magnicornis* showing two-stage ontogeny. *Journal of Vertebrate Paleontology* **21**, 803–806.

Rudolf, V.H.W. and Rödel, M.O. (2007) Phentoypic plasticity and optimal timing of metamorphosis under certain time constraints. *Evolution and Ecology* **21**, 121–143.

Sanchez, S., Klembara, J., Castanet, J., & Steyer, J.S. (2008) Salamander-like development in a seymouriamorph revealed by palaeohistology. *Biology Letters* **4**, 411–414.

Sanchez, S., de Ricqlès, A., Schoch, R.R., & Steyer, J.S. (2010) Developmental plasticity of limb bone microstructural organization in *Apateon*: histological evidence of paedomorphic conditions in branchiosurs. *Evolution and Development* **12**, 315–328.

Schlichting, C. & Pigliucci, M. (1998) *Phenotypic Plasticity: a Reaction Norm Perspective*. Sunderland: Sinauer.

Schmalhausen, I.I. (1949) *Factors of Evolution*. Chicago: University of Chicago Press.

Schoch, R.R. (2002) The early formation of the skull in extant and Paleozoic amphibians. *Paleobiology* **28**, 378–396.

Schoch, R.R. (2009a) The evolution of life cycles in early amphibians. *Annual Review of Earth and Planetary Sciences* **37**, 135–162.

Schoch, R.R. (2009b) Developmental evolution as a response to diverse lake habitats in Paleozoic amphibians. *Evolution* **63**, 2738–2749.

Schoch, R.R. (2010) Heterochrony: the interplay between development and ecology in an extinct amphibian clade. *Paleobiology* **36**, 318–334.

Schoch, R. R. & Fröbisch, N.B. (2006) Metamorphosis and neoteny: alternative developmental pathways in an extinct amphibian clade. *Evolution* **60**, 1467–1475.

Schoch, R.R. & Witzmann, F. (2009) Osteology and relationships of the temnospondyl *Sclerocephalus*. *Zoological Journal of the Linnean Society London* **157**, 135–168.

Semlitsch, R.D., Harris, R.N., & Wilbur, H.M (1990) Paedomorphosis in Ambystoma talpoideum: maintenance of population variation and alternative life-history pathways. *Evolution* **44**, 1604–1613.

Stearns, S.C. (1992) *The Evolution of Life Histories*. Oxford: Oxford University Press.

Vallin, G. & Laurin, M. (2004) Cranial morphology and affinities of *Microbrachis*, and a reappraisal of the phylogeny and lifestyle of the first amphibians. *Journal of Vertebrate Paleontology* **24**, 56–72.

van Valen, L. (1973) Festschrift. *Science* **180**, 488.

Waddington, C.H. (1942) Canalization of development and the inheritance of acquired characters. *Nature* **150**, 563–565.

Werner, E.E. (1986) Amphibian metamorphosis: growth rate, predation risk, and the optimal size at transformation. *American Naturalist* **128**, 319–341.

West-Eberhard, M.J. (1989) Phenotypic plasticity and the origins of diversity. *Annual Review of Ecology and Systematics* **20**, 249–278.

West-Eberhard, M.J. (2003) *Developmental Plasticity and Evolution*. New York: Oxford University Press.

White, B.A. & Nicoll, C.S. (1981) Hormonal control of amphibian metamorphosis. In: L.I. Gilbert & E. Frieden (eds.), *Metamorphosis: a Problem Of*

Developmental Biology. New York: Plenum Press, pp. 363–396.

Wilder, I.W. (1925) *The Morphology of Amphibian Metamorphosis*. Northampton, MA: Smith College Fiftieth Anniversary Publication.

Wildy, E.L., Chivers, D.P., Kiesecker, J.M., & Blaustein, A.R. (2001) The effects of food level and conspecific density on biting and cannibalism in larval long-toed salamanders, *Ambystoma macrodactylum*. Oecologia **128**, 202–209.

Witzmann, F. (2006). Developmental patterns and ossification sequence in the Permo-Carboniferous temnospondyl *Archegosaurus decheni* (Saar-Nahe Basin, Germany). *Journal of Vertebrate Paleontology* **26**, 7–17.

Woltereck, R. (1909) Weitere experimentelle Untersuchungen über Artveränderung, speziell über das Wesen quantitativer Artunterschiede der Daphniden. *Verhandlungsbericht der deutschen zoologischen Gesellschaft* **1909**, 110–172.

9 Phylogeny

Phylogenetic systematics has cast much light on the field of amphibian evolution and the initial diversification of tetrapods. Although major questions remain unanswered, the cladistic approach has forced workers in the field to lay open their own reasoning, address the arguments of others, and refine their own research questions. Several widely used software packages have contributed greatly to this line of research (Laurin and Reisz 1997; Laurin 1998; Anderson 2001, 2007; Ruta *et al.* 2003a). Even though some of the currently discussed hypotheses were first proposed many decades ago, cladistics has made it easier to formulate them and elucidate their stratigraphic, functional, and evolutionary implications. The "total evidence" approach has accumulated a large quantity of data that are hard to review. However, it has also established a very useful platform (database) for subsequent analyses. New analyses do not have to start from scratch, and authors can build on previous work by others. This saves a lot of time, but also means hard work to understand every aspect of morphology in a wide range of taxa. First-hand examination of material is often obligatory, which may require extensive travel and work in numerous scientific collections. The major challenge for contemporary researchers is to understand the composition of data sets, evaluate the reliability of characters and character states considered by previous authors, decide on the treatment of missing data, and choose the methods and algorithms of analysis that best fit the problem under study.

Amphibian Evolution: The Life of Early Land Vertebrates, First Edition. Rainer R. Schoch.
© 2014 Rainer R. Schoch. Published 2014 by John Wiley & Sons, Ltd.

9.1 Phylogeny of amphibians

Two schools of thought are apparent when cladistic analyses of amphibians over the last 25 years are compared: (1) a *morphological school* focused on the significance of the most important features that support monophyletic clades and (2) an *analytical school* which highlights the analytical tools and seeks to maximize congruence between large data sets. Both schools have their strengths and shortcomings, and the way forward is to endorse both perspectives. The morphological school is uneasy with the total evidence approach because it highlights the different significance and robustness of characters. It is a strength that it evaluates character definitions, their biological context, and the limitations and problems of single characters. For instance, features rich in anatomical structures (e.g., pedicely) are viewed as more significant than "absence characters" by this school (Hecht 1976; Boy 1981; Milner 1988). However, potential synapomorphies are often discussed in more detail than homoplasies, which is a shortcoming of this approach.

The analytical school holds that the biological aspects of characters are mostly obscure, and that they should therefore be treated (and weighted) equally. This is why congruence is more important in the analytical approach than the robustness of characters. This perspective puts more emphasis on the different ways to analyze large data sets (> 20 taxa, > 100 characters), and usually asks broader questions (Laurin and Reisz 1997; Ruta *et al.* 2003a). While there are good reasons to employ "total evidence," an overly relaxed attitude towards character selection and definition can be problematic (Schoch and Milner 2004; Ruta and Coates 2007; Anderson 2008). Many analyses based on large data sets are notorious for their inclusion of doubtful characters (which are usually recognized by the fact that they cannot be verified by subsequent authors), and often nodes are diagnosed on the basis of long lists of rather obvious homoplasies.

The first phylogenetic analyses of early tetrapod and amphibian relationships were conducted "by hand" (without the use of computer programs). These pioneering studies

often formed the basis of all subsequent work in collecting large numbers of characters (Bolt 1969; Boy 1972; Bolt and Lombard 1985; Milner 1988, 1993; Panchen and Smithson 1988; Trueb and Cloutier 1991). In the last decade, large data sets have been put together by several teams of authors, and analytical aspects have been discussed in detail. An influential set of papers was published by Michel Laurin and colleagues (Laurin and Reisz 1997; Laurin 1998; Vallin and Laurin 2004), who were the first to analyze a large data set of all major tetrapod clades the software PAUP (phylogenetic analysis using parsimony). Laurin formulated an entirely new hypothesis on that basis, in which lysorophians were proposed for the first time as a sister taxon of all Lissamphibia (the lepospondyl hypothesis, LH). This was followed by a still larger analysis by Ruta *et al.* (2003a), who expanded the frame to cover a wider range of stem-tetrapods, considered more ingroup taxa, and by that means found support for the temnospondyl hypothesis (TH). Most recently, Anderson *et al.* (2008) have described a new potential lissamphibian relative from the Early Permian (*Gerobatrachus*), finding support for a diphyletic origin of lissamphibians (the diphyly hypothesis, DH).

9.2 The big picture: tetrapod diversification

The early evolution of tetrapods has formed a controversial topic for many years (Panchen and Smithson 1988; Ahlberg and Milner 1994). In recent years, a consensus on several major questions has emerged (Laurin and Reisz 1997; Clack 2001; Anderson 2001, 2007; Ruta *et al.* 2003a, 2007; Coates *et al.* 2008).

Limbed tetrapodomorphs. The relationships between the four-legged tetrapodomorphs are still mysterious, and their clarification will shed more light on the origin of tetrapods. The supertree approach of Ruta *et al.* (2003b) highlighted the consensus that all four mentioned groups rank below crown tetrapods. This is probably the most honest approach, as the faint evidence for relationships with a particular branch (amniote or lissamphibian) is not convincing in most groups.

Most authors agree that whatcheeriids and *Crassigyrinus* are more basal than colosteids and baphetids, but the phylogenetic topologies mainly reflect the primary assumptions of authors (polarity of characters). Baphetids and colosteids share a few characters with stem-amphibians, which seems to "attract" them towards temnospondyls; it is an open question whether this is a real phylogenetic signal. Likewise, *Crassigyrinus* shares features with stem-amniotes, especially anthracosaurs, but these may well be plesiomorphic states not recognized as such because of problems with identifying polarity. One reason for these problems is that both *Acanthostega* and *Ichthyostega* are clearly derived in some features, such as in the absence of an intertemporal or the highly autapomorphic ear. Only when the limbed tetrapodomorphs have been studied in more detail can their diversity be appreciated, which in turn will be the key to polarize the characters that may place the groups discussed here more safely into the tetrapod tree.

Crown Tetrapoda. Most recent authors agree that temnospondyls and "reptiliomorphs" form a basal dichotomy in early tetrapod phylogeny (Laurin and Reisz 1997; Anderson 2001; Ruta *et al.* 2003a, 2003b). This means that anthracosaurs, gephyrostegids, seymouriamorphs, lepospondyls, and amniotes form a monophyletic group, with temnospondyls forming their sister taxon. An alternative was presented by Clack (2001), who found seymouriamorphs to nest with temnospondyls and baphetids, but this node has not been confirmed by other studies. The old name "Reptiliomorpha" is here used only as a label for this large embolomere-to-amniote grade. The major question here is how crown Tetrapoda fits into the phylogeny of all these Paleozoic groups. There are two alternatives: (1) either temnospondyls are related to one or more lissamphibian clades (resulting in a "large crown"), or (2) all lissamphibians evolved from lepospondyls (the lepospondyl hypothesis), which would result in a "small crown." These two divergent concepts have important implications for the status of Paleozoic groups: in the "small crown" version, anthracosaurs and seymouriamorphs would be stem-tetrapods along with temnospondyls, in contrast to embolomeres and seymouriamorphs

forming part of the amniote stem in the "large crown" alternative. This is why describing phylogenetic patterns and naming nodes is still so difficult in this field: researchers need to define their preferred hypothesis of tetrapod relationships before using names such as "crown tetrapod," "stem-amniote," or "stem-amphibian."

9.3 The origin of lissamphibians

Few issues have been more controversial in vertebrate phylogeny than the Paleozoic ancestry of the Lissamphibia (Figure 9.1). This topic involves a set of related questions that require analysis of large data sets. It is a field in which integration of fossil and extant data is obligatory, and where the consideration of molecular data has attained increasing importance (Ruta and Coates 2007; Anderson 2008; Marjanović and Laurin 2009; Pyron 2011).

In the last few decades, three fossil taxa have played a pivotal role in the debate on lissamphibian origins: (1) the Early Permian amphibamid *Doleserpeton* was the first Paleozoic tetrapod with pedicellate teeth (Bolt 1969; Sigurdsen and Bolt 2010); (2) the Early Jurassic apodan *Eocaecilia* provided new evidence on the caecilian stem-group (Jenkins and Walsh 1993; Jenkins *et al.* 2007); and (3) the amphibamid *Gerobatrachus* shed new light on the stem-group of Batrachia (Anderson *et al.* 2008).

Temnospondyl hypothesis (Figure 9.1A). Traditionally, temnospondyls were considered as ancestors of anurans because of their temporal notches and the structure of the stapes (Quenstedt 1850; Watson 1940; Bolt and Lombard 1985). They also share the large palatal openings, double occipital condyles, broad vomer plates, and short trunk ribs with lissamphibians. Bolt (1969, 1977, 1979) then identified pedicellate teeth in the amphibamids *Doleserpeton* and *Amphibamus*. Closer examination of other amphibamids showed that their vertebrae are similar to the lissamphibian condition (Bolt 1969; Daly 1994) – this was an important discovery, as the spool-shaped vertebrae of lissamphibians differ substantially from the rhachitomous vertebrae in most temnospondyls. The lightly built palate of amphibamids and the

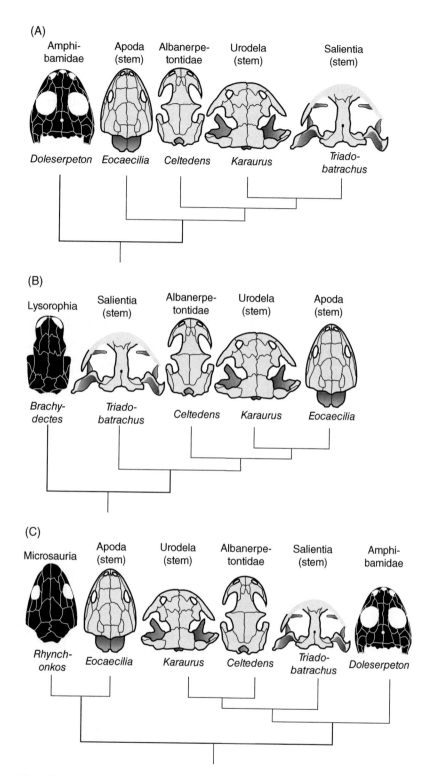

Figure 9.1 The origin of lissamphibians is still a matter of controversy. The three disputed alternatives are: (A) the temnospondyl, (B) the lepospondyl (B), and (C) the diphyly hypothesis. The three differ not only in which Paleozoic taxa form the stem-group, but also in the internal relationships between Apoda, Urodela, and Salientia. Finally, whereas (A) and (B) are consistent with the monophyly of Lissamphibia, (C) holds that apodans are more closely related to amniotes than to batrachians.

reduced number of presacral vertebrae (19–21 in some taxa) were further highlighted as shared features with extant amphibians. Together, these characters form the core of the temnospondyl hypothesis (TH), which was most comprehensively formulated by Milner (1988, 1993). All variants of the TH agree that the lissamphibian characters were acquired successively in temnospondyl phylogeny (e.g., the large palatal openings early, the double occipital condyle later), but that amphibamids are the single clade that shares the bulk of lissamphibian autapomorphies (Figure 9.2).

Authors proposing or supporting the TH were Moodie (1916), Watson (1919, 1940), Reig (1964), Bolt and Lombard (1985), Milner (1988, 1993), Bolt (1991), Trueb and Cloutier (1991), Ruta *et al.* (2003a), Schoch and Milner (2004), Ruta and Coates (2007), Sigurdsen and Green (2011), and Maddin *et al.* (2012).

Lepospondyl hypothesis (Figure 9.1B). Some microsaurs are remarkably similar to modern salamanders in body proportions, while others resemble caecilians in the massive structure of their skulls, elongate trunks, and miniature limbs.

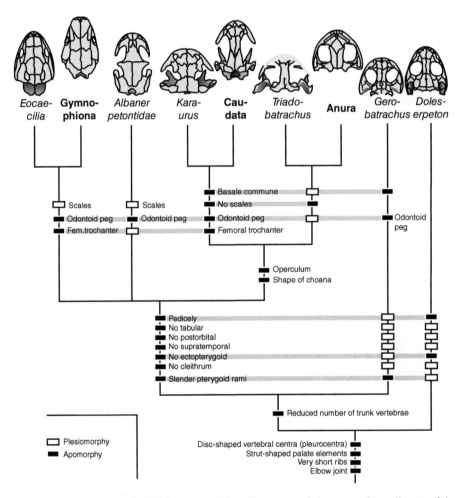

Figure 9.2 The temnospondyl hypothesis (TH) is supported by a diverse set of characters from all parts of the skeleton. Here, the most important characters are mapped onto a cladogram in which the TH is the preferred topology. Note that *Gerobatrachus* may be viewed either as a stem-lissamphibian (as shown here) or as a stem-batrachian (see supporting characters marked in black). Either topology has to live with a range of homoplasies.

As with temnospondyls and frogs, lepospondyls have repeatedly been associated with salamanders and caecilians (Carroll and Currie 1975; Carroll and Holmes 1980), but these authors usually accepted anurans as closely related to temnospondyls. Apart from some rather idiosyncratic ideas (Cox 1967), Laurin and Reisz (1997) were the first to propose the lepospondyl hypothesis (LH) based on a cladistic analysis. Their version holds that all lissamphibians were derived from lysorophians, and that this clade is nested within microsaurs. Lepospondyls and microsaurs would consequently be paraphyletic with respect to the Lissamphibia.

The relatively few characters uniquely shared by lissamphibians and lysorophians are the lack of the postorbital, ectopterygoid, cleithrum, and pineal foramen. A problem has been the homology of a dermal bone at the posterolateral end of the skull table in lysorophians, which might be the supratemporal, the tabular, or a compound element (Marjanović and Laurin 2008a). The LH has been proposed and developed by Laurin (1998), Vallin and Laurin (2004), Marjanović and Laurin (2007, 2008a, 2008b), and Pyron (2011), all using the original data set of Laurin and Reisz (1997).

Diphyly hypothesis (Figure 9.1C) (Carroll and Currie 1975; Carroll and Holmes 1980; Carroll 2009; Anderson *et al.* 2008). In many respects, the diphyly hypothesis (DH) forms a compromise between the other two concepts. It does so by accepting the shared characters of anurans and dissorophoids on the one hand and those of caecilians and microsaurs on the other hand. It forms an attractive option for paleontologists, but neotologists hold that the monophyly of Lissamphibia is firmly supported by morphological and molecular data (Mickoleit 2004; San Mauro *et al.* 2005).

The central question asked by the DH is whether salamanders nest with caecilians or frogs, known as the Batrachia versus Procera hypotheses (Schoch and Milner 2004). Carroll and Holmes (1980) originally suggested that microsaurs were the stem-group of both salamanders and caecilians, whereas frogs were derived from dissorophoid temnospondyls. Later, Carroll (2009) accepted the Batrachia concept, arguing for a temnospondyl origin of batrachians, but upheld the microsaur origin for caecilians. Anderson (2001) and Anderson *et al.* (2008) have further elaborated Carroll's view,

finding caecilians to nest within a certain clade of microsaurs, being the sister taxon of *Rhynchonkos*. However, in a recent analysis that included new μCT-derived anatomical data of *Eocaecilia*, Anderson and colleagues found gymnophionans to nest with batrachians within the amphibamid temnospondyls (Maddin *et al.* 2012).

Salamanders and *Eocaecilia* share the presence of the odontoid peg on the atlas vertebra with microsaurs and lysorophians. This structure is also present in some temnospondyls (e.g., *Sclerocephalus*, plagiosaurids), but not well known in amphibamids. The DH elegantly explains the divergent structure of the ear region: impedance-matching ear with tympanum, middle ear cavity and anuran-like stapes in temnospondyls, in contrast to the "primitive" stapes of salamanders and caecilians. Likewise, the broad parasphenoid is shared by salamanders, caecilians, microsaurs, and lysorophians, whereas the abbreviated basal plate of anurans is also found in the temnospondyl *Amphibamus*. A problem for the diphyly hypothesis is that the aforementioned features all support the Procera version, whereas the recently described *Gerobatrachus* and *Eocaecilia* strongly support the Batrachia hypothesis.

Carroll (2009) has enthusiastically developed his own version of the diphyly hypothesis, proposing separate origins of anurans and salamanders from within the dissorophoid temnospondyls. Based largely on symplesiomorphies such as gill rakers, external gills, and cranial ossification sequences, he proposed branchiosaurids as a sister taxon of salamanders, in contrast to frogs having evolved from amphibamids. Cladistically, branchiosaurids were shown to form part of the amphibamid clade and therefore come almost as near to the anuran/batrachian/lissamphibian condition as *Doleserpeton* or *Amphibamus* (Milner 1988; Schoch and Milner 2008; Fröbisch and Schoch 2009b). Apart from that, there are no synapomorphies that would support a separate origin of salamanders from branchiosaurids.

Features supporting several hypotheses. There is a range of lissamphibian characters found in both temnospondyls and lepospondyls. These include the cylindrical vertebrae, the absence of palatal tusks, and the lack of plicidentine (labyrinthodonty). In addition, both temnospondyls and lepospondyls share the reduced set of four fingers in the hand, as found in Batrachia (Mickoleit 2004). These features had first been ranked as supporting the LH, but their appearance in a few amphibamid temnospondyls now renders their significance equivocal.

Comparison of hypotheses (Figure 9.3). Although each of the three hypotheses finds good support in the studies proposing it, they rest on rather different bodies of evidence. This was analyzed in some detail by Schoch and Milner (2004), Ruta and Coates (2007), and Anderson (2008). The TH rests on a large quantity of characters from all parts of the skeleton, whereas the LH is supported mainly by "absence characters." On closer inspection, only the smallest set of bones is really absent in all lissamphibians, whereas, for instance, the jugal, prefrontal, postfrontal, and tabular are retained in *Eocaecilia*. If Anderson *et al.* (2008) are right in that *Gerobatrachus* is a stem-batrachian, then most "absence characters" were acquired independently by caecilians and batrachians.

The three hypotheses also differ with respect to the monophyly of Lissamphibia itself: whereas TH and LH are consistent with the Lissamphibia–Amniota dichotomy, DH requires at least caecilians to form the sister group of Amniota, with Lissamphibia being paraphyletic. This is a major problem in communication between neotologists and adherents of the DH, as the molecular and soft-anatomical evidence for lissamphibian monophyly are usually considered robust (Parsons and

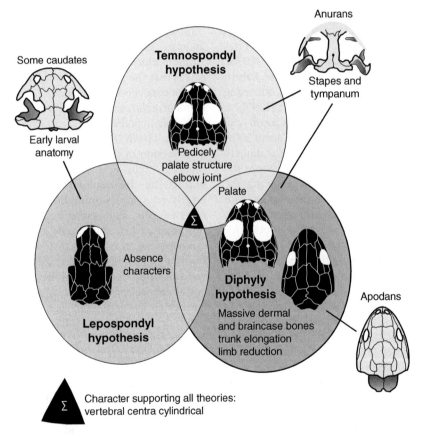

Figure 9.3 Which characters support which lissamphibian hypothesis? Whereas the temnospondyl and diphyly hypotheses are based on various skeletal features, the lepospondyl hypothesis employs various absence characters but little else. The cylindrical centra, a feature long emphasized, turned out to be indecisive.

Williams 1963; Milner 1988; San Mauro *et al.* 2005; Marjanović and Laurin 2007).

Finally, the three hypotheses also differ in the main evolutionary scenarios explaining their topologies of branching. In the temnospondyl version, miniaturization has been proposed as a major factor, accounting for the loss of bones and the immature appearance of lissamphibians when compared to Paleozoic tetrapods (Milner 1988; Fröbisch and Schoch 2009a). This scenario may also apply for the LH, although this has not been explicitly formulated (but see Carroll 1990). The topology of the lysorophian–lissamphibian hypothesis implies that neoteny played an important role in the evolution of lissamphibians: the close resemblance of lysorophians and larval/neotenic proteids and sirenids is striking, and the shared features between these groups contribute disproportionately to supporting the LH. The diphyly alternative suggests the evolution of burrowing adaptations in gymnarthrid microsaurs as a prerequisite for caecilian origins, apart from batrachian derivation from amphibamids.

Of course, all these scenarios can be used to criticize the hypotheses: miniaturization, burrowing, and neoteny are such recurrent phenomena in amphibian evolution that they might easily have evolved iteratively (Schoch and Milner 2004). This means that characters affected by these evolutionary factors are likely to be homoplastic and may not be the most reliable phylogenetic signals.

Albanerpetontidae. It is sometimes overlooked that there was a fourth clade of lissamphibians, which unfortunately died out only very recently – the Albanerpetontidae. These salamander-like forms were small and had a reduced skeleton, as in batrachians, but retained a high number of presacral vertebrae and bony scales, as in caecilians. This pattern suggests that albanerpetontids nest between Gymnophiona and Batrachia (Gardner 2001). Others proposed that they formed a clade with Gymnophiona (Ruta *et al.* 2007) or Procera (Gymnophiona+Caudata) (Laurin and Reisz 1997). Albanerpetontids are therefore critical for the morphological analysis of lissamphibian intrarelationships, although most molecular phylogenies favor the Batrachia hypothesis (Anderson 2008).

Lissamphibian monophyly. Since the influential paper by Parsons and Williams (1963), the monophyly of Lissamphibia has been confirmed by neontological studies, with skeletal, soft-anatomical, and molecular data all supporting a lissamphibian–amniote dichotomy (Kumar and Hedges 1998; Mickoleit 2004; San Mauro *et al.* 2005; Zhang *et al.* 2005; Marjanović and Laurin 2007). Conversely, based on most phylogenetic scenarios of Paleozoic tetrapods, the diphyly hypothesis requires caecilians to be closer to amniotes than to batrachians (Anderson *et al.* 2008). Interestingly, the most recent molecular clock data are more consistent with the diphyly hypothesis than with either of the monophyletic scenarios (Anderson *et al.* 2008). One may argue that the estimates still vary greatly, and that molecular data of lissamphibians are particularly prone to long branch attraction. However, it is also true that not all of the morphological characters supporting lissamphibian monophyly are equally convincing or unequivocal. For instance, the existence of two sensory papillae in the inner ear (p. basilaris and p. amphibiorum) is potentially a very informative trait, but the absence of papillae in dipnoans and the presence of a basilar papilla in *Latimeria* (Fritzsch 1987) render outgroup comparison equivocal. Conversely, potential lissamphibian synapomorphies such as the operculum and opercularis muscle were possibly lost in limbless caecilians, with the effect that they appear to be apomorphic for Batrachia only. One of the most convincing lissamphibian characters is pedicely, despite its absence in larval morphs and some anurans with firmly attached teeth. After all, the mounting molecular data put lissamphibian monophyly on increasingly firm ground.

Conjectures and refutations. The recent debate has revealed some remarkable differences between authors working on the lissamphibian origin problem. The morphological and analytical schools have already been mentioned, and although the two sides have grown closer, aspects of their divergent perspectives are still found in current debates. However, most authors now accept cladistic principles, employ the same software, and apply the same procedures in analyzing their data sets. Differences are more subtle today, but still have an impact on the results. For instance, first-hand examination of specimens is not universally practiced as an important first step in defining characters. Some authors rely almost entirely on the literature, and in publications in which reconstructions rather than original specimens are figured, this often leads to misunderstandings. These studies usually give relatively short character definitions, involving many "yes" or "no" states, such as "absence" versus "presence" of bones or their parts. Other authors focus very much on shape characters, and have been criticized for over-splitting morphological features. Thus, although character weighting is generally avoided in cladistic analyses, an implicit weighting is often practiced by lumping or splitting characters. This is not only a facet of subjectivity, but also accounts for major differences between data sets. Sigurdsen and Green (2011) have recently analyzed this problem for studies focusing on the lissamphibian origin.

Another source of disagreement is the inclusion of ontogeny. Steyer (2000) emphasized the problem that the distinction of larval and adult morphs is important for the proper analysis of temnospondyl phylogeny. However, this was only a first approach to the problem of comparing ontogenies across taxa. The truth is, unfortunately, that the identification of standardized stages has been elusive. As Wiens et al. (2005) put it, "ontogeny discombobulates phylogeny," not only in salamanders but in early tetrapod evolution as well.

Molecular clock, stratigraphy, and fossil calibration. It was already clear to Darwin (1859) that the fossil record cannot be read literally. In vertebrate paleontology, there are huge gaps in the fossil record, and in view of the small number of productive tetrapod deposits in the Devonian and Carboniferous, one may almost lose confidence. Stratigraphy and phylogeny are especially hard to calibrate in this field. Recently, help has come from an unexpected quarter – molecular data from extant taxa. Nucleotide and amino acid sequences of living species not only provide a rich source of data for phylogenetic analysis, but may also contribute to measuring the age of a particular node. This is possible because the rate of molecular evolution (by nucleotide substitution) usually correlates with the time passed since the split between two studied taxa (Margoliash 1963). However, this *molecular clock* requires calibration by fossil data. Paleontological data do not provide direct age estimates for divergence events, but may give precise minimum constraints on the

calibration of molecular clocks (Benton and Donoghue 2006). Recently, Parham et al. (2012) have formulated a suite of best practices for fossil calibrations, showing how important the integration of fossil and recent data has become.

In recent years, molecular clocks have been used to date the relatively long branches of the three lissamphibian clades, as well as the age of crown tetrapods (lissamphibian–amniote split). In extant tetrapods, these rank among the longest branches, dating back to the Late Carboniferous or earlier according to most studies. Still, these studies vary greatly in dating the origins of Lissamphibia, Batrachia, and the three extant clades (San Mauro et al. 2005; Zhang et al. 2005; Marjanović and Laurin 2007, 2008a; Anderson 2008; Pyron 2011; Sigurdsen and Green 2011). This is no wonder, as it depends on the preferred phylogeny chosen as basis for calibration.

Are we closer to the answer now? I believe we are, with morphology, molecular data, and cladistic analyses having played important roles here. The large-scale analyses by Laurin (1998), Ruta and Coates (2007), and Anderson et al. (2008) have accumulated a huge database, which has recently been merged (Sigurdsen and Green 2011). Detailed anatomical work on many temnospondyls, seymouriamorphs, and lepospondyls has established a much broader morphological platform for further work on characters. Focused studies of amphibamids and *Eocaecilia* have revealed new autapomorphies of lissamphibians, such as Sigurdsen and Bolt's (2009) data on the structure of the forelimb. CT scanning has also contributed important observations in the last decade (e.g., Maddin et al. 2011).

Sigurdsen and Green (2011) analyzed the three hypotheses and the data matrices on which they rest. They created a supermatrix from the three main data sets and recoded characters whose states had to be reformulated after recent fossil discoveries. Their result was clear-cut, favoring the temnospondyl hypothesis with *Doleserpeton* as a sister taxon of Lissamphibia. The discovery of *Doleserpeton*, *Eocaecilia*, and *Gerobatrachus* has had a huge impact on the lissamphibian origin problem, and authors have increasingly come to view amphibamid temnospondyls as convincing sister taxa of frogs and salamanders. Whether having pedicellate teeth or not, it is difficult not to view *Gerobatrachus* as a stem-batrachian: its mix of anuran, salamander, batrachian, and lissamphibian characters is just too seductive. Although the case is still not settled, Anderson (2008) is probably right in his statement that the main question now has shifted to the origin of caecilians. In the most recent paper focusing on caecilian origins, a study that was co-authored by Anderson, Lissamphibia has been found to nest within Amphibamidae (Maddin *et al.* 2012).

Although there are many homoplasies and reversals to be accepted in all three hypotheses, the temnospondyl hypothesis holds the pivotal position. It has now accumulated the largest suite of apomorphies that are rich in homologous structures, whereas the lepospondyl hypothesis suffers from the problem that (1) most of its supporting characters are not apomorphic but homoplastic, (2) that these characters involve the reappearance of a large number of lost bones throughout the skeleton, and (3) that lissamphibians lack numerous apomorphies shared by lepospondyls and amniotes (Sigurdsen and Green 2011).

After all, amphibamids come closest to lissamphibians not only in their dermal skull morphology, braincase organization, hyobranchial apparatus, and forelimb structure, but they also share the complex life cycle, including a drastic metamorphosis, with lissamphibians. The successive origin of this life cycle is now well documented in temnospondyls, and finds no parallel in other early tetrapod clades. At any rate, future research will depend on further detailed analysis of morphology, and on diversified techniques of cladistics and other methods of analysis. Perhaps most importantly, it will depend on the critical eye of authors being cast not only on the data of colleagues, but on their own as well. Although a consensus is very much to be desired, the persistence of alternative hypotheses may be the best spur to critical reflection.

References

Ahlberg, P.E. & Milner, A.R. (1994) The origin and early diversification of tetrapods. *Nature* **368**, 507–514.

Anderson, J.S. (2001) The phylogenetic trunk: maximal inclusion of taxa with missing data in an analysis of the Lepospondyli (Vertebrata, Tetrapoda). *Systematic Biology* **50**, 170–193.

Anderson, J.S. (2007) Incorporating ontogeny into the matrix: a phylogenetic evaluation of developmental evidence for the origin of modern amphibians. In: J.S. Anderson & H.-D. Sues (eds.), *Major Transitions in Vertebrate Evolution*. Bloomington: Indiana University Press, pp. 182–212.

Anderson, J.S. (2008) Focal review. The origin(s) of modern amphibians. *Evolutionary Biology* **35**, 231–247.

Anderson, J.S., Reisz, R.R., Scott, D., Fröbisch, N.B., & Sumida, S.S. (2008) A stem batrachian from the Early Permian of Texas and the origin of frogs and salamanders. *Nature* **453**, 515–518.

Benton, M.J. & Donoghue, P.C.J. (2006). Paleontological evidence to date the Tree of Life. *Molecular Biology and Evolution* **24**, 26–53.

Bolt, J.R. (1969) Lissamphibian origins: possible protolissamphibian from the Lower Permian of Oklahoma. *Science* **166**, 888–891.

Bolt, J.R. (1977) Dissorophoid relationships and ontogeny, and the origin of the Lissamphibia. *Journal of Paleontology* **51**, 235–249.

Bolt, J.R. (1979) *Amphibamus grandiceps* as a juvenile dissorophid: evidence and implication. In: M.H. Nitecki (ed.), *Mazon Creek Fossils*. New York: Academic Press, pp. 529–563.

Bolt, J.R. (1991) Lissamphibian origins. In: H-P. Schultze & L. Trueb (eds.), *Origin of the Higher Groups of Tetrapods: Controversy and Consensus*. Ithaca: Cornell University Press, pp. 194–222.

Bolt, J.R. & Lombard, R.E. (1985) Evolution of the amphibian tympanic ear and the origin of frogs. *Biological Journal of the Linnean Society* **24**, 83–99.

Boy, J.A. (1972) Die Branchiosaurier (Amphibia) des saarpfälzischen Rotliegenden (Perm, SW-Deutschland). *Abhandlungen des Hessischen Landesamts für Bodenforschung* **65**, 1–137.

Boy, J.A. (1981) Zur Anwendung der Hennigschen Methode in der Wirbeltierpaläontologie. *Paläontologische Zeitschrift* **55**, 87–107.

Carroll, R.L. (1990) A tiny microsaur from the Lower Permian of Texas: size constraints in Palaeozoic tetrapods. *Palaeontology* **33**, 893–909.

Carroll, R.L. (2009) *The Rise of Amphibians*. Baltimore: Johns Hopkins University Press.

Carroll, R.L. & Currie, P. (1975) Microsaurs as possible apodan ancestors. *Zoological Journal of the Linnean Society* **57**, 229–247.

Carroll, R.L. & Holmes, R. (1980) The skull and jaw musculature as guides to the ancestry of salamanders. *Zoological Journal of the Linnean Society* **68**, 1–40.

Clack, J.A. (2001) *Eucritta melanolimnetes* from the Early Carboniferous of Scotland: a stem tetrapod showing a mosaic of characteristics. *Transactions of the Royal Society of Edinburgh Earth Sciences* **92**, 75–95.

Coates, M.I., Ruta, M., & Friedman, M. (2008) Ever since Owen: changing perspectives on the early evolution of tetrapods. *Annual Review of Ecology, Evolution and Systematics* **39**, 571–592.

Cox, C.B. (1967) Cutaneous respiration and the origin of the modern Amphibia. *Proceedings of the Linnean Society* **178**, 37–47.

Daly, E. (1994) The Amphibamidae (Amphibia: Temnospondyli), with a description of a new genus from the Upper Pennsylvanian of Kansas. *Miscellaneous Publications of the University of Kansas Museum of Natural History* **85**, 1–59.

Darwin, C. (1859) *On the Origin of Species by Means of Natural Selection*. London: John Murray.

Fritzsch, B. (1987) Inner ear of the coelacanth fish *Latimeria* has tetrapod affinities. *Nature* **327**, 153–154.

Fröbisch, N.B. & Schoch, R.R. (2009a) The largest specimen of *Apateon*, and the life history pathway of neoteny in the Palaeozoic family Branchiosauridae. *Fossil Record* **12**, 83–90.

Fröbisch, N.B. & Schoch, R.R. (2009b) Testing the impact of miniaturization on phylogeny: Paleozoic dissorophoid amphibians. *Systematic Biology* **58**, 312–327.

Gardner, J.D. (2001) Monophyly and affinities of albanerpetontid amphibians (Temnospondyli; Lissamphibia). *Zoological Journal of the Linnean Society* **131**, 309–352.

Hecht, M. (1976) Phylogenetic inference and methodology as applied to the vertebrate record. *Evolutionary Biology* **9**, 335–363.

Jenkins, F.A., Jr. & Walsh, D.M. (1993) An Early Jurassic caecilian with limbs. *Nature* **365**, 246–250.

Jenkins, F.A., Jr., Walsh, D.M., & Carroll, R.L. (2007) Anatomy of *Eocaecilia micropodia*, a limbed caecilian of the Early Jurassic. *Bulletin of the Museum of Comparative Zoology Harvard* **158**, 285–366.

Kumar, S. & Hedges, S.B. (1998) A molecular time-scale for vertebrate evolution. *Nature* **392**, 917–920.

Laurin, M. (1998) The importance of global parsimony and historical bias in understanding tetrapod evolution. Part I. Systematics, middle ear evolution, and jaw suspension. *Annales des Sciences naturelles* **19**, 1–42.

Laurin, M. & Reisz, R.R. (1997) A new perspective on tetrapod phylogeny. In: S.S. Sumida & K.L.M. Martin (eds.), *Amniote Origins: Completing The Transition To Land*. London: Academic Press, pp. 9–59.

Maddin, H., Olori, J.C., & Anderson, J.S. (2011) A redescription of *Carrolla craddocki* (Lepospondyli: Brachystelechidae) based on high-resolution CT, and the impacts of miniaturization and fossoriality on morphology. *Journal of Morphology* **272**, 722–743.

Maddin, H., Jenkins, F.A., & Anderson, J.S. (2012) The braincase of *Eocaecilia micropodia* (Lissamphibia, Gymnophiona) and the origin of caecilians. *PloS ONE* **7**, e50743.

Margoliash, E. (1963) Primary structure and evolution of cytochrome C. *Proceedings of the National Academy of Sciences* **50**, 672–679.

Marjanović, D & Laurin, M. (2007) Fossils, molecules, divergence times, and the origin of lissamphibians. *Systematic Biology* **56**, 369–388.

Marjanović, D. & Laurin, M. (2008a) A reevaluation of the evidence supporting an unorthodox hypothesis on the origin of extant amphibians. *Contributions to Zoology* **77**, 149–199.

Marjanović, D. & Laurin, M. (2008b) Assessing confidence intervals for stratigraphic ranges of higher taxa: the case of Lissamphibia. *Acta Palaeontologica Polonica* **53**, 413–432.

Marjanović, D. & Laurin, M. (2009) The origin(s) of modern amphibians: a commentary. *Evolutionary Biology* **36**, 336–336.

Mickoleit, G. (2004) *Phylogenetische Systematik der Wirbeltiere*. Munich: Pfeil.

Milner, A.R. (1988) The relationships and origin of living amphibians. In: M.J. Benton (ed.), *The Phylogeny and Classification of the Tetrapods*. Oxford: Clarendon Press, Vol. 1, pp. 59–102.

Milner, A.R. (1993) The Paleozoic relatives of lissamphibians. In: D. Cannatella & D. Hillis (eds.),

Amphibian relationships: phylogenetic analysis of morphology and molecules. *Herpetological Monographs* 7, 8–27.

Moodie, R.L. (1916) *The Coal Measures Amphibia of North America.* Publications of the Carnegie Institution 238. Washington, DC: Carnegie Institution.

Panchen, A.L. & Smithson, T.R. (1988) The relationships of early tetrapods. In: M.J. Benton (ed.), *The Phylogeny and Classification of the Tetrapods.* Oxford: Clarendon Press, Vol. 1, pp. 1–32.

Parham, J.F., Donoghue, P.C.J., Bell, C.J., *et al.* (2012) Best practices for justifying fossil calibrations. *Systematic Biology* **61**, 346–359.

Parsons, T.S. & Williams, E.E. (1963) The relationships of the modern Amphibia: a re-examination. *Quarterly Review of Biology* **38**, 26–53.

Pyron, R.A. (2011) Divergence time estimation using fossils as terminal taxa and the origins of Lissamphibia. *Systematic Biology* **60**, 466–481.

Quenstedt, F.A. v. (1850) *Die Mastodonsaurier aus dem grünen Keupersandsteine Württembergs sind Batrachier.* Tübingen: Laupp.

Reig, O. (1964) El problema del origen monofilético o polifilético de los anfibios, con consideraciones sobre las relaciones entre Anuros, Urodelos y Ápodos. *Ameghiniana* 3, 191–211.

Ruta, M. & Coates, M.I. (2007) Dates, nodes and character conflict: addressing the lissamphibian origin problem. *Journal of Systematic Palaeontology* 5, 69–122.

Ruta, M., Coates, M.I., & Quicke, D.L.J. (2003a) Early tetrapod relationships revisited. *Biological Reviews* 78, 251–345.

Ruta, M., Jeffrey, E., & Coates, M.I. (2003b) A supertree of early tetrapods. *Proceedings of the Royal Society of London B* **270**, 2507–2516.

Ruta, M., Pisani, D., Lloyd, G.T., & Benton, M.J. (2007) A supertree of Temnospondyli: cladogenetic patterns in the most species-rich group of early tetrapods. *Proceedings of the Royal Society of London, Series B* **274**, 3087–3095.

San Mauro, D., Vences, M., Alcobendas, M., Zardoya, R., Meyer, A. (2005) Initial diversification of living amphibians predated the breakup of Pangea. *American Naturalist* **165**, 590–599.

Schoch, R.R. & Milner, A.R. (2004) Structure and implications of theories on the origin of lissamphibians. In: G. Arratia, M.H.V. Wilson, & R. Cloutier (eds.), *Recent Advances in the Origin and Early Radiation of Vertebrates.* Munich: Pfeil, pp. 347–377.

Schoch, R.R. & Milner, A.R. (2008) The intrarelationships and evolutionary history of the temnospondyl family Branchiosauridae. *Journal of Systematic Palaeontology* **6**, 409–431.

Sigurdsen, T. & Bolt, J.R. (2009) The lissamphibian humerus and elbow joint, and the origins of modern amphibians. *Journal of Morphology* **270**, 1443–1453.

Sigurdsen, T. & Bolt, J.R. (2010) The Lower Permian amphibamid *Doleserpeton* (Temnospondyli: Dissorophoidea), the interrelationships of amphibamids, and the origin of modern amphibians. *Journal of Vertebrate Paleontology* **30**, 1360–1377.

Sigurdsen, T. & Green, D.M. (2011) The origin of modern amphibians: a re-evaluation. *Zoological Journal of the Linnean Society* **162**, 457–469.

Steyer, J.S. (2000) Ontogeny and phylogeny in temnospondyls: a new method of analysis. *Zoological Journal of the Linnean Society* **130**, 449–467.

Trueb, L. & Cloutier, R. (1991) A phylogenetic investigation of the inter- and intrarelationships of the Lissamphibia (Amphibia, Temnospondyli). In: H.-P. Schultze & L. Trueb (eds.), *Origin of the Higher Groups of Tetrapods: Controversy and Consensus.* Ithaca: Cornell University Press, pp. 223–313.

Vallin, G. & Laurin, M. (2004) Cranial morphology and affinities of *Microbrachis*, and a reappraisal of the phylogeny and lifestyle of the first amphibians. *Journal of Vertebrate Paleontology* **24**, 56–72.

Watson, D.M.S. (1919) The structure, evolution and origin of the Amphibia. The "Orders" Rachitomi and Stereospondyli. *Philosophical Transactions of the Royal Society London B* **209**, 1–73.

Watson, D.M.S. (1940) The origin of frogs. *Transactions of the Royal Society of Edinburgh* **60**, 195–231.

Wiens, J.J., Bonnet, R.M., & Chippindale, P.T. (2005) Ontogeny discombobulates phylogeny. *Systematic Biology* **54**, 91–110.

Zhang, P., Zhou, H., Chen, Y.Q., Liu, Y.F., & Qu, L.H. (2005) Mitogenomic perspectives on the origin and phylogeny of living amphibians. *Systematic Biology* **54**, 391–400.

10 Macroevolution

The 330 myr history of amphibian evolution is full of interesting patterns, ranging from species-level changes to major evolutionary events and extinction. This is the domain of macroevolution, which is studied by the integration of fossil and extant data. The field includes a diverse set of questions. Which *evolutionary novelties* originated in stem-tetrapods and how did they contribute to their conquest of land? Were there *key innovations* that permitted early tetrapods and amphibians to construct new niches and diversify? Which modes of *speciation* are known in modern amphibians, and are there patterns from the fossil record matching those? What evidence is there for evolutionary stasis and its opposite, rapid punctuational change? How do evolutionary constraints emerge: for instance, how can genome size affect the volume of cells, metabolism, and development? Can re-evolution of lost features – such as suggested by many phylogenetic hypotheses – be made plausible? And finally, how were early tetrapods affected by extinction events, both minor and major, in the fossil record? This chapter also provides an opportunity to look back on the origin of tetrapods and terrestriality, and to ask how far we have progressed on the path towards understanding the fish–tetrapod transition.

Amphibian Evolution: The Life of Early Land Vertebrates, First Edition. Rainer R. Schoch.
© 2014 Rainer R. Schoch. Published 2014 by John Wiley & Sons, Ltd.

10.1 What is macroevolution?

Macroevolution is defined here as an evolutionary pattern revealed by the fossil record – it is *not* a process. Overwhelming evidence indicates that the same factors of evolution were active throughout Earth history and across all phyla (Simpson 1944; Futuyma 2005). Earlier claims that paleontology demanded additional, macroevolutionary processes (Schindewolf 1950; Stanley 1979) have not been successful in delivering the evidence (Futuyma 2005). There are no laws of evolution, just as there are no laws of human history (Popper 1957).

However, it is not sufficient to explain patterns of macroevolution by adaptations alone – there are various additional aspects to be considered. Let us recall that adaptations are features that enhance the survival and reproductive success of individuals (Mayr 1983). However, not each and every organismic feature does so (Gould and Lewontin 1979). Evolutionary change is often not a process of pure optimization, but rather a tinkering constrained by multiple trade-offs (Jacob 1977). When numerous selection pressures act on a feature, compromises are formed and constraints emerge that limit the effectivness of adaptations (Schwenk and Wagner 2004). In macroevolution, this results in patterns that often appear counterintuitive under a strictly adaptationist perspective. This is why dialectical thinking is more appropriate (Wake 1991). For instance, the study of developmental evolution or directional change in body size often reveals limitations to body plans not caused by selection.

In this chapter, I will first discuss some common patterns of macroevolution in fossil amphibians and then focus on some of the major factors that have been identified in amphibian evolution. Although all these evolutionary patterns and processes occur in many other groups as well, their interaction produces specific patterns that are typical of amphibians, both exinct and extant.

10.2 Patterns of early tetrapod evolution

The 330 myr fossil record of amphibians and the initial diversification of tetrapods provide interesting patterns of macroevolution. Most of these span tens of millions of years, but the incomplete understanding of many taxa makes identification of evolutionary factors difficult. However, documenting these patterns is more than an end in itself, because they add to the greater picture of tetrapod evolution and provide insight into the long-term evolution of amphibians, which the study of extant taxa cannot.

Evolutionary novelties. Is evolution only a tinkerer with pre-existing parts, or does it also "invent" new things? At any rate, the origin and successful evolution of entirely new structures is a rather rare phenomenon. Although it already puzzled Darwin (1859), this field is still in an early phase of study (Love 2003; Wagner and Larsson 2007). Not only are the factors producing novelties still obscure, but the identification of novelties is also not easily accomplished. In fossil tetrapods, skeletal remains are the major source of morphological information. New skeletal elements may appear suddenly in the fossil record, but their soft-anatomical correlates (associated muscles, tendons, cartilages, and connective tissues) are often unknown. In recent times, the genetic and developmental mechanisms behind novelties have been inferred from the study of modern vertebrates (Wagner and Larsson 2007), but the map is still largely blank.

1. The middle ear of tetrapods forms an evolutionary novelty, although all of its components have precursors in bony fishes. In this case, it is the *novel combination* of the parts that matters (middle ear cavity, skeletal stapes element, stapedial muscles and nerves). The enclosure of the stapes within the spiracle, the filling of the spiracle exclusively with air, and the appearance of the ear drum acted together to form the tetrapod middle ear. Here, a new combination of parts permitted the origin of a new function: terrestrial hearing evolved from air breathing.

2. The origin of the autopodium – the hand and foot skeleton – is an often-cited example of evolutionary novelty (Clack 2009). Irrespective of whether primordial digits were present in *Panderichthys* (Boisvert *et al.* 2008), the origin of the autopodium involved the *novel gene expression* of *Hox* genes that are present in

bony fishes (Wagner and Larsson 2007). The resulting digits have not only made an essential contribution to the tetrapod body plan, but they have given rise to a multitude of locomotions: crawling, high walk, erect gait, swimming in secondarily aquatic taxa, arboreal escalation, gliding in some lizards and mammals, and flying in pterosaurs, birds, and bats.

3. A further example is provided by the external gills of tetrapods, which are only retained in larval lissamphibians. In contrast to internal gills, these structures are derived from gill septa and must be regarded as a novelty, because juveniles of *Eusthenopteron* and other tetrapodomorph fishes lacked external gills (Witzmann 2004; Schoch and Witzmann 2011). External gills permitted early tetrapod larvae to populate diverse water bodies, and the ready resorption of external gills during metamorphosis probably contributed to the success of lissamphibian larvae.

4. The tadpole morphotype forms a highly derived organism that involves several novelties. As such, new components, the labial cartilages, keratinized "teeth," and the repatterned hyo-branchial skeleton contribute to its unique morphology. In this case, not only morphology, but also the function of the whole buccal pumping apparatus, constitutes a novelty.

5. Evolutionary novelties are not restricted to morphology, but may also appear in development. The novel "rewiring" of gene regulatory networks (GRNs) has been much highlighted recently (Davidson 2006) – it may or may not have an impact also on morphology. The drastic metamorphosis of lissamphibians may be another example, because although it does not lead to novel structures, it forms a developmental novelty in itself, and it divides the life of amphibians into two separate existences, which form separate foci for selection to act upon.

Key innovations. A key innovation need not be an evolutionary novelty (Lauder 1981), but it involves a new adaptive feature that contributes essentially to the evolutionary success of a clade (Lauder and Liem 1989). Key innovations enable the construction of new niches, permitting a sustained breakthrough for many millions of years. Subsequent morphological diversification usually allows the newly formed clade to play a dominant role in many ecosystems.

The exaptation of digits – originally fin-like structures for some unknown kind of locomotion under water – for terrestrial locomotion led to a most successful key innovation. The restructuring of elbow and knee joints, reorientation of the autopodium, and altered limb posture permitted tetrapods to colonize dry land, which permitted an extensive evolutionary radiation in Late Devonian and Mississippian times. This radiation has resulted in ~24000 extant species of tetrapods, among numerous extinct lineages.

Temnospondyls provide an example of how developmental evolution and a key innovation might have contributed to the radiation of lissamphibian ancestors (Schoch and Milner 2008). In dissorophoid temnospondyls, the larval period was modified to produce filter-feeding morphs. Delayed consolidation of the skull roof, resulting from slower rates of ossification, retained a mobile skull well into adulthood, and a modification in the branchial denticles permitted them to act as a filter for plankton. This permitted branchiosaurids and amphibamids to exploit niches not previously occupied by other temnospondyl larvae. The vast number of species described from Pennsylvanian and Permian lake deposits suggests that branchiosaurids were able to populate a wide range of aquatic paleoecosystems.

Tadpoles form a further and much better-understood example of larval innovations, with their remodeled skull, sucking mouth, and body shape having produced a larval morph that not only constructed its own niche (detritus-feeder in ephemeral ponds) but also established a platform for a tremendous evolutionary diversification (Haas 2003). The different types of tadpoles and their fine-tuned morphologies permitted anurans to breed in an extremely wide range of aquatic habitats (Orton 1953).

A well-studied more recent key innovation is the tongue-projection apparatus in plethodontid salamanders (Wake and Deban 2000). In these miniature caudates, a pair of hyobranchial elements was greatly extended to permit a fast and far projection of the tongue in capturing insect prey (Wake 1966). This mechanism requires stereoscopic vision, which in turn led to the modification of eye and head structure, as well as properties of the brain. The new feeding apparatus has paved the way for the most remarkable evolutionary radiation in salamanders, with more than half the extant species of salamanders belonging to this clade (Frost *et al.* 2006).

Atavisms and re-evolution. A common phenomenon in amphibian macroevolution is the re-appearance of structures or elements that have been absent for many million years (Smirnov 1997; Wiens 2011). These structures have traditionally been called atavisms (evolutionary sense), but are now more often referred to as reversals (cladistic sense). In the strict sense, an atavism forms an extreme case of a reversal, such as when a long-lost bone re-appears, usually in a new anatomical context, producing a highly aberrant morphotype.

Reversals cause little trouble for the cladist, as long as large data sets are dealt with – they form just one "odd" character, after all. For the evolutionary biologist, they are more than an oddity, because sometimes they provide insight into developmental processes or evolutionary transformations that are otherwise inaccessible. The re-appearance of a long-lost structure may shed light on how this structure became reduced in evolution, and its appearance in a novel anatomical context might reveal how the ancient structure formed in ontogeny and by which factors it was influenced.

1. The intertemporal disappeared and re-evolved various times in early tetrapods. Present in *Eusthenopteron* and *Panderichthys*, it is absent in *Tiktaalik*, present in *Ventastega*, absent again in *Acanthostega* and *Ichthyostega*, and its presence and size in crown tetrapods is highly variable. This confusing pattern offers little evidence for any mechanism by which the loss and re-appearance of an intertemporal might have proceeded.

2. Several elements between the cheek and skull table (lost in all extant lissamphibians) re-appear in large and exceptionally old specimens of some anuran species (Smirnov 1997). The homology of these elements is not entirely clear, but by their position they match the supratemporal, postorbital, and tabular of stem-amphibians. If this holds true, then the three elements might have never been lost entirely, but simply hidden in a late phase of the developmental trajectory, which is usually not realized. Only in the case of an extended life span, accomplished by hormonal treatment by Smirnov, are the three ancient elements produced. However, their morphology departs markedly from the ancient pattern, because their late ontogenetic formation means that they have to fit into the fully established anuran skull roof. This exemplifies why atavisms often generate unusual ("odd") morphologies, because their expression in a new context leads to further changes in the existing phenotype.

3. The mandibular dentition, which is almost universally absent in more than 5400 species of modern frogs, was reported to appear in a single extant genus, *Gastrotheca* (Wiens 2011). Whereas many potential reversals remain ambiguous because of homology assessment, the regained mandible teeth form a rather clear-cut case and may serve best to illustrate the reality of atavisms. Wiens (2011) argued that structures of which some serial homolog exists (e.g., teeth in the upper and lower jaws) might be easier to re-evolve in places where they became lost, even if this loss occurred hundreds of million years ago, as in the frog mandibular dentition.

Skeletal reduction. A major pattern in both lissamphibians and amniotes is the reduction of skeletal elements. Both crown groups lack a wide range of bones, especially in the dermal skull and pectoral girdle, which apparently occurred in parallel several times. The early tetrapod body plan included numerous skull bones (prefrontal, postfrontal, postorbital, jugal, intertemporal, supratemporal, tabular, postparietal, ectopterygoid) and elements of the pectoral girdle (cleithrum, interclavicle) that were reduced to rudiments or entirely lost in lissamphibians. These reductions must have occurred convergently in caecilians and batrachians, as the stem-taxa *Eocaecilia* and *Gerobatrachus* indicate. The opposite pattern, the addition of new elements, has occurred very rarely (e.g., the origin of the second ear ossicle, the operulum). In lissamphibians, the loss of elements has been referred to evolutionary size reduction, beyond a threshold which triggered reorganization and simplification of skeletal components (Milner 1988; Fröbisch and Schoch 2009; Schoch 2013). Two aspects come into play here: (1) heterochrony, specifically the observation that some of the bones that are missing in lissamphibians were the last to form in temnospondyls (Schoch 2002), and (2) miniaturization itself, combined with the pattern that lissamphibians have rather large cell volumes. The heterochrony hypothesis holds that ontogeny was truncated and some late-ossifying elements simply failed to form. In contrast, the miniaturization hypothesis suggests that size reduction combined with large cells resulted in reduced stem-cell populations, which failed to form skeletal elements in cases where they did not reach a given minimum number required to differentiate into osteoblasts (Atchley and Hall 1991).

Tempo and mode of evolution. In addition to asking *what* evolves and *how*, one may also study *how fast* evolutionary change proceeds, and whether the *rate* is constant or variable. Do evolutionary changes come about gradually, as Darwin (1859) implied, or are there phases of rapid change interspersed with long-term stability? A second essential problem is how directed evolutionary change may be, and how frequent cases of trends are, as opposed to random fluctuations. If viewed from extreme perspectives, both questions may lead to non-Darwinian concepts: saltation and orthogenesis. Because both of these concepts have long been abandoned by evolutionary biologists, I will focus on moderate cases of tempo and mode in evolution, for which there is unequivocal evidence and which are consistent with the evolutionary synthesis of Simpson (1944) and Mayr (1963).

Punctuated equilibria. Contrary to gradual changes, the punctuated equilibria model was suggested to be a combination of geologically short phases of rapid change (punctuation), followed by longer periods of morphological stasis (Eldredge and Gould 1972). This model appeared to match evolutionary patterns in the fossil record much better than gradualism (Stanley 1979). Based on Wright's (1938) founder effect hypothesis,

elaborated by Mayr (1963), Eldredge and Gould (1972) suggested that speciation occurred in short-term events within small peripheral populations, driven by random genetic drift as decisive factor in evolutionary "bottlenecks" with reduced variation. However, recent studies have revealed that this largely theoretical model does not hold, as selection was found to be much more significant than drift, with variation not significantly reduced in small populations, and hence no strict founder effect could be identified (Coyne and Orr 2004). Thus, the originally proposed mechanism of the punctuated equilibria model is no longer realistic – so what about the pattern?

In a recent review of the fossil evidence across a wide range of animal groups, Hunt (2007) found that only 5% of the analyzed evolutionary sequences were consistent with directional change, whereas the remaining 95% fell almost equally into random walk and stasis. Both random walk and stasis (in the 10^5–10^7 year range) are also the most common patterns in the amphibian fossil record, whereas rapid short-term directional evolution ("punctuations" in the 10^3–10^5 year range) is simply too rapid to be recorded paleontologically. Thus, whereas one component of the punctuated equilibria pattern (stasis) appears to be rather widespread, the second one (punctuation) falls beyond the range of paleontological study, which altogether gives the model a rather metaphysical status. Accordingly, there are no clear-cut examples of punctuation from the fossil record of amphibians. In section 10.3, new data on speciation in modern amphibians will be discussed that contribute to a more diversified picture.

Stasis. More apparent than speciation and punctuation are cases of evolutionary stasis, which is when morphology is conserved over many millions of years in a given species (Hunt 2007). As Gould (2002) put it, "stasis is data," challenging the conventional view that only morphological change documents evolution in the fossil record (Figure 10.1). In fact, stasis may result from evolutionary forces (stabilizing selection) as much as from the failure to evolve (e.g., due to insufficient genetic variation). Stasis may equally result from adaptation to a specific habitat, as well as from resistance against environmental changes. As such, it probably involves

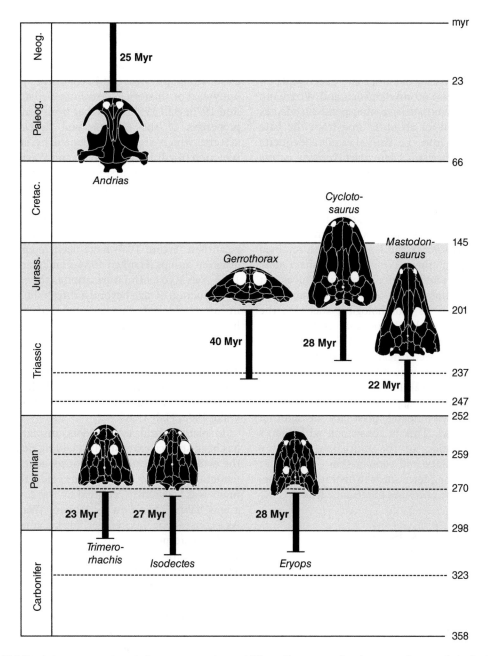

Figure 10.1 Stasis is a recurrent evolutionary pattern in amphibians. Here, taxa showing extensive morphological stasis are mapped onto a stratigraphic range chart. The most extreme case forms the plagiosaurid *Gerrothorax*, which survived at least 40 myr almost unaltered (adapted from Schoch and Witzmann 2012).

various different processes that converge in similar macroevolutionary patterns.

In the fossil record of amphibians, several clear-cut examples of long-term morphological stasis have been reported: (1) the plagiosaurid stereospondyl *Gerrothorax pulcherrimus* from the Middle–Late Triassic of Europe and Greenland persisted for at least 40 myr (Schoch and Witzmann 2012), (2) the dvinosaurian temnospondyl *Isodectes obtusus* existed for about 27 myr from the Late Carboniferous into the Early Permian (Sequeira 1998), and (3) the giant salamander *Andrias*, native to South China and Japan today, has a history of 25 myr, as finds from the Oligocene of Rott near Bonn indicate (Westphal 1958). Despite many differences, these three taxa all share a fully aquatic mode of life. Whereas *Andrias* is a skin-breathing neotene, *Gerrothorax* and *Isodectes* both retained gills throughout their lives; all three share slow rates of growth and development. Stasis does not necessarily imply more stable environments: notwithstanding its morphological stasis on the macroscopical level, *Gerrothorax pulcherrimus* reveals a remarkable variation at the histological level, and paleoecological data show that it managed to inhabit very diverse habitats (Schoch and Witzmann 2012). In this case, stasis on one level (morphology) appears to result from an enhanced developmental plasticity at another level (histology). This is consistent with Flatt's (2005) description of the mutual relationship between plasticity and canalization – the latter may produce long-term stasis in evolutionary lineages such as the examples mentioned here.

10.3 Major factors of amphibian evolution

Factors of evolution can only be fully analyzed in extant species, and lissamphibians provide a rich field for such work. However, the patterns associated with identified factors may also be found in the fossil record. In the following I give some examples of well-studied factors in modern amphibian evolution that have also been discussed for Paleozoic and Mesozoic taxa (Figure 10.2). Here, "factors" are understood as causes of major evolutionary changes, but they may result from

various different processes. Selection forms the most important, but by no means the only process behind evolutionary factors. Constraints form another set of factors, but contrary to the conventional view constraints need not be opposed to selection (Schwenk and Wagner 2004); they may arise from physical and chemical properties (universal or ahistoric constraints: Seilacher 1970; Reif 1975; Hall 1999) as well as from conservative properties of the genome and developmental system, which are largely upheld by stabilizing selection (Riedl 1978; Schwenk and Wagner 2004).

Body size. Size traits, such as body length or volume, form important foci of selection pressures, and they contribute to ecological divergence as well as adaptive radiation (Bonett *et al.* 2009). In extant amphibians, both miniaturization and gigantism (Figure 10.3) have been reported from different groups (Hanken 1984; Hanken and Wake 1993; Yeh 2002). Miniaturization is the evolutionary reduction of size beyond a threshold, resulting in morphological changes imposed by design limitations (Hanken and Wake 1993). Miniaturized taxa of different clades often resemble one another closely because of these limitations (Wake 2009). This is an example of evolutionary constraint imposed not by selection but simply by physical necessity (Maynard Smith *et al.* 1985; Hanken and Wake 1993; Hall 1999).

In plethodontid salamanders, miniaturization has been studied extensively, yielding rich data on the evolutionary causes and consequences of size reduction (Hanken 1983). In this clade, the loss of lungs was apparently required by the conquest of a new habitat: fast-flowing streams (Wake 1966). As a consequence, skin breathing became the only mode of respiration in plethodontids, causing strong selection for small body size (Wake 2009). The resulting miniaturization had profound effects on the functionality of organs, most importantly the brain. Here a specific factor comes into play, to be fully discussed below under *Cell volume and genome size*: many amphibians have disproportionately large cell volumes. Consequently, a salamander has fewer body cells than a lizard, bird, or mouse of similar size. Miniature animals with large cells face a serious problem: crucial organs may run out of cells when body size is reduced. In order to maintain function, many

Figure 10.2 The macroevolution of early tetrapods involved two major factors that formed recurrent themes: body size evolution and the change of developmental rates. X, developmental evolution; C, constraints imposed by evolution of body size.

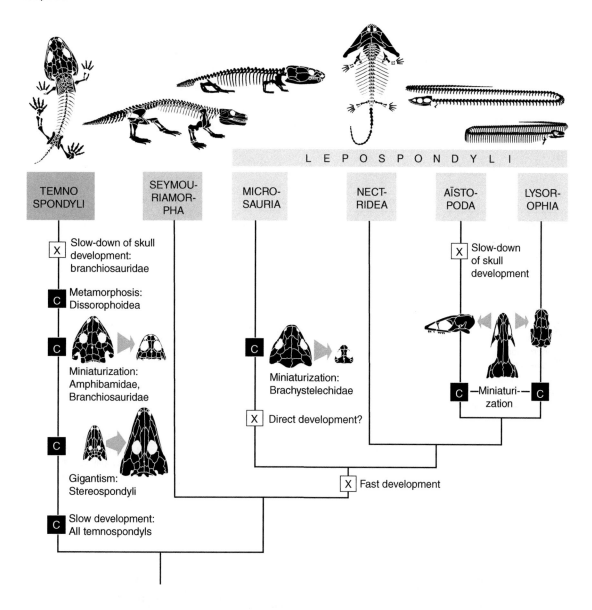

organs require a minimal number of constituent cells. In the brain and optical system of plethodontids, this problem is apparent when the number of cells in the relevant brain region is compared with other amphibians: 500,000 cells in a frog compared to only 30,000 in a plethodontid salamander (Roth *et al*. 1988). The functionality problem was solved by compensatory processes: an increase of parts required for the optical system at the expense of the forebrain and an increase in the packing density of neurons (Roth *et al*. 1988). Miniaturization may well explain why skeletal

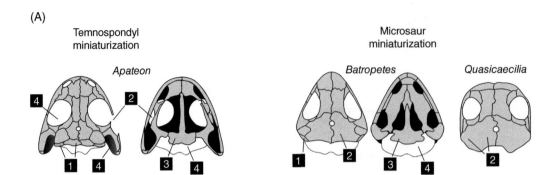

(A)

Temnospondyl miniaturization

Apateon

Microsaur miniaturization

Batropetes   *Quasicaecilia*

(B)

Morphological consequences of miniaturization

		Temnospondyli	Microsauria	Aistopoda
1	Rudimentary elements	Postparietal, tabular, prefrontal, postfronal	Prefrontal, postfrontal	Prefrontal, postfrontal Lacrimal
2	Loss of elements	Jugal, ectopterygoid	Postparietal, tabular	Postparietal, tabular, supratemporal
3	Novel morphology	Posterior skull table, parasphenoid	Posterior skull table, parasphenoid, braincase fusion	Frontal unpaired, posterior skull table
4	Disproportionate regions	Sensory capsules, eyes	Sensory capsules, eyes	Sensory capsules, eyes
5	Increased variation	Suture morphology, atavistic bones	Unknown	Unknown

(C)

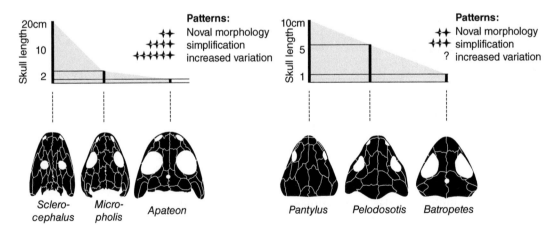

Size range and patterns

Patterns:
- ++ Noval morphology
- ++++ simplification
- ++++++ increased variation

Patterns:
- ++ Noval morphology
- +++ simplification
- ? increased variation

Sclero-cephalus   *Micro-pholis*   *Apateon*

Pantylus   *Pelodosotis*   *Batropetes*

Figure 10.3 (A) Miniaturization is a common macroevolutionary phenomenon in both temnospondyls and lepospondyls. (B, C) Body size reduction often correlates with additional patterns, which are listed here (adapted from Schoch 2013, reproduced with permission of Taylor & Francis).

elements were lost in lissamphibians: the number of stem cells might have failed to reach the critical threshold required to form a condensation (Atchley and Hall 1991). At any rate, miniaturization may provide an essential component of the answer to the question of why the skeletons of Paleozoic tetrapods and lissamphibians differ so profoundly.

A common result of miniaturization is simplification of anatomical structures (Schoch 2013). This poses a problem for the systematist, because lost characters and reduced complexity enhance the level of homoplasy (Wake 1991). Modern amphibians are particularly affected when body size decreases (Hanken 1983, 1984; Yeh 2002). Salamander phylogeny in particular is affected by parallelism and convergence (Wake 2009). Miniaturized salamanders and anurans have lost similar sets of bones, and other elements have a more rudimentary morphology than in outgroups (Hanken 1983; Yeh 2002). Incomplete ossification results from truncation of development combined with a reduced developmental rate. On the basis of his observations in salamanders, Hanken (1983) proposed a null hypothesis according to which miniaturization results in structural simplification, novel morphological structures, and increased intraspecific variation.

Miniaturization has been reported from different Paleozoic tetrapods, especially amphibamid temnospondyls (Milner 1988; Fröbisch and Schoch 2009) and microsaurs (Carroll 1990). Despite their disparate morphologies, tiny members of both clades share reduced circumorbital and parasphenoid elements and disproportionately large otic capsules (Carroll 1990). Fusion of braincase elements is also a common result of miniaturization (Schoch 2013). Similarly, the formation of cylindrical vertebrae has occurred convergently in lepospondyls and the smallest amphibamid temnospondyls, a further example of simplification of the ancestral rhachitomous vertebral centrum. Aïstopods and lysorophians are two other clades with high levels of miniaturization, in which the tiny skull was lightly built and lacked similar sets of bones. Furthermore, aïstopods show patterns of bone rudimentation and loss similar to those seen in microsaurs and temnospondyls.

It is likely that miniaturization played a role in the origin of lissamphibians (Milner 1988; Schoch 2013), but as mentioned earlier it remains unclear whether the loss of bones is a direct result of drastic size decrease (insufficient cell number to produce bones) or an indirect effect caused by the truncation of the ontogenetic trajectory (Schoch 2002). Miniaturization often results from resource or habitat limitations (Bonett et al. 2009).

Conversely, gigantism has been studied only occasionally in lissamphibians, with the most extreme case being large aquatic salamanders (cryptobranchids and amphiumids). In the cryptobranchid Andrias japonicus, gigantism is accompanied by delayed sexual maturity, an incomplete metamorphosis, longevity, and a stable habitat (Matsui et al. 2008). The largest cryptobranchid, indeed the largest known lissamphibian, was Aviturus exsecratus from the Paleocene of Mongolia, which appears to have been a more terrestrial form that evolved during a climatic optimum (Vasilyan and Böhme 2012).

Recently, Bonett et al. (2009) highlighted that selection for giant body size may be favored by (1) increased resource abundance, (2) ecological release from predators, or (3) the necessity for long-distance dispersal. In Mesozoic temnospondyls, the first of these may have been the decisive factor, permitting capitosaurs, trematosaurs, and metoposaurs to repeatedly evolve 3–6 m long top predators in the rich tropical–subtropical freshwater ecosystems of the Triassic. Apart from that, gigantism appears to correlate with slow developmental rates in both salamanders and Mesozoic temnospondyls.

Body size evolution has traditionally played a big role in the discussion of directed evolution, highlighted by "Cope's rule." This holds that size tends to increase within clades, with basal taxa being often small (Cope 1880; McKinney 1990). On a very large scale, Payne et al. (2009) found that maximum size of life had increased at least in two major pulses in the Proterozoic and early Paleozoic, each time associated with a significant increase in body-plan complexity. McKinney (1990) emphasized the evolutionary–ecologic reasoning that many clades which radiated after an extinction event started with small species. Based on a microevolutionary model of selection on size, Kingsolver and Pfennig (2004) suggested that the dominance of directional selection within

(A)

Ontogenetic trajectories

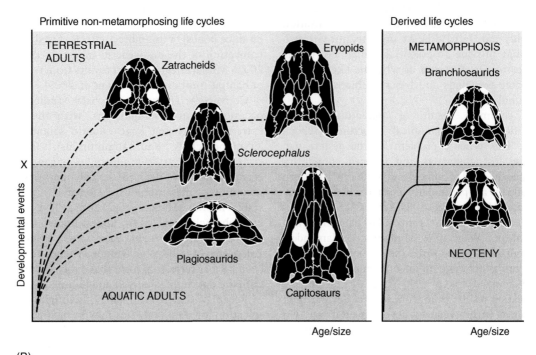

(B)

Modes of trajectory evolution

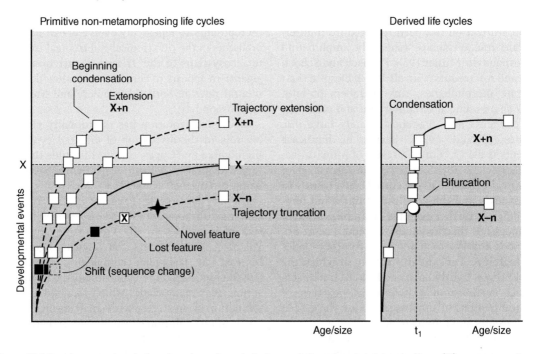

Figure 10.4 Developmental evolution played a major role in the evolution of early tetrapods. Here, (A) processes of heterochrony are exemplified by temnospondyls, illustrating (B) different modes by which ontogenetic trajectories evolved (adapted from Schoch 2013, reproduced with permission of Taylor & Francis).

populations could translate into macroevolutionary trends towards increased size within clades. In a recent analysis of the stem-lineages of tetrapods and amniotes, Laurin (2004) found some evidence for phyletic size increase in the amniote stem, but not in early tetrapods.

Cell volume and genome size. The biological body size of an organism (assessed by the number of its constituent cells) may differ profoundly from its metric body size. In lungfishes and many amphibians, the two scales diverge substantially, because both groups have much larger cells than other vertebrates (Hanken and Wake 1993). In many metazoans, cell volume correlates with the genome size in the nucleus, for reasons that are still not entirely understood (Cavalier-Smith 1978). Hence the large cells of salamanders probably result from their huge content of DNA in the nucleus. In turn, large genomes set maximum limits to developmental rates, and in aquatic salamanders that have large cells this often results in pedomorphosis (Sessions and Larson 1987). Thus, increase in genome size increases cell size and slows down metabolism and rate of development (Hanken and Wake 1993). In salamanders, metamorphosing taxa usually have smaller cell volumes than neotenic ones (D.B. Wake, personal communication 2012).

Recently, histology has yielded the first data on cell volume size of some Paleozoic and Mesozoic tetrapods (Sanchez *et al.* 2010b). Among the lepospondyls, lysorophians, microsaurs, and nectrideans had small cell volumes (as assessed by osteocyte lacunae), consistent with those of most amniotes (Organ *et al.* 2010). Lepospondyl cell volumes ranged between 100 and 220 μm^3, contrasting with the much larger cells of lissamphibians, notably salamanders (300–400 μm^3). Constraints imposed by cell size were therefore probably not as severe in lepospondyls as they are in extant salamanders. In temnospondyls, few samples have so far been analyzed, but they apparently had a wide range of cell volume sizes, with branchiosaurids comparing well with salamanders (350–400 μm^3). The analysis of cell volume in fossil taxa has just started, but it forms a promising field for future research.

Developmental evolution. Following the "evo-devo" approach, paleontology has recently contributed data to the study of developmental evolution on various levels. Early amphibians, especially temnospondyls, have provided rich data on heterochronic changes (patterns), but more recently also on the extent of plasticity and canalization (processes).

Heterochrony is a very common pattern in lissamphibians and Paleozoic temnospondyls, and it has been identified on various taxonomic scales. In salamanders it has been identified as a major factor (Wake and Roth 1989; Wake 2009). The different modes of developmental evolution and their relation to metamorphosis have already been analyzed, and it has been argued that evolution proceeded on different pathways in biphasic and uniphasic life cycles (Schoch 2010). Here, it may be sufficient to stress that macroevolutionary patterns of heterochrony (Figure 10.4) are abundant but their adaptive context often remains obscure. The slow-down of development in neotenes (salamanders, branchiosaurids) often forms an adaptive strategy to exploit favorable aquatic conditions or avoid harmful terrestrial habitats.

The ancient life cycle of tetrapods and stem-amphibians was either a simple uniphasic one (aquatic adults), or biphasic as in seymouriamorphs with a long transformational period (terrestrial adults). On a small scale, the evolutionary "fine-tuning" of ontogenetic trajectories appears to have been an adaptive strategy of early temnospondyls, permitting them to cope with unstable environments or broaden the range of habitats they could occupy. In digited stem-tetrapods, which were aquatic gill-breathers (Coates and Clack 1991), a certain amount of flexibility is indicated by the presence of tracks (Clack 1997a, 2012). However, this flexibility was much less pronounced than in temnospondyls, as the rather inflexible ontogenies of seymouriamorphs reveal: these were not capable of evolving neoteny, reaching sexual maturity only before or during transformation to a terrestrial existence (Sanchez *et al.* 2008). Heterochrony probably formed a widespread pattern in the developmental evolution of taxa inhabiting the water–land interface. In stem-amphibians, the terrestrialization of dissorophoids apparently necessitated the evolution of a more radical transformation, producing metamorphosis and the biphasic life cycle of lissamphibians.

Turning from pattern to process, two general adaptive strategies of developmental evolution can be identified. These are plasticity and canalization, two apparently contradictory properties of organisms, which are nevertheless often linked. Some temnospondyls had wide reaction norms, especially branchiosaurids and the genus *Sclerocephalus*. These responded to diverse lake habitats by (1) extending or truncating life span and the ontogenetic trajectory, (2) adjusting growth rate, and (3) fine-tuning the timing of sexual maturity (Schoch 2009a; Sanchez *et al.* 2010a). More subtle cases of plasticity are found in Triassic stereospondyls, such as *Gerrothorax* and *Mastodonsaurus*, in which histology reveals a wide range of annual growth rates, resorption, and the deposition of ballast in the skeleton, depending on the properties of lake habitats. In these taxa, plasticity affected the microanatomical level, whereas gross morphology was especially constrained by canalization.

The origin of a drastic, lissamphibian-like metamorphosis involved a different role for plasticity and canalization. The wide range of plastic responses exemplified by temnospondyls was here restricted to two well-constrained pathways: neoteny and metamorphosis. Within each of these pathways, plasticity permits much variation, but the restriction to two alternatives forms a strong case of canalization. This constraint can be broken by evolving direct development (skipping the larval stage), which produces miniature adults and removes neoteny as an option.

Reproduction. Like few other vertebrates, extant amphibians have remarkably diverse modes of reproduction (Wake 1982; Wake and Hanken 1996). The ability to evolve live-bearing taxa or ones that lay eggs on dry land forms a major advantage in colonizing new habitats or responding to environmental perturbations. Both random fluctuations and gradual changes of external parameters are readily addressed by this flexibility, either by means of plasticity or by an evolutionary change from one mode of reproduction to another one. This contributed to the success of lissamphibians and permitted adaptive radiations in many environments (Wake 1982).

Whereas all three modern clades of amphibians have evolved viviparity, it is unknown whether Paleozoic or Mesozoic groups managed to do the same. Lepospondyls are likely to have abandoned the larval stage and thus either laid eggs on land or bore live young. In contrast, temnospondyls and seymouriamorphs appear to have had aquatic larvae throughout. Each of these Paleozoic clades appears to have been bound to a particular reproductive strategy. Therefore, the large extent of evolutionary flexibility in lissamphibian reproductive strategies might well be restricted to that clade.

Speciation. Do species really exist, or are they just categories invented by humans to classify organisms? Are species merely metaphors invented to describe in simple terms what is really much more complicated? There is increasing support for the pragmatic notion that species are segments of population-level evolutionary lineages (de Queiroz 1998). Often, such lineages form highly complicated strings of populations that separate and reunite, characterized by limited gene flow and resulting in genetic disparity. The point at which two evolutionary lineages have irretrievably diverged is therefore difficult to identify. It seems that the case will remain fuzzy, highlighted by numerous disagreements about the total number of species in a clade (Vences and Wake 2007).

How do new species form? Species are sets of populations whose unity is maintained by gene flow. Any interruption of gene flow may eventually lead to speciation, although short-term fluctuations in and interruption of gene flow are very common. In modern amphibians, gene flow is often restricted to small populations, and species-wide flow remains exceptional (Vences and Wake 2007), which in theory should be a common cause of species multiplication. However, even in extant species, identification of the main factors of speciation is a difficult task. Molecular data have recently contributed crucial information to the speciation of amphibian taxa. In lissamphibians, most studied species formed by allopatry (dichopatry), the geographic separation of a formerly single habitat. Others originated by peripatry, which is when a small fraction of a population invades a new area at the periphery of the main habitat. Finally, cases of sympatric speciation have also been reported, but these form a small minority (Vences and Wake 2007). Lissamphibians are notorious for hybridization even between relatively distant taxa, such as anuran species of the genus *Rana*; this forms a nightmare for the paleontologist, as there are almost no criteria by which to prove this phenomenon in the fossil record. Much more severely, it is the crux of paleontology that new species often form (1) within very small populations (bottlenecks) whose preservation potential is minimal and (2) within remarkably short periods of time. In addition, speciation does not necessarily involve morphological change, as is clear from the many cryptic species that have been identified only after molecular analysis has delivered the necessary genome data.

Tracing the origin of new species in the fossil record is therefore a difficult and controversial task. Incipient species may often not be recognized as such, and thus rates of speciation and diversification are not necessarily correlated. Identification of speciation events in the fossil record requires a continuous sequence of sediments, the stratigraphic succession of large samples of the study taxon, and complete preservation throughout these samples. These requirements are not easily met, considering that this means the close succession of conservation *Lagerstätten*. In Paleozoic tetrapods, the lake deposits of the central European *Rotliegend* facies are as close as one may get to analyzing patterns of speciation. These finely laminated mudstones formed in the deeper parts of the water body and under continued conditions of deposition. However, the preserved time intervals usually fall in the 10^2–10^3 year range, according to annual deposition cycles (varve lamination: see Boy and Sues 2000). Speciation events most probably lie in the gaps between the successive lake deposits, which cover longer time intervals than the lake deposits themselves (~10^4–10^5 years). In the Saar–Nahe basin, two different patterns of speciation appear to have occurred:

1. The large fish-eater *Sclerocephalus haeuseri* usually formed the single top predator in Saar–Nahe lakes (Schoch 2009b), but in the Kappeln lake a bimodal "population" pattern has been found in which two morphs or nascent species of *Sclerocephalus* co-occurred. In later deposits, two separate species are recognized that are usually grouped into the genera *Sclerocephalus* (*S. haeuseri*) and

Glanochthon (*G. angusta*). These differ not only in morphology (snout length, dentition, developmental traits), but also ecologically (actinopterygian versus acanthodian prey). The two coeval "populations" of the Kappeln lake are very similar to the two separate species and genera of later deposits. Conversely, in all lake deposits older than the Kappeln lake, only a single "population" of *Sclerocephalus* was present, albeit with considerable morphological variation; in the Jeckenbach lake, single specimens resembling the incipient species of the Kappeln lake occasionally occur. These observations suggest that between the deposition in the Jeckenbach and Kappeln lakes, the Saar–Nahe population of *Sclerocephalus haeuseri* underwent a speciation. Morphology appears to have taken the lead in this process, because both morphs or incipient species in the Kappeln lake fed on the same species of fish. The autochthonous co-occurrence of both morphs in the Kappeln lake is indicated by the abundance of larvae and juveniles in the same deposit. Although the Kappeln sample forms only a segment of the two lineages, it is possible that the mode of speciation was sympatric.

2. The small branchiosaurid *Apateon pedestris* forms an abundant vertebrate in many lake deposits of the European *Rotliegend* basins. In small or seasonally unstable water bodies, this plankton-feeder was often the only vertebrate (Boy 2003). In adapting to local conditions, *Apateon* was apparently more plastic than other temnospondyls, and this plasticity affected both development and morphology. Many lake deposits in France and Germany housed their own *Apateon* "populations." Mean adult size and development ranged broadly from lake to lake, suggesting that developmental plasticity played a major role. This recalls the remarkable radiation of cichlid fishes in the East African lakes (Liem 1974; Coyne and Orr 2004). Because neoteny constrained *Apateon* populations to be fully aquatic dwellers, local populations were more or less isolated from each other. This suggests that vicariant speciation was a common mode of evolution in branchiosaurids.

10.4 Clades, space, and time

Species multiply, diversify, and become disparate by adaptive radiations. These processes require not only much time, but also a lot of space. Evolving clades usually disperse over large areas, often continent-wide, sometimes globally. Amphibians, both extant and extinct, are no exception. This is the main problem of *vicariance*: how does the hierarchy of phylogenetic branching map on geography, and how did this pattern change through time? Imagine drawing a cladogram on a map and tracing the evolving cladogram through the successively subdividing supercontinent. Vicariance can be especially exciting when sub-clades correlate with newly separated continents.

The relation between phylogeny and biogeography is best exemplified by salamanders, in which dispersal has been studied by Milner (1983), who proposed a vicariance model based on phylogeny, paleogeography, and paleoclimatology. He suggested that salamander distribution is well explained by a single major dispersal in the early Mesozoic resultuing in a cosmopolitan Laurasian fauna. Further evolution followed a sequence of vicariance events that correlate with geographic isolation events in the northern hemisphere, which happened during the later Mesozoic and early Cenozoic. He concluded that urodeles originated after the separation from the saliential lineage and migrated into the northern temperate humid belt of northern Euramerica. Subsequent separation of continental blocks by the successively developing Turgai Sea, Mid-Continental Sea, and nascent Atlantic Ocean isolated the salamanders, resulting in hynobiids in east Asia, dicamptodontids in western North America, and salamandrids in Europe. These groups could later expand when seas withdrew and the Bering land bridge formed. Zhang *et al.* (2005) have recently confirmed this on a general scale, concluding from molecular data that salamanders most likely originated in east Asia (and indeed the oldest crown-group salamanders are from China), the neobatrachian anurans had an African–Indian origin, and caecilians most likely formed in the tropical belt of Pangaea.

Patterns of early tetrapod distribution are more difficult to trace, because many time slices are preserved only at Euramerican localities and adjacent regions. The available evidence indicates that global distribution of clades was not necessarily the rule, despite the existence of Pangaea in the Permian and Triassic.

In the Devonian and Early Carboniferous, stem-tetrapods were not restricted to Euramerica (their main fossil-bearing region), because some skeletal material and tracks have also been reported from Australia (Clack 1997b; Warren 2007). Anthracosaurs and baphetids were apparently tropical forest dwellers in Euramerica, while lepospondyls were more widespread over the whole Euramerican subcontinent, as were early temnospondyls (Milner and Panchen 1973). More clear-cut patterns emerge in the Permian, where dissorophoids and eryopids have been found in North America, Europe, and Russia, whereas surviving edopoids (a Late Carboniferous Euramerican clade) were reported from an unexpected fauna in Niger (Steyer *et al.* 2006). Even branchiosaurids, which had long been considered endemic to Europe, have also been identified in North America (Milner 1982) and Siberia (Werneburg 2009). In seymouriamorphs, seymouriids occur throughout Euramerica, whereas discosauriscids are confined to Europe and Inner Asia (Klembara and Ruta 2005). In the Triassic, temnospondyl finds are abundant from all regions of Pangaea, with trematosaurs having dispersed most widely, at a fast rate, and apparently also using marine passages for their migration (Schoch and Milner 2000). Plagiosaurids and brachyopoids are the only stereospondyls that appear to have excluded one another: in the Triassic, plagiosaurids were confined to Greenland, Europe, and China, whereas brachyopoids occurred in North America, South America, South Africa, and Australia; only in the Jurassic did brachyopids invade Inner Asia (Shishkin 2000). Finally, chroniosuchians were long thought to be confined to European Russia, but have recently been reported from Germany and Inner Asia, thus having had at least a European–Asian distribution (Witzmann *et al.* 2008).

10.5 Diversity, disparity, and extinction

Modern amphibians provide alarming examples of the global decline of species (Wake 2009). The rate at which new species are discovered, especially among anurans, appears to be exceeded only by the fast pace at which lissamphibian species disappear. This is extinction in progress, and its extent indicates that it will probably form the beginning of a modern mass extinction event. Especially in the tropics, amphibian hot-spot habitats are destroyed by large-scale deforestation and the expansion of human settlements. These threats are imminent and occur at a very small time scale (10^1–10^3 years). At much larger time scales, spanning 10^5–10^7 years, patterns of speciation, diversity decline, and extinction are usually analyzed in the fossil record. Such analyses require robust phylogenetic hypotheses for the groups under consideration and stratigraphic calibration. Depending on the preferred cladistic hypothesis, more or less long-range extensions (for stratigraphically younger sister taxa) and ghost lineages (for stem-groups) have to be postulated. These must be considered in assessments of how many lineages managed to cross particular stratigraphic boundaries or became extinct.

How can such macroevolutionary changes in the fossil record be measured? Diversity is a widely used label, applied by different authors to various types of phenomena. In vertebrate paleontology, it usually refers to the macroevolutionary metric for species number in a given clade, time window, or geographic region. Thus, diversity measures taxonomic variety. Apart from that, morphological variety within a clade is also often measured, referred to as disparity.

Disparity. Morphological variety and its relation to phylogeny has been studied in temnospondyls, the most diverse (speciose) group of early tetrapods – which is also known for its uniformity compared with lepospondyls, for instance (Ruta 2009). Stayton and Ruta (2006) analyzed geometric morphometrics of Permo-Triassic temnospondyl skulls, based on phylogenetic hypotheses by Yates and Warren (2000) and Schoch and Milner

(2000). All studied taxa belong to the Stereospondylomorpha, a clade that originated in the Late Carboniferous and became the dominant group of Mesozoic temnospondyls. They found that morphological and phylogenetic distances are not concordant in most stereospondyl groups – this means that some clades diversified without major morphological change (e.g., capitosaurs), whereas some small clades were highly disparate (e.g., plagiosaurids, trematosaurs). In a disparity analysis of temnospondyls, Ruta (2009) found that all major clades were widely separated in morphospace. Similar levels of disparity were found throughout temnospondyls, with dvinosaurs and dissorophoids each being more disparate than edopoids and eryopoids. What, after all, does disparity measure here? In the cited example, morphological variation correlates with paleoecological diversification: stream-, lake- and sea-dwelling trematosaurs were more disparate *and* ecologically diverse than the uniform capitosaur top predators. Likewise, dissorophoids were small terrestrial carnivores and insectivores and therefore more disparate than the large fish-eating eryopoids or the presumably tetrapod-eating seymouriamorphs. Future studies might want to focus on the analysis of paleoecology and disparity, based on phylogenetic scenarios.

Biases of the fossil record. The assessment of diversity in the fossil record requires consideration of how incomplete it may be for the studied group. Much more severely than the marine record, terrestrial faunas are subject to taphonomic biases. Many habitats on dry land are unlikely to preserve even fragments of bone, let alone articulated skeletons or larger samples: intense weathering, soil formation, and scavengers/detritivores will destroy almost everything. Thus, inhabitants of dry plains and dense forests have very little potential to become preserved. Probably only those terrestrial taxa occasionally crossing streams will be buried in river sediments, but these are usually deposited as isolated and water-worn bone fragments. In such rocks, only skulls are normally diagnostic enough to establish the presence of a particular taxon. Many Late Permian and Triassic deposits are of this type, and probably only temnospondyls were large enough to permit preservation and identification – this may explain why lissamphibians are so extremely rare in the early Mesozoic, although long ghost lineages have been inferred from almost any recent phylogeny. The same may account for lepospondyl diversity in the Permian. The very unlikely discovery of a fissure fill can change the picture, as evidenced by the Early Permian cave deposits at Fort Sill (full of small stem-amphibians and tiny lepospondyls), or Middle Jurassic marls in Britain (rich in early lissamphibians). Even in the Cenozoic record of lissamphibians, fissure-fill deposits contribute greatly to knowledge of past diversity (Böhme 2008).

Higher-quality preservation occurs in floodplain deposits, which often yield complete skeletons. However, "often" is a relative term here: collecting in such monotonous mudstones can be utterly time-consuming. Despite the good preservation, fossils cannot be predicted in such deposits. Rich tetrapod *Lagerstätten* form in slow-flowing streams, oxbow and peat lakes, larger lakes and lagoons, and extensive river deltas. Not surprisingly, such deposits contribute disproportionately to the knowledge of the amphibian fossil record. Judging by modern habitats, such places were not necessarily the richest in species, and the most diverse tropical rainforests usually have a very poor fossilization potential. It is fortunate that parts of the Late Carboniferous ecosystems were preserved in lakes and bogs within such rainforests, providing the bulk of data on early tetrapod evolution. In Saxony (eastern Germany), an Early Permian forest was preserved after its destruction by a fire; very unexpectedly, fossil collectors discovered molds of burned skeletons from various forest dwellers, among them temnospondyls and early amniotes (R. Werneburg, personal communication 2012). The amphibian fossil record is neither exceptionally good nor poor, but simply extremely heterogeneous. There is a clear bias towards freshwater habitats and their inhabitants, whereas more terrestrial taxa are disproportionately rare. Nevertheless, true patterns of diversity emerge when similar deposits in different regions or periods produced widely divergent numbers of species. Even when this restricted set of samples is studied, numerous patterns are found that indicate changes in amphibian diversity and extinction.

Diversity and extinction. On a gross scale, early tetrapod diversity falls into several phases, with the transition from the Paleozoic to the Mesozoic era involving the most fundamental changes. Although the prominent mass extinction events (P–T, T–J, K–Pg) appear to have affected fewer early amphibian taxa than in many other tetrapod clades, there were still phases of increased extinction rates, especially during the Late Permian, when many Paleozoic groups declined. In the end, only temnospondyls and chroniosuchians survived the Permo-Triassic (P–T) boundary. Stem-tetrapods, embolomeres, seymouriamorphs, and probably all lepospondyls disappeared before or during the P–T boundary (Milner 1990, 1996).

Although falling within the same 10–15 myr interval (Late Permian), the decline and extinction of these divergent groups appears to have had very different reasons. Embolomeres were confined to lakes and coal seams in tropical rainforests of the Carboniferous, which had vanished in the Early Permian (except for China, from where no such finds have yet been reported); the stream-dwelling *Archeria* is the only known Permian anthracosaur to date. Stem-tetrapods of unclear phylogenetic position (baphetids, colosteids, whatcheeriids) were probably also affected by the disappearance of equatorial forests.

In contrast to embolomeres, seymouriamorphs and lepospondyls survived well into the Permian. Like the dissorophid and trematopid temnospondyls, seymouriamorphs inhabited floodplains and dry upland habitats that were also populated by the rapidly diversifying amniotes. These regions were evidently more affected by the increasing hothouse climate of the Late Permian than the freshwater habitats of most temnospondyls, and it is conceivable that both dissorophoids and seymouriamorphs rank among the victims of the end-Permian extinction. However, both clades vanished or substantially declined well before the P–T boundary, suggesting that their disappearance correlates with gradually increasing aridity of terrestrial habitats in wide regions of Pangaea rather than with the end-Permian extinction itself. Dissorophoids and seymouriamorphs were probably more vulnerable to aridity than amniotes because they still relied on freshwater habitats for reproduction and the larval phase.

Indeed, both parareptiles and archosauromorphs were much less affected by Permian extinction (both within and at the end of the period) than were early tetrapods or synapsids (Dilkes 1998; Modesto *et al.* 2001).

Among the stem-amniotes, chroniosuchians were the only clade to persist well into the Middle Triassic, with chroniosuchids in Kyrgyzstan (Schoch *et al.* 2010) and bystrowianids in Russia and Germany (Golubev 2000; Witzmann *et al.* 2008). It is impossible to interpret this pattern until reliable data become available on the life history and adult lifestyle of these stem-amniotes. Their general lack of lateral-line grooves and the-bone profiler data presented by Laurin *et al.* (2004) suggest that chroniosuchians were terrestrial rather than aquatic.

As the largest early tetrapod clade, temnospondyls provide rich data for the analysis of diversity and extinction through the Paleozoic and Mesozoic. The mass extinction at the end of the Permian did not have such a catastrophic impact on most temnospondyls as on other tetrapods. Ruta and Benton (2008) analyzed temnospondyl ranges based on several alternative phylogenies, finding two peaks of diversity in the Early Permian and Early Triassic. Terrestrial and aquatic groups contributed in very different ways to these peaks in the Permian and Triassic samples. Whereas the Early Permian was a time of ecologically diverse temnospondyl faunas with large numbers of amphibious and terrestrial taxa (dissorophids, trematopids, amphibamids, zatracheids, eryopids), the Late Permian hothouse climate appears to have favored a trend away from terrestrial and amphibious taxa towards fully aquatic ones, which then contributed disproportionately to the Early Triassic peak (rhinesuchids, lydekkerinids, rhytidosteids, trematosaurids). Particularly around the P–T boundary, a remarkable increase in the number of genera and species has been recorded, despite the extinction of others (Milner 1990). Ruta and Benton (2008) found that the recovery after the end-Permian extinction occurred earlier for temnospondyl families than for genera and species. Conversely, an extinction rate peak was identified in the Late Carboniferous (Moscovian–Kasimovian). After the Triassic, diversity dropped substantially within stereospondyls, the only survivors of the vast temnospondyl clade.

Lissamphibians were the only amphibian clade to survive the end of the Cretaceous, with albanerpetontids as a fourth lineage to cross the Cretaceous–Paleogene (K–Pg) boundary (Milner 1993, 1994). These four lineages appear to have survived the K–Pg extinction event without major changes in diversity. Based on a detailed analysis of lissamphibian diversity in the Jurassic–Eocene interval, Fara (2004) found a "virtual extinction-free gradual rise" of the group. He further concluded that the K–Pg event did not have such a major impact on small terrestrial vertebrates as it had on dinosaurs. Amphibian diversity figures have only very recently started to dwindle. Albanerpetontids died out in the early Pliocene, only ~3 myr ago. Within the Pleistocene, starting ~1 myr ago, climatic oscillations accompanied alternating phases of glaciations and melting (Barnosky 2008). These fluctuations affected numerous taxa, and after the end of the last glaciation (~11 000 years ago), large terrestrial vertebrates became extinct at a faster pace, with climate change and increasing human impact named as major factors (Martin 2005). Whereas the causes of the early Holocene extinction are still controversial, the current global disappearance of species is undoubtedly related to human activities. The modern extinction of animal species has already been referred to as a "sixth mass extinction" (Novacek 2007).

How severe is the present-day disappearance of lissamphibian species? Recently, Wake and Vredenburg (2008) have reviewed the body of evidence for the current decline of amphibian populations. They reported a rate of extinction 211 times higher than that of normal background extinction. They further attested that human activities – both direct and indirect – are involved in almost every aspect of the current amphibian extinction spasm. Habitat modification and destruction are often accompanied by the application of fertilizers and pesticides, which kill amphibians or lead to their sterility. The introduction of exotic species that feed on native amphibians poses a further threat. Finally, current climate change with global warming and increased climatic variability affects many amphibian species that have specialized or small habitats and low fecundity (Wake and Vredenburg 2008).

A second major problem (probably not caused by human influence) is an infectious disease caused by a fungus (chytridiomycosis), which has led to the massive decline and extinction of many frogs in Central America and Australia (Pounds *et al.* 1997). Wake and Vredenburg (2008) concluded that the eventual survival of robust frog, salamander, and caecilian species is likely, but that the losses in amphibian diversity are already heavy and will have manifold effects on the biosphere. Clearly, the study of amphibian biology and paleobiology will be essential to tackle the problems caused by harmful human influence now and in the future.

10.6 The evolution of terrestriality

It was no small surprise when the limbed tetrapodomorphs *Acanthostega* and *Ichthyostega* turned out to have been essentially aquatic animals (Coates and Clack 1990, 1991). In particular, the possession of internal gills prompted a reconsideration of life habits in these digited taxa. As Clack (2012) put it, there has been a recent separation of two formerly connected questions: (1) why did stem-tetrapods leave the water and (2) why did they evolve hands and feet? Since the work of Clack, Shubin, and colleagues we have learned that digits evolved long before stem-tetrapods left the water. It is therefore interesting to review the alternative hypotheses that have been proposed for the origin of tetrapods (Figure 10.5), and compare these with the now-available evidence. A detailed discussion has recently been given by Clack (2012).

These hypotheses do not all deal with the same basic question, but vary between explaining the origin of digits (D), the origin of terrestrial locomotion (L), or the origin of terrestriality or a land-dwelling life (T). A general prerequisite is that stem-tetrapods possessed both internal gills and lungs, which is inferred by means of the extant phylogenetic bracket, as well as from anatomical evidence in some taxa.

1. A central dogma in theories of tetrapod origin was the ecological scenario that tetrapodomorphs were attracted by food resources outside the water (T). This would have been invertebrates, particularly arthropods, which formed abundant and diverse land dwellers by Late Devonian times. The classic picture thus

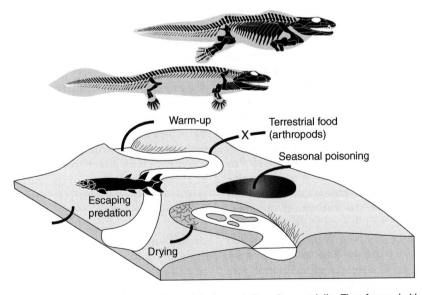

Figure 10.5 Various scenarios have been proposed to explain the evolution of terrestriality. They focused either on pressures to *leave* the water (seasonal poisoning, predation, drying) or pressure to *invade* the dry land (warm-up, terrestrial food). These hypotheses are not equally likely; e.g., evidence for the exploitation of terrestrial food sources has been elusive.

includes an *Ichthyostega*-like animal feeding on insects and millipedes. However, evidence for such feeding habits can only be indirect, as stomach contents are unknown in these taxa, with putative feeding inferred from dentitions and jaw mechanics. These do not support the central dogma, but instead suggest predation on larger fishes rather than invertebrates. Thus, most well-known stem-tetrapods and early crown tetrapods were probably fish-eaters, and potential land excursions were not made in search of food.

2. An influential hypothesis holds that drying ponds forced stem-tetrapods to migrate to persisting pools (L). This scenario was derived from the observation that many Late Devonian red beds show features of seasonal drying. Romer (1958, 1966) expanded on this scenario, suggesting that those taxa with the strongest limbs were the most successful in dispersing across land bridges. Romer (1958) thus presented an early version of the modern interpretation that digits evolved in aquatic animals, long before tetrapods became fully terrestrial.

3. Instead, Orton (1954) proposed that limbs might have evolved to bury in mud, not to crawl on land (D). This would be in line with the drying theory, yet with the variant hypothesis that the drying pools were not left, but used to aestivate droughts. This hypothesis could be tested if stem-tetrapods were found in burrows, but this has never been reported.

4. A conceivable scenario is a response to selection pressures generated by increased competition for food (Clack 2012). This may form the first step of scenario (1), initially forcing stem-tetrapods to leave a particular water body (L) and potentially explore new food resources along the shoreline or even on land (T). However, this scenario falls short of naming such food resources.

5. A further suggestion holds that predators (bony fishes, large arthropods) might have pressured stem-tetrapods more and more towards refugia around the shore, eventually forcing them to leave the water (McNamara and Selden 1993). This could imply either the search for new, less dangerous water bodies (L), or the transition to feeding on land (T).

6. Recently, Carroll *et al.* (2005) have suggested an explanation employing physiology. They argued that a raised body temperature would have been favorable to large predators such as stem-tetrapods. In preferring shallower and warmer water, such taxa might have been ultimately attracted by the land as a place to warm up for fish-hunting in the water (T). Although difficult to test, this scenario is at least consistent with the dentition of tetrapodomorphs and their inferred fish-eating habits, at the same time providing an explanation of how the land might have formed an attraction to leave the water.

7. Lately, Clack (2007, 2012) has worked on a more inclusive scenario that considers atmospheric oxygen levels, increased plant cover of the land surface, and anatomical changes in the spiracular and pectoral regions of stem-tetrapods. She suggested that air breathing was enhanced by using the spiracular chamber (Brazeau and Ahlberg 2006), which increased in size continuously between *Tiktaalik* and *Acanthostega*. Loss of the opercular bones might have permitted stem-tetrapods to raise the head for engulfing air more readily or frequently, such as during phases of air shortage in the water. This is generally consistent with the data of Berner (2006), who found low atmospheric oxygen levels in the earliest Carboniferous (Mississippian). The increased density of forests and size of tree stems indicates that more complex ecosystems were established on land, leading to increased plant decay. This, in turn, must have had an impact on many pools and lagoons, leading to low oxygen levels that might have triggered the above-mentioned evolutionary adjustments in stem-tetrapods. This hypothesis has the strength that it combines data from different fields (geochemistry, paleobotany, vertebrate morphology), but it cannot provide a model for the transition to land *per se*.

8. Alternatively, stem-tetrapods may have exploited the niche for predating on stranded fish (or invertebrates) in the tidal zone (Clack 2012). This hypothesis was prompted by the discovery of digited tracks in marine rocks of the Polish Middle Devonian (Niedźwiedzki

et al. 2010), although the referral of these tracks to stem-tetrapods has been questioned (King *et al.* 2011). Again, this hypothesis would only explain the first step, during which tetrapodomorphs approached the shore, explaining the evolution of digits in the aquatic environment (D).

9. Here, I propose a variant of Romer's drying-pond hypothesis, involving seasonal water poisoning by plant growth (algal blooms). This occurs in many modern pools, and has been identified as a factor of stressed ecosystems in Paleozoic environments (Boy 2003). Unlike in drying ponds, aestivation is not an option in such habitats, as oxygen may be entirely absent, killing all lake dwellers. It evidently often leads to mass mortality in small or shallow lakes (Boy 1998). In such environments, stem-tetrapods with digits might have had a huge advantage because they could escape (L). Fleeing water poisoning would thus have triggered exaptation of hands and feet after they had evolved under water. This hypothesis thus names a specific selective pressure for leaving the water. Clack (2012) and Pierce *et al.* (2012) have emphasized that polydactyl autopodia were not used for locomotion on land, but rather formed a different type of strong fins that were exapted for terrestrial locomotion only in a second step.

The preceding discussion has revealed that we might be dealing with two different variants of selective pressures when discussing the origin of terrestriality: (a) pressure to *leave the water* (drying, oxygen shortage, poisoning, escaping predation), or alternatively (b) pressure to *invade land* (attraction by food, temperature). The first pressure would have minimized the time spent on land, the second should have maximized it. At present, selection for the ability to leave dangerous water bodies appears to be the more plausible scenario for the initial steps on land. It is not ruled out that the "conquest" of land occurred in different clades in parallel, and in different ways: stem-amphibians developed a more active role for their limbs in locomotion, whereas stem-amniotes probably applied undulation of their elongate trunk instead.

References

Atchley, W.R. & Hall, B.K. (1991) A model for development and evolution of complex morphological structures. *Biological Reviews* **66**, 101–157.

Barnosky, A. (2008) Megafauna biomass tradeoff as a driver of Quarternary and future extinctions. *Proceedings of the National Academy of Sciences* **105**, 11543–11548.

Berner, R.A. (2006) GEOCARBSULF: a combined model for Phanerozoic atmospheric O_2 and CO_2. *Geochimica et Cosmochemica Acta* **70**, 5653–5665.

Böhme, M. (2008) The Miocene Climatic Optimum: evidence from ectothermic vertebrates of Central Europe. *Palaeogeography, Palaeoclimatology, Palaeoecology* **195**, 389–340.

Boisvert, C., Mark-Kurik, E., & Ahlberg, P. (2008) The pectoral fin of *Panderichthys* and the origin of digits. *Nature* **456**, 636–638.

Bonett, R.M., Chippindale, P.T., Moler, P.E., van Devender, R.W., & Wake, D.B. (2009) Evolution of gigantism in amphiumid salamanders. *Plos One* **4**, e5615.

Boy, J.A. (1998) Möglichkeiten und Grenzen einer Ökosystem-Rekonstruktion am Beispiel des spätpaläozoischen lakustrinen Paläo-Ökosystems. 1. Theoretische Grundlagen. *Paläontologische Zeitschrift* **72**, 207–240.

Boy, J.A. (2003) Paläoökologische Rekonstruktion von Wirbeltieren: Möglichkeiten und Grenzen. *Paläontologische Zeitschrift* **77**, 123–152.

Boy, J.A. & Sues, H.-D. (2000) Branchiosaurs: larvae, metamorphosis and heterochrony in temnospondyls and seymouriamorphs. In: H. Heatwole & R.L. Carroll (eds.), *Amphibian Biology. Volume 4. Palaeontology.* Chipping Norton, NSW: Surrey Beatty, pp. 973–1496.

Brazeau, M. D. & Ahlberg, P. E. (2006) Tetrapod-like middle ear architecture in a Devonian fish. *Nature* **439**, 318–321.

Carroll, R.L. (1990) A tiny microsaur from the Lower Permian of Texas: size constraints in Palaeozoic tetrapods. *Palaeontology* **33**, 893–909.

Carroll, R.L., Irwin, J., & Green, D.M. (2005) Thermal physiology and the origin of terrestriality in vertebrates. *Zoological Journal of the Linnean Society* **143**, 345–358.

Cavalier-Smith, T. (1978) Nuclear volume control by nuclear skeletal DNA, selection for cell volume and cell growth and the solution of the DNA C-value paradox. *Journal of Cell Science* **34**, 247–278.

Clack, J.A. (1997a) Devonian tetrapod trackways and trackmakers; a review of the fossils and footprints. *Palaeogeography, Palaeoclimatology, Palaeoecology* **130**, 227–250.

Clack, J.A. (1997b) The evolution of tetrapod ears and the fossil record. *Brain, Behavior and Evolution* **50**, 198–212.

Clack, J.A. (2007) Devonian climate change, breathing, and the origin of the tetrapod stem group. *Integrative and Comparative Biology* **47**, 510–523.

Clack, J. A. (2009) The fin to limb transition: new data, intepretations, and hypotheses from paleontology and developmental biology. *Annual Reviews of Earth and Planetary Sciences* **37**, 163–179.

Clack, J.A. (2012) *Gaining Ground: the Origin and Evolution of Tetrapods*, 2nd edition. Bloomington: Indiana University Press.

Coates, M.I. & Clack, J.A. (1990) Polydactyly in the earliest known tetrapod limbs. *Nature* **347**, 66–69.

Coates, M.I. & Clack, J.A. (1991) Fish-like gills and breathing in the earliest known tetrapod. *Nature* **352**, 234–236.

Cope, E. D. (1880) Extinct Batrachia. *American Naturalist* **14**, 609–610.

Coyne, J.A. & Orr, H.A. (2004) *Speciation*. Sunderland: Sinauer.

Darwin, C. (1859) *On the Origin of Species by Means of Natural Selection*. London: John Murray.

Davidson, E.H. (2006) *The Regulatory Genome: Gene Regulatory Networks in Development and Evolution*. New York: Academic Press.

de Queiroz, K. (1998) The general lineage concept of species, species criteria, and the process of speciation. In: D.J. Howard & H. Berlocher (eds.), *Endless Forms: Species and Speciation*. Oxford: Oxford University Press, pp. 57–77.

Dilkes, D.W. (1998) The Early Triassic rhynchosaur *Mesosuchus browni* and the interrelationships of basal archosauromorph reptiles. *Philosophical Transactions of the Royal Society London B* **353**, 501–541.

Eldredge, N. & Gould, S.J. (1972) Punctuated equilibria: an alternative to phyletic gradualism. In: T.J.M. Schopf (ed.), *Models in Paleobiology*. San Francisco: Freeman, Cooper, pp. 82–115.

Fara, E. (2004) Estimating minimum global species diversity for groups with a poor fossil record: a case study of Late Jurassic–Eocene lissamphibians. *Palaeogeography, Palaeoclimatology, Palaeoecology* **207**, 59–82.

Flatt, T. (2005) The evolutionary genetics of canalization. *Quarterly Review of Biology* **80**, 287–316.

Fröbisch, N.B. & Schoch, R.R. (2009) Testing the impact of miniaturization on phylogeny: Paleozoic dissorophoid amphibians. *Systematic Biology* **58**, 312–327.

Frost, D.R., Grant, T., Faivovich, J., *et al.* (2006) The amphibian tree of life. *Bulletin of the American Museum of Natural History* **297**, 1–370.

Futuyma, D.J. (2005) *Evolution*. Sunderland: Sinauer.

Golubev, V.K. (2000) [Permian and Triassic chroniosuchians and biostratigraphy of the Upper Tatarian series in Eastern Europe]. *Trudy Paleontologiceskogo Instituta RAS* **276**, 1–172. (In Russian.)

Gould, S.J. (2002) *The Structure of Evolutionary Theory*. Cambridge, MA: Belknap Press.

Gould, S.J. & Lewontin, R.C. (1979) The spandrels of San Marco and the Panglossian Paradigm: a critique of the adaptationist programme. *Proceedings of the Royal Society of London B* **205**, 581–598.

Haas, A. (2003) Phylogeny of frogs as inferred from primarily larval characters (Amphibia: Anura). *Cladistics* **19**, 23–89.

Hall, B.K. (1999) *Evolutionary Developmental Biology*, 2nd edition. Dordrecht: Kluwer.

Hanken, J. (1983) Miniaturization and its effects on cranial morphology in plethodontid salamanders, genus *Thorius* (Amphibia: Plethodontidae). II. The fate of the brain and sense organs and their role in skull morphogenesis and evolution. *Journal of Morphology* **177**, 155–268.

Hanken, J. (1984) Miniaturization and its effects on cranial morphology in plethodontid salamanders, genus *Thorius* (Amphibia: Plethodontidae). I. Osteological variation. *Biological Journal of the Linnean Society* **23**, 55–75.

Hanken, J. & Wake, D.B. (1993) Miniaturization of body size: organismal consequences and evolutionary significance. *Annual Review of Ecology and Systematics* **24**, 501–519.

Hunt, G. (2007) The relative importance of directional change, random walks, and stasis in the evolution of fossil lineages. *Proceedings of the National Academy of Sciences* **104**, 18404–18408.

King, H.M., Shubin, N.H., Coates, M.I., & Hale, M.E. (2011) Behavioral evidence for the evolution of walking and bounding before terrestriality in sarcopterygian fishes. *Proceedings of the National Academy of Sciences* **108**, 21146–21151.

Kingsolver, J.G. & Pfennig, D.W. (2004) Individual-level selection as a cause of Cope's rule of phyletic size increase. *Evolution* **58**, 1608–1612.

Klembara, J. & Ruta, M. (2005) The seymouriamorph tetrapod Utegenia shpinari from the ?Upper Carboniferous-Lower Permian of Kazakhstan. Part I: Cranial anatomy and ontogeny. *Transactions of the Royal Society of Edinburgh: Earth Sciences* **94**, 45–74.

Lauder, G. & Liem, K. (1989) The role of historical factors in the evolution of complex organismal functions. In: D.B. Wake & G.V. Roth (eds.), *Complex Organismal Functions: Integration and Evolution in Vertebrates*. Dahlem Konferenzen. New York: Wiley, pp. 63–78.

Lauder, G.V. (1981) Form and function: structural analysis in evolutionary morphology. *Paleobiology* **7**, 430–442.

Laurin, M. (2004) The evolution of body size, Cope's rule and the origin of amniotes. *Systematic Biology* **53**, 594–622.

Laurin, M., Girondot, M., & Loth, M.L. (2004) The evolution of long bone microstructure and lifestyle of lissamphibians. *Paleobiology* **30**, 589–613.

Liem, K. (1974) Evolutionary strategies and morphological innovations: cichlid pharyngeal jaws. *Systematic Zoology* **22**, 425–441.

Love, A.C. (2003) Evolutionary morphology, innovation, and the synthesis of evolutionary and developmental biology. *Biology and Philosophy* **18**, 309–345.

Martin, P.S. (2005) *Twilight of the Mammoths: Ice Age Extinctions and the Rewilding of America*. Berkeley: University of California Press.

Matsui, M., Tominaga, A., Liu, W.Z., & Tanaka-Ueno, T. (2008) Reduced genetic variation in the Japanese giant salamander, *Andrias japonicus* (Amphibia: Caudata). *Molecular Phylogenetics and Evolution* **49**, 318–326.

Maynard Smith, J., Burian, R., Kauffman, S., *et al.* (1985) Developmental constraints and evolution. *Quarterly Review of Biology* **60**, 265–287.

Mayr, E. (1963) *Animal Species and Evolution*. Cambridge, MA:Harvard University Press.

Mayr, E. (1983) How to carry out the adaptationist program? *American Naturalist* **121**, 324–334.

McKinney, M.L. (1990) Trends in body-size evolution. In: K. McNamara (ed.), *Evolutionary Trends*. Chichester: Wiley, pp. 75–118.

McNamara, K. & Selden, P. (1993) Strangers on the shore. *New Scientist* **139**, 23–27.

Milner, A.R. (1982) Small temnospondyl amphibians from the Middle Late Carboniferous of Illinois. *Palaeontology* **25**, 635–664.

Milner, A.R. (1983) The biogeography of salamanders in the Mesozoic and early Cenozoic: a ladistic vicariance model. In: R.W. Sims, J.H. Price, & P.E.S. Whalley (eds.), *Evolution, Time and Space: the Emergence of the Biosphere*. London: Academic Press, pp. 431–468.

Milner, A.R. (1988) The relationships and origin of living amphibians. In: M.J. Benton (ed.), *The Phylogeny and Classification of the Tetrapods*. Oxford: Clarendon Press, Vol. 1, pp. 59–102.

Milner, A.R. (1990) The radiations of temnospondyl amphibians. In: P.D. Taylor & G.P. Larwood (eds.), *Major Evolutionary Radiations*. Oxford: Clarendon Press, pp. 321–349.

Milner, A.R. (1993) Amphibian-grade Tetrapoda. In: M.J. Benton (ed.), *The Fossil Record 2*. London: Chapman & Hall, pp. 665–680.

Milner, A.R. (1994) Late Triassic and Jurassic amphibians. In: N.C. Fraser & H.-D. Sues (eds.), *In the Shadow of the Dinosaurs: Early Mesozoic Tetrapods*. New York: Cambridge University Press, pp. 5–22.

Milner, A.R. (1996) A revision of the temnospondyl amphibians from the Upper Carboniferous of Joggins, Nova Scotia. *Special Papers in Palaeontology* **52**, 81–103.

Milner, A.R. & Panchen, A.L. (1973) Geographical variation in the tetrapod faunas of the Upper Carboniferous and Lower Permian. In: D.H. Tarling & S.K. Runcorn (eds.), *Implications of Continental Drift to the Earth Sciences*. New York: Academic Press, pp. 353–368.

Modesto, S., Sues, H.-D., & Damiani, R. (2001) A new Triassic procolophonoid reptile and its implications for procolophonoid survivorship during the Permo-Triassic extinction event. *Proceedings of the Royal Society of London B* **268**, 2047–2052.

Niedźwiedzki, G., Szrek, P., Narkiewicz, K., Narkiewicz, M., & Ahlberg, P.E. (2010) Tetrapod trackways from the Middle Devonian Period of Poland. *Nature* **463**, 43–48.

Novacek, M. (2007) *Terra*. New York: Farrar Straus Giroux.

Organ, C.L., Canoville, A., Reisz, R.R., & Laurin, M. (2010) Paleogenomic data suggest mammal-like genome size in the ancestral amniote and derived large genome size in amphibians. *Journal of Evolutionary Biology* **24**, 372–380.

Orton, G.L. (1953) The systematics of vertebrate larvae. *Systematic Zoology* **2**, 63–75.

Orton, G.L. (1954) Original adaptive significance of the tetrapod limb. *Science* **120**, 1042–1043.

Payne, J.L. Boyer, A.G., Brown, J.H., *et al.* (2009) Two-phase increase in the maximum size of life over 3.5 billion years reflects biological innovation and environmental opportunity. *Proceedings of the National Academy of Sciences* **106**, 24–27.

Pierce, S.E., Clack, J.A., & Hutchinson, J.R. (2012) Three-dimensional limb joint mobility in the early tetrapod *Ichthyostega*. *Nature* **486**, 523–526.

Popper, K.R. (1957) *The Poverty of Historicism*. London: Routledge & Kegan Paul.

Pounds, J.A., Fogden, M.P.L., Savage, J.M., & Gorman, G.C. (1997) Test of null models for amphibian declines on a tropical mountain. *Conservation Biology* **11**, 1307–1322.

Reif, W.E. (1975) Lenkende und limitierende Faktoren in der Evolution. *Acta Biotheoretica* **24**, 136–162.

Riedl, R. (1978) *Order in Living Organisms*. New York: Wiley.

Romer, A.S. (1958) Tetrapod limbs and early tetrapod life. *Evolution* **12**, 365–369.

Romer, A.S. (1966) *Vertebrate Paleontology*. Chicago: University of Chicago Press.

Roth, G., Rottluff, B., & Linke, R. (1988) Miniaturization, genome size and the origin of functional constraints in the visual system of salamanders. *Naturwissenschaften* **75**, 297–304.

Ruta, M. (2009) Patterns of morphological evolution in major groups of Palaeozoic Temnospondyli (Amphibia: Tetrapoda). *Special Papers in Palaeontology* **81**, 91–120.

Ruta, M. & Benton, M.J. (2008) Calibrated diversity, tree topology and the mother of mass extinctions: the lesson of temnospondyls. *Palaeontology* **51**, 1261–1288.

Sanchez, S., Klembara, J., Castanet, J., & Steyer, J.S. (2008) Salamander-like development in a seymouriamorph revealed by palaeohistology. *Biology Letters* **4**, 411–414.

Sanchez, S., Germain, D., de Ricqlès, A., *et al.* (2010a) Limb-bone histology of temnospondyls: implications for understanding the diversification of palaeoecologies and patterns of locomotion of Permo-Triassic tetrapods. *Journal of Evolutionary Biology* **23**, 2076–2090.

Sanchez, S., de Ricqlès, A., Schoch, R.R., & Steyer, J.S. (2010b) Developmental plasticity of limb bone microstructural organization in *Apateon*: histological evidence of paedomorphic conditions in branchiosaurs. *Evolution and Development* **12**, 315–328.

Schindewolf, O.H. (1950) *Grundfragen der Paläontologie*. Stuttgart: Schweizerbart.

Schoch, R.R. (2002) The early formation of the skull in extant and Paleozoic amphibians. *Paleobiology* **28**, 378–396.

Schoch, R.R. (2009a) The evolution of life cycles in early amphibians. *Annual Review of Earth and Planetary Sciences* **37**, 135–162.

Schoch, R.R. (2009b) Developmental evolution as a response to diverse lake habitats in Paleozoic amphibians. *Evolution* **63**, 2738–2749.

Schoch, R.R. (2010) Heterochrony: the interplay between development and ecology in an extinct amphibian clade. *Paleobiology* **36**, 318–334.

Schoch, R.R. (2013) How body size and development biased the direction of evolution in early amphibians. *Historical Biology* **25**, 155–165.

Schoch, R.R. & Milner, A.R. (2000) Stereospondyli. In: P. Wellnhofer (ed.), *Handbuch der Paläontologie*. Munich: Pfeil, Vol. 3B.

Schoch, R.R. & Milner, A.R. (2008) The intrarelationships and evolutionary history of the temnospondyl family Branchiosauridae. *Journal of Systematic Palaeontology* **6**, 409–431.

Schoch, R.R. & Witzmann, F. (2011) Bystrow's paradox: gills, forssils, and the fish-to-tetrapod transition. *Acta Zoologica* **92**, 251–265.

Schoch, R.R. & Witzmann, F. (2012) Cranial morphology of the plagiosaurid *Gerrothorax pulcherrimus* as an extreme example of evolutionary stasis. *Acta Zoologica* **92**, 251–265.

Schoch, R.R., Voigt, S., & Buchwitz, M. (2010) A chroniosuchid from the Triassic of Kyrgyzstan and analysis of chroniosuchian relationships. *Zoological Journal of the Linnean Society* **160**, 515–530.

Schwenk, K. & Wagner, G.P. (2004) The relativism of constraints on phenotypic evolution. In: M. Pigliucci & K. Preston (eds.), *Phenotypic Integration: Studying the Ecology and Evolution of Complex Phenotypes*. Oxford: Oxford University Press, pp. 390–408.

Seilacher, A. (1970) Arbeitskonzept zur Konstruktionsmorphologie. *Lethaia* **3**, 393–396.

Sequeira, S.E.K. (1998) The cranial morphology and taxonomy of the saurerpetontid *Isodectes obtusus* comb. nov., (Amphibia: Temnospondyli) from the Lower Permian of Texas. *Zoological Journal of the Linnean Society* **122**, 237–259.

Sessions, S.K. & Larson, A. (1987) Developmental correlates of genome size in plethodontid salamanders and their implications for genome evolution. *Evolution* **41**, 1239–1251.

Shishkin, M.A. (2000) Mesozoic amphibians from Mongolia and the Central Asian republics. In: M.J. Benton, M.A. Shishkin, D.M. Unwin, & E.N. Kurochkin (eds.), *The Age of Dinosaurs in Russia and Mongolia*. Cambridge: Cambridge University Press, pp. 297–308.

Simpson, G.G. (1944) *Tempo and Mode in Evolution*. New York: Columbia University Press.

Smirnov, S.V. (1997) Additional dermal ossifications in the anuran skull: morphological novelties or archaic elements? *Russian Journal of Herpetology* **4**, 17–27.

Stanley, S.M. (1979) *Macroevolution: Pattern and Process*. San Francisco: Freeman.

Stayton, T. & Ruta, M. (2006) Geometric morphometrics of the skull roof of stereospondyls (Amphibia: Temnospondyli). *Palaeontology* **49**, 307–337.

Steyer, J.S., Damiani, R.J., Sidor, C.A., *et al.* (2006) The vertebrate fauna of the Upper Permian of Niger. IV. *Nigerpeton ricqlesi* (Temnospondyli: Cochleosauridae), and the edopoid colonization of Gondwana. *Journal of Vertebrate Paleontology* **26**, 18–28.

Vasilyan, D. & Böhme, M. (2012) Pronounced peramorphosis in lissamphibians: *Aviturus exsecratus* (Urodela, Cryptobranchidae) from the Paleocene–Eocene Thermal Maximum of Mongolia. *PLoS ONE* **7**, e40665.

Vences, M. & Wake, D.B. (2007) Speciation, species boundaries and phylogeography of amphibians. In: H. Heatwole & R.L. Carroll (eds.), *Amphibian Biology. Volume 6. Systematics*. Chipping Norton, NSW: Surrey Beatty, pp. 2613–2669.

Wagner, G.P. & Larsson, H.C.E. (2007) Fins and limbs in the study of evolutionary novelties. In: B.K. Hall (ed.), *Fins into Limbs: Evolution, Development, and Transformation*. Chicago: University of Chicago Press, pp. 49–61.

Wake, D.B. (1966) Comparative osteology and evolution of the lungless salamanders, family Plethodontidae. *Memoirs of the South California Academy of Science* **4**, 1–111.

Wake, D.B. (1991) Homoplasy: the result of natural selection, or evidence of design limitations? *American Naturalist* **138**, 543–567.

Wake, D.B. (2009) What salamanders have taught us about evolution. *Annual Review of Ecology and Systematics* **40**, 333–352.

Wake, D.B. & Deban, S.M. (2000) Terrestrial feeding in salamanders. In: K. Schwenk (ed.), *Feeding: Form, Function, and Evolution in Tetrapod Vertebrates*. Boston: Academic Press, pp. 95–116.

Wake, D.B. & Hanken, J. (1996) Direct development in the lungless salamanders: what are the consequences for developmental biology, evolution and phylogenesis? *International Journal of Developmental Biology* **40**, 859–869.

Wake, D.B. & Roth, G. (1989) The linkage between ontogeny and phylogeny in the evolution of complex systems. In: D.B. Wake & G. Roth (eds.), *Complex Organismal Function: Integration and Evolution in Vertebrates*. Chichester: Wiley, pp. 361–377.

Wake, D.B. & Vredenburg, V.T. (2008) Are we in the midst of the sixth mass extinction? A view from the world of amphibians. *Proceedings of the National Academy of Sciences* **105**, 11466–11473.

Wake, M.H. (1982) Diversity within a framework of constraints. Amphibian reproductive modes. In: D. Mossakowski & G. Roth (eds.), *Environmental Adaptation and Evolution*. Stuttgart: Gustav Fischer, pp. 87–106.

Warren, A.A. (2007) New data on *Ossinodus pueri*, a stem-tetrapod from the Early Carboniferous of Australia. *Journal of Vertebrate Paleontology* **27**, 850–862.

Werneburg, R. (2009) The Permotriassic branchiosaurid *Tungussogyrinus* Efremov, 1939 (Temnospondyli, Dissorophoidea) from Siberia restudied. *Fossil Record* **12**, 105–120.

Westphal, F. (1958) Die tertiären und eurasiatischen Riesensalamander (Genus *Andrias*, Urodela, Amphibia). *Palaeontographica A* **110**, 20–92.

Wiens, J.J. (2011) Re-evolution of lost mandibular teeth in frogs after more than 200 million years, and re-evaluating Dollo's Law. *Evolution* **65**, 1283–1296.

Witzmann, F. (2004) The external gills of Palaeozoic amphibians. *Neues Jahrbuch für Geologie und Paläontologie Abhandlungen* **232**, 375–401.

Witzmann, F., Schoch, R.R., & Maisch, M.W. (2008) A relic basal tetrapod from the Middle Triassic of Germany. *Naturwissenschaften* **95**, 67–72.

Wright, S. (1938) Size of population and breeding structure in relation to evolution. *Science* **87**, 430–431.

Yates, A.M. & Warren, A.A. (2000). The phylogeny of the 'higher' temnospondyls (Vertebrata: Choanata) and its implications for the monophyly and origins of the Stereospondyli. *Zoological Journal of the Linnean Society* **128**, 77–121.

Yeh, J. (2002) The effect of miniaturized body size on skeletal morphology in frogs. *Evolution* **56**, 628–641.

Zhang, P., Zhou, H., Chen, Y.Q., Liu, Y.F., & Qu, L.H. (2005) Mitogenomic perspectives on the origin and phylogeny of living amphibians. *Systematic Biology* **54**, 391–400

Index

Amphibian Evolution: The Life of Early Land Vertebrates, First Edition. Rainer R. Schoch.
© 2014 Rainer R. Schoch. Published 2014 by John Wiley & Sons, Ltd.

Plate 1.1 Field work. One of the most fruitful expeditions was the 1987 field campaign to the Devonian of Greenland, which yielded rich material of the tetrapodomorph *Acanthostega*. (A) Skull of *Acanthostega* in lateral view ("Grace"). (B) Skeleton of "Boris," a further *Acanthostega* specimen. (C) Jenny Clack, head of the expedition. (D) Stensiö Bjerg (locality). (E) Excavation. (F) Hindlimb of *Ichthyostega*. Photos: University Museum of Zoology, Cambridge.

Plate 1.2 Stem-amniote fossils. (A) Chroniosuchian *Chroniosaurus* from the Late Permian of Russia. (B) Anthracosaur *Archeria* from the Early Permian of Texas. (C) Juvenile of seymouriamorph *Discosauricus* from the Early Permian of the Czech Republic. (D) Two adults of *Seymouria* from the Early Permian of Germany ("Romeo and Juliet of Tambach"). (D) courtesy of Thomas Martens (Gotha).

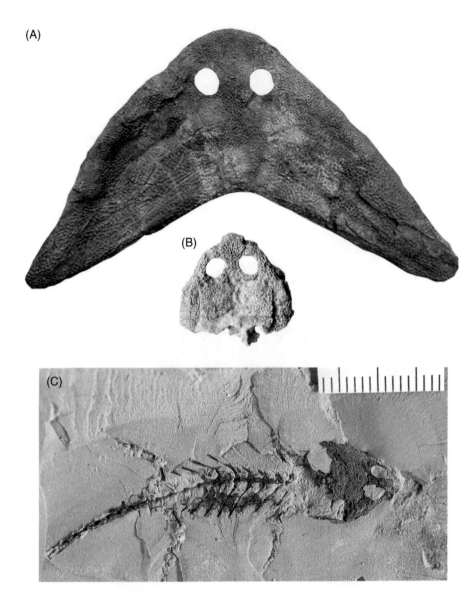

(A)

(B)

(C)

Plate 1.3 Stem-amniote fossils. Growth series of nectridean lepospondyl *Diplocaulus*, Early Permian of Texas. (A) Adult with typical boomerang shape. (B) Juvenile with narrow cheeks. (C) Tiny skeleton, showing the much narrower skull and fully ossified postcranium.

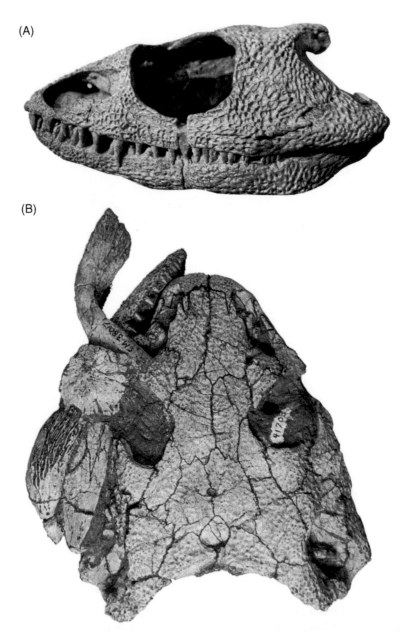

Plate 1.4 Stem-amphibian fossils. Dissorophoid temnospondyls (Early Permian). (A) Skull of trematopid *Phonerpeton* in side view, with enlarged naris (Texas). (B) Skull of trematopid *Ecolsonia* (New Mexico).

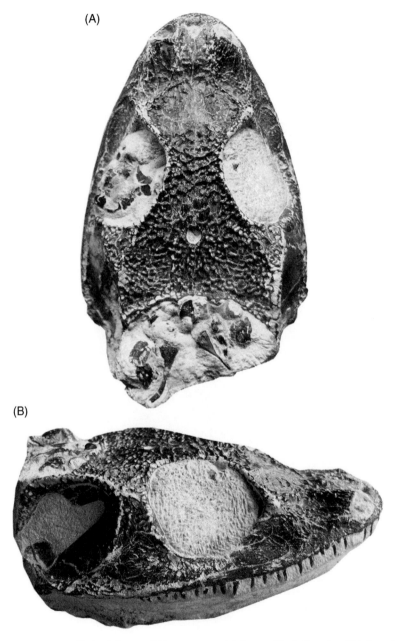

Plate 1.5 Stem-amphibian fossil. Skull of dissorophid temnospondyl *Cacops* (Early Permian, Oklahoma): (A) dorsal view; (B) side view.

Plate 1.6 Stem-amphibian fossil. Skeleton of temnospondyl *Sclerocephalus* (Pennsylvanian, Germany).

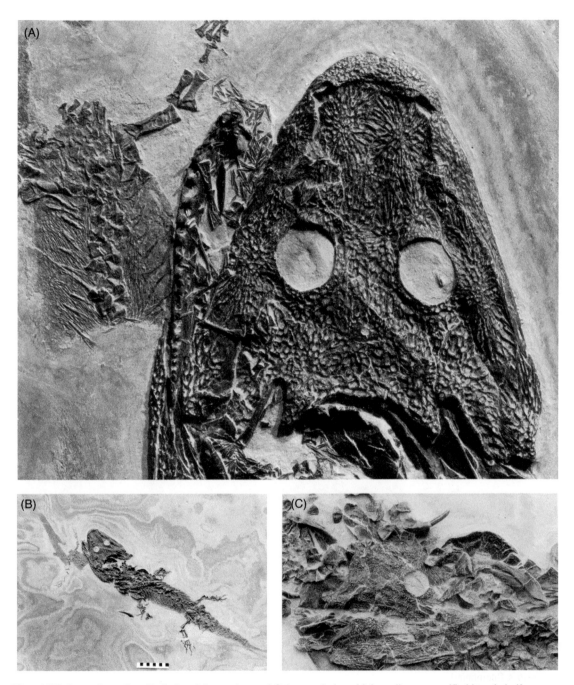

Plate 1.7 Paleoecology. Cannibalistic adult specimen of *Sclerocephalus* with juvenile conspecific bitten in half (Pennsylvanian, Germany): (A) skull of predator, with posterior part of prey; (B) complete specimen; (C) skull of prey.

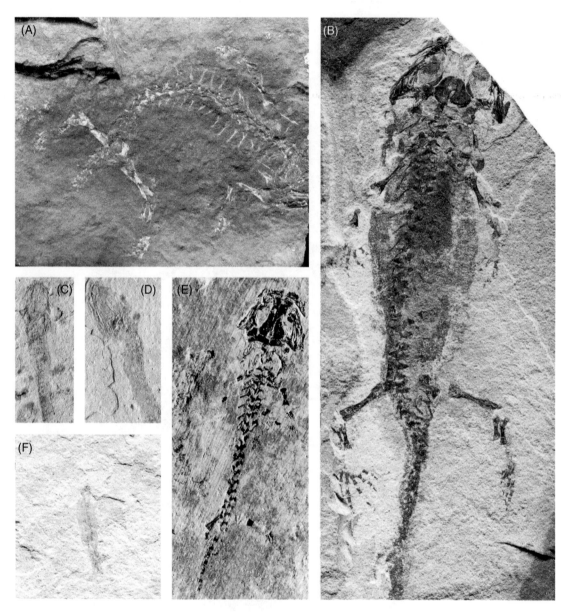

Plate 1.8 Ontogeny. Larvae of branchiosaurid *Apateon* (Pennsylvanian–Permian, Germany). (A) Metamorphosing specimen of *A. gracilis* (Niederhäslich). (B) Adult neotene of *A. pedestris* (Odernheim). (C–F) larvae of different sizes (*A. pedestris*, Odernheim).

Plate 1.9 Gigantism. Mesozoic aquatic top predator *Mastodonsaurus* (Middle Triassic, Germany): (A–C) mounted juvenile skeleton (3.5 m long); (D) skull in dorsal view, with pronounced lateral-line grooves; (E) palate, with differentiated fangs and marginal teeth.

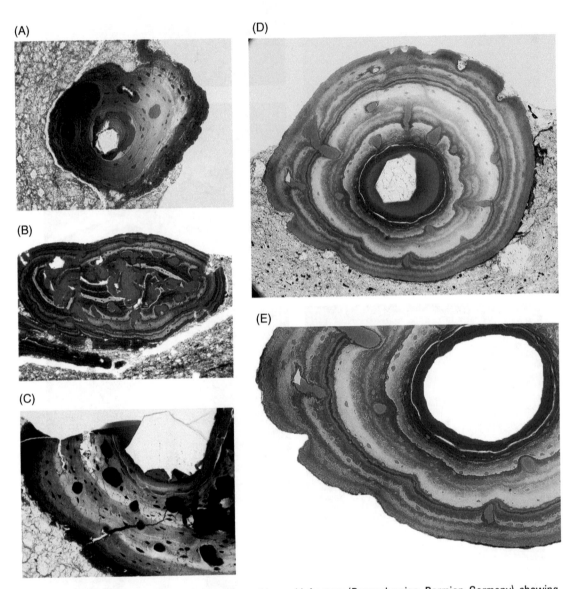

Plate 1.10 Paleohistology. (A–E) Long bones of branchiosaurid *Apateon* (Pennsylvanian–Permian, Germany), showing lines of arrested growth (LAGs) and bone cells.

Plate 1.11 Fossil preservation. (A) Articulated skeleton of *Gerrothorax* with excellent bone preservation in a limestone concretion (Kupferzell, Germany). (B) Crushed skull of *Archegosaurus* in a siderite nodule (Lebach, Germany). (C) Three-dimensional preservation of an *Amphibamus* skull in an ironstone nodule without bone matrix (Mazon Creek, Illinois). (D) Crushed skeleton of microsaur *Batropetes*, with pelvis and tail displaced by tectonic activity (Niederkirchen, Germany). (E) Disarticulated skull and mandibles of *Kupferzellia* (Vellberg, Germany).

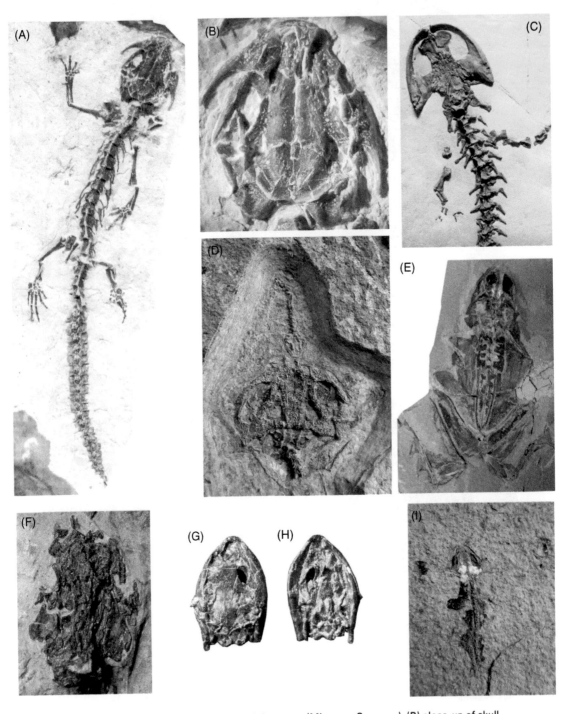

Plate 1.12 Lissamphibian fossils. (A) Salamandrid *Ichthyosaura* (Miocene, Germany); (B) close-up of skull. (C) Cryptobranchid *Andrias* (Oligocene, Germany). (D) Skull of salamandrid *Chelotriton* with larval skeleton near its mouth (Miocene, Germany). (E) Anuran *Rana strausi* (Pliocene, Germany). (F–H) Skull of *Eocaecilia* (Arizona). (I) Larval salamander with skin impression and calcareous ear capsules (Miocene, Germany).

Plate 1.13 Lissamphibians. (A) African Tree Frog *Afrixalus fornasini*. (B) Gray Tree Frog *Chiromantis xerampelina*. (C) Dwarf Squeaker *Arthroleptis xenodactyloides*. (D) Black Mountain Dusky Salamander *Desmognathus welteri*. (E) Three-lined Salamander *Eurycea guttolineata*. (F) Red-legged Salamander *Plethodon shermani*. (G, H) Kirk's Caecilian *Scolecomorphus kirkii*. Reproduced with permission of Hendrik Müller (Jena).

Plate 1.14 Field work. Excavations in the Triassic (southwest Germany). (A) Lower Keuper section (Vellberg). (B, C) Excavations in temnospondyl-rich mudstones (Vellberg). (D–G) Discovery and plaster-jacketing of a *Mastodonsaurus* skull (Kupferzell). (H) Large temnospondyl fangs.

Plate 1.15 Reconstruction of Paleozoic environments and tetrapodomorphs. (A) Pennsylvanian forest (reproduced with permission of Mary Parrish). (B) *Sclerocephalus* with its preferred prey, actinopterygian fish *Paramblypterus*. (C) *Eusthenopteron*. (D) *Acanthostega* (left) and *Ichthyostega* (right).

Plate 1.16 Reconstruction of Mesozoic environments and stem-amphibians. (A) Early Triassic river-bank environment (Buntsandstein facies, Germany and France). (B) Swampy deltaic environment with large horsetails (Lower Keuper, Germany). (C) *Mastodonsaurus*. (D) *Trematolestes*.

Printed and bound by CPI Group (UK) Ltd, Croydon, CR0 4YY

27/10/2024

14580288-0004